W9-BXC-964

Complete Web Monitoring

Complete Web Monitoring

Alistair Croll and Sean Power

JERICHO PUBLIC LIBRARY

O'REILLY®

Beijing · Cambridge · Farnham · Köln · Sebastopol · Taipei · Tokyo

Complete Web Monitoring
by Alistair Croll and Sean Power

Copyright © 2009 Alistair Croll and Sean Power. All rights reserved.
Printed in the United States of America.

Published by O'Reilly Media, Inc., 1005 Gravenstein Highway North, Sebastopol, CA 95472.

O'Reilly books may be purchased for educational, business, or sales promotional use. Online editions are also available for most titles (*http://my.safaribooksonline.com*). For more information, contact our corporate/institutional sales department: (800) 998-9938 or *corporate@oreilly.com*.

Editor: Simon St.Laurent		**Indexer:** Lucie Haskins	
Production Editor: Sumita Mukherji		**Cover Designer:** Karen Montgomery	
Copyeditor: Amy Thomson		**Interior Designer:** David Futato	
Proofreader: Sada Preisch		**Illustrator:** Robert Romano	

Printing History:

June 2009:   First Edition.

Nutshell Handbook, the Nutshell Handbook logo, and the O'Reilly logo are registered trademarks of O'Reilly Media, Inc. *Complete Web Monitoring*, the image of a raven, and related trade dress are trademarks of O'Reilly Media, Inc.

Many of the designations used by manufacturers and sellers to distinguish their products are claimed as trademarks. Where those designations appear in this book, and O'Reilly Media, Inc., was aware of a trademark claim, the designations have been printed in caps or initial caps.

While every precaution has been taken in the preparation of this book, the publisher and authors assume no responsibility for errors or omissions, or for damages resulting from the use of the information contained herein.

RepKover. This book uses RepKover™, a durable and flexible lay-flat binding.

ISBN: 978-0-596-15513-1

[M]

1211567705

For Kirsten, who is by equal measures patient, smart, understanding, and mischievous, and has sacrificed a year of evenings and weekends without complaint or recrimination.

—Alistair

For Dad, Mom, and Mark. Your undying support and love is the reason that I find myself here today. "Thank you" doesn't even begin to describe the respect, gratitude, and love I feel for you.

—Sean

Table of Contents

Part IV. Online Communities, Internal Communities, and Competitors

Preface

This is a book about achieving complete web visibility. If you're involved in something web-based—and these days, everything is web-based—you need to understand the things on the Web that affect you. Of course, today, everything affects you, which means looking far beyond what visitors did on your site to the communities and competitors with which you're involved.

While the Web is a part of our daily lives, web visibility lags far behind. Web operators seldom have a complete understanding of how visitors interact with their sites, how healthy their web properties are, what online communities are saying about their organizations, or what their competitors are up to. What little visibility they *do* have is fragmented and imperfect.

It doesn't have to be this way. In recent years, advances in monitoring, search, data collection, and reporting have made it possible to build a far more comprehensive picture of your web presence.

This book will show you how to answer some fundamental questions about that presence:

- What did visitors do on my site?
- How did they go about doing it?
- Why did they do it in the first place?
- Could they do it, or was it broken or slow?
- What are people saying about my organization elsewhere?
- How do I compare to my competitors?

Answering these questions means integrating many disciplines: web analytics, user surveys, usability testing, performance monitoring, community management, and competitive analysis.

At first, this might seem like a daunting task. There are hundreds of tools from which to choose, ranging from free to extremely expensive. There are millions of sites to track and more visitors to watch. Many stakeholders—each with their own opinions—probably care about your website, including investors, executives, marketers, designers, developers, and support teams.

We hope to arm you with a good understanding of the various technologies at your disposal so that you can craft and implement a strategy for complete web monitoring.

How to Use This Book

This book is divided into five sections.

- Part one—Chapters 1 through 4—looks at the need for complete web monitoring. Because every business is different, we look at how the business you're in affects the metrics and performance indicators that matter to you. *We suggest you read through this section of the book to get an understanding of the scope of web visibility.*

- Part two—Chapters 5 through 7—shows you how to understand what people did on your website. This goes beyond simple analytics, encompassing things like usability, recording interactions, and surveying visitors to better understand their needs. Each chapter offers an introduction and history of the technology, a framework for thinking about it, implementation details, and a look at how to share the resulting data within your organization. *You can consult each chapter when it's time to roll out that kind of visibility.*

- In part three—Chapters 8 through 10—we cover web performance monitoring. This is the complement to part two, looking at the question of whether visitors *could* do things, rather than what they did. After reviewing the elements of web performance, we look at synthetic testing and real user monitoring (RUM). *For an understanding of web performance, read* Chapter 8; *when it's time to deploy performance monitoring, review* Chapters 9 and 10.

- Part four—Chapters 11 through 16—looks outward toward the rest of the Internet. This includes communities, both external and internal, as well as competitors. We look at the emergence of online communities and how companies are using them successfully today. We offer a framework for thinking about communities, and look at eight common community models and ways of monitoring them. Then we take a quick look at the internally facing communities. Finally, we show how to track your competitors with many of the tools we've covered for watching your own sites. *If you're involved in community monitoring or competitive analysis, you should read through this part of the book.*

- Part five of the book—Chapters 17 and 18—discusses how to integrate and consolidate multiple kinds of monitoring data so you have a holistic view. We then consider what lies ahead for web monitoring and invite you to continue the discussion with us online.

What Will and Won't Be Covered

This book covers an extremely wide range of topics. We've tried to provide a foundation and a summary of important facts about each kind of monitoring, as well as concrete examples and practical implementation suggestions.

There are other books that delve more deeply than we do into the individual subjects covered here. Where this is the case, we'll point you to what we consider to be the definitive texts and websites on the topic.

We're not going to cover how to build a website, or how to make it faster and more usable, or how to get people to talk about you. What we *will* do, however, is show you how to measure whether your site is fast, effective, and usable, and determine whether people are talking about you on the Web.

Who You Are

If you're responsible for reporting and improving the online part of your business, this book is for you. If that sounds like a broad audience, you're right: comprehensive web visibility is by definition multidisciplinary. That doesn't mean, however, that the disciplines should be disconnected from one another. Consider the following:

- If you're responsible for web analytics, you need to know how performance, community referrals, and usability affect your visitors.
- If you're in charge of operating the web infrastructure, you need to know how to monitor sudden bursts of traffic or how to set up testing to mimic where real users are coming from.
- If you're managing communities, you need to know how activity within those communities leads to business outcomes on your website.
- If you're a competitive analyst, you need to know how your site and online presence compare to others.

The information in this book applies to entrepreneurs and marketing executives who need to define success metrics for their organizations, but also to the product managers, web operators, developers, and systems administrators responsible for achieving those metrics.

What You Know

We assume that you understand the basic elements of the Web—URLs, search engines, web navigation, and social networks. You don't necessarily know how web applications function beneath the covers, but you understand that they break in unpredictable ways. You know that your site could be faster, but you're not exactly sure how slow it is or whether your users care.

Ultimately, you know that the Web drives your business.

Conventions Used in This Book

Items appearing in the book are sometimes given a special appearance to set them apart from the regular text. Here's how they look:

Italic
> Indicates new terms, URLs, email addresses, filenames, file extensions, pathnames, and directories.

`Constant width`
> Indicates commands, options, switches, variables, attributes, keys, functions, types, classes, namespaces, methods, modules, properties, parameters, values, objects, events, event handlers, XML tags, HTML tags, macros, the contents of files, or the output from commands.

`Constant width bold`
> Shows commands or other text that should be typed literally by the user.

`Constant width italic`
> Shows text that should be replaced with user-supplied values.

 This icon signifies a tip, suggestion, or general note.

Using Code Examples

This book is here to help you get your job done. In general, you may use the code in this book in your programs and documentation. You do not need to contact us for permission unless you're reproducing a significant portion of the code. For example, writing a program that uses several chunks of code from this book does not require permission. Selling or distributing a CD-ROM of examples from O'Reilly books does require permission. Answering a question by citing this book and quoting example code does not require permission. Incorporating a significant amount of example code from this book into your product's documentation does require permission.

We appreciate, but do not require, attribution. An attribution usually includes the title, author, publisher, and ISBN. For example: "*Complete Web Monitoring*, by Alistair Croll and Sean Power. Copyright 2009 Alistair Croll and Sean Power, 978-0-596-15513-1."

If you feel your use of code examples falls outside fair use or the permission given above, feel free to contact us at *permissions@oreilly.com*.

How to Contact Us

Web monitoring is changing fast. As you'll see in our concluding chapter, many factors—including mobility, client-side applications, new browsers, video, and a flurry of social network platforms—make it hard to maintain good web visibility.

While this book serves as a foundation for complete web monitoring, we've got a companion site at *www.watchingwebsites.com*, where we're providing parallel content, links to tools, and a forum for you to help us grow and extend the book you have in your hands. You can reach both of us by emailing *authors@watchingwebsites.com*.

Vendor Policy

This book is full of examples and screenshots. Many vendors were kind enough to provide us with access to their systems, and we made copious use of the free tools that abound online. Don't treat illustration as endorsement, however. In many cases, we used a free tool because we had access to it. Also, some companies—particularly Google—are so pervasive and have such a breadth of tools that they figure prominently in the book. Other firms have alternate solutions.

One of the reasons we're maintaining the *www.watchingwebsites.com* website is to keep track of the various newcomers and the inevitable failures as the web monitoring industry matures and consolidates.

Safari® Books Online

When you see a Safari® Books Online icon on the cover of your favorite technology book, that means the book is available online through the O'Reilly Network Safari Bookshelf.

Safari offers a solution that's better than e-books. It's a virtual library that lets you easily search thousands of top tech books, cut and paste code samples, download chapters, and find quick answers when you need the most accurate, current information. Try it for free at *http://my.safaribooksonline.com*.

Reviewers

This book was a group effort. Web monitoring is a complex subject, and we sought the advice of many industry experts. We were especially eager to get feedback from vendors, since both of us have worked for vendors and we needed to make sure the book was nonpartisan and even-handed.

Inaccuracies and omissions herein are our fault; smart ideas and clever quips are probably those of our reviewers. We'd have been lost without:

- Imad Mouline, who reviewed the first parts of the book thoroughly while taking his vendor hat off entirely.
- Bob Page, who saved us from certain embarrassment on more than one occasion and provided excellent feedback.
- Hooman Beheshti, whose encyclopedic knowledge of web performance and willingness to break out a sniffer at a moment's notice were invaluable.
- Matt Langie and Sean Hammons, who provided excellent feedback on the web analytics chapter.
- Bryan Eisenberg, who provided a history lesson and great feedback on A/B testing and experimentation.
- Avinash Kaushik, who was always able to quietly ask the central question that's obvious in hindsight and makes the pieces fall into place.
- Stephanie Troeth, who could see beyond the sentences to the structure and convinced us to rewrite and restructure the WIA and VOC chapters.
- Tal Schwartz at Clicktale and Ariel Finklestein at Kampyle, who gave us great material and feedback, and continue their ongoing innovation within the industry.
- Johnathan Levitt at iPerceptions, who provided us with a review and insights into VOC.
- The Foresee team, notably Lee Pavach and Larry Freed through Eric Feinberg, who brought them to us.
- Robert Wenig and Geoff Galat at Tealeaf, who checked our impartiality and pointed us at great resources on their blogs.
- Fred Dumoulin at Coradiant, who provided us with a review of the web performance and RUM sections, and continues to work in opening up web monitoring.
- Doug McClure, who lent his expertise on ITIL, BSM, and IT monitoring in general.
- Joe Hsy at Symphoniq, who added his voice to the RUM section and rounded out collection methodologies.
- Tim Knudsen at Akamai, who gave us feedback on RUM and CDNs.
- Daniel Schrijver of Oracle/Moniforce, who provided us with great last-minute feedback (and jumped through corporate hoops to get it to us).
- The team at Gomez, in particular Imad Mouline and Samantha McGarry.
- The team at Keynote, in particular Vik Chaudhary and David Karow.
- The team at Webmetrics, in particular Peter Kirwan, Arthur Meadows, and Lenny Rachitsky.
- The team at AlertSite, in particular Richard Merrill and Ken Godskind.
- David Alston of Radian6, who reviewed the entire community section, gave us great feedback, and opened the kimono.
- The Scout Labs team, who brought us their candor, insight, and humor.

- Steve Bendt and Gary Koellig from Best Buy, who gave us great feedback.
- Stephen Pierzchala, who had excellent comments and observations.
- Lauren Moores from Compete.com, who had great input in the competitive monitoring chapter and let us use Compete's tools and data.
- Beth Kanter, who tore through the entire community section over the course of an afternoon and evening, and gave us excellent and concrete comments upon which to act.

We haven't listed all of you, and for this we owe you beers and apologies.

Acknowledgments

This book wouldn't have been possible without the patience and input of many people. We'd especially like to thank:

- The people at Networkshop, Coradiant, Bitcurrent, GigaOm, Techweb, FarmsReach, Syntenic, and Akoha (especially Austin Hill) for putting up with our constant distraction and midnight oil and for sharing monitoring data with us when needed.
- Technology conferences, blogs, and industry events where much of the material in this text was first presented and tweaked, including Interop, Web2Expo, Mesh, Enterprise 2.0, eMetrics, Podcamp, Democamp, and Structure.
- The restaurants, pubs, and coffee shops that let us linger and put up with heated discussions, particularly the staff at Lili & Oli, The Burgundy Lion, Brutopia, the Wired Monk, JavaU, Toi, Moi et Café, and Laïka.
- All the people who write free tools that help us build the Web. There are hundreds of plug-ins, sites, clients, search engines, and reporting platforms. We've covered many of them in this text. Without these tools, written and released into the wild or offered as free entry-level services, the Web would be a slower, less usable, less reliable place; these developers and companies have our gratitude.
- Many of the industry heavyweights who have given us their time and their thoughts, and who paved the way for a book like this, including Avinash Kaushik, Paul Mockapetris, Lenny Heymann, Steve Souders, Guy Kawasaki, Ross Mayfield, Hooman Beheshti, Imad Mouline, Bob Page, Cal Henderson, Jim Sterne, Bryan Eisenberg, Eric Reis, Dave McClure, Eric T. Peterson, Charlene Li, and Jeremiah Owyang.
- The artists and DJs who have kept us sane along the way. Notably, Underworld, Simon Harrison, Pezzner and others at Freerange, and Dirk Rumpff. You speak our language and have made it possible for us to remain sane during long nights of writing and editing.

- The entire team at O'Reilly, including Simon, Lauren, Amy, Marlowe, Sumita, Rob, Rachel, and Mike, all the way up to Tim. Thanks for believing in us (and being patient).

Alistair thanks Sean, Ian, Hooman, Dan, Marc, Oliver, and the rest of the Interop posse; everyone at Bitnorth; Eric and JFD; Jesse, Brady, Jen, and everyone else at O'Reilly; three Phils, Cheryl, Paulette, Michelle, Gordon, and the rest of the Kempton Resort; Brett, Paul, and Randy; and of course Doreen, Becky, and Adam.

Sean thanks Alistair, A-K, Brig, Deww, Ian, Hazel, Jayme, Kirsten, Marc, Pablo, Scott, Simon, and Wes for just being there and never asking anything in return.

Cary Goldwax, Dan Koffler, Steve Hunt, Scott Shinn, Guy Pellerin, Tara Hunt, James Harbottle, Lily, Oliver & Angela, Brad & Am, Chris & Melissa, Claudine, Steve Fink, CR, the Houghs and Foucault, Phil Dean, Dennis, Naomi, the shade in between, Adam & the choopa team, the blessed team and the folks in #ud & #colosol.

Finally, the group we really want to thank is you. Our Twitter followers, our Facebook friends, we're strangers now, but hope to know you soon. Find us online (@sean power and @acroll on Twitter), and come talk to us. These connections are the ones that bring the book and our shared knowledge alive! Thank you.

PART I

The Business Of Web Monitoring

While all businesses need web visibility, the nature of each business directly affects what metrics you need to track. In this section, we look at why you need a complete web monitoring strategy, how to tell what business you're in, and which performance indicators you need to track. Part I contains the following chapters:

- Chapter 1, *Why Watch Websites?*
- Chapter 2, *What Business Are You In?*
- Chapter 3, *What Could We Watch?*
- Chapter 4, *The Four Big Questions*

Why Watch Websites?

Smart companies make mistakes faster.

When it comes to figuring out what business you're in, there's only one rule: nobody gets it right the first time. Whether you're a fledgling startup or a global corporation launching a new brand, you're going to tweak your strategy before you find the best way to go to market.

There's no secret recipe for this. It takes time and resources to try something out, and the sooner you know it's wrong, the sooner you can try something different, wasting less time and money.

The changes you try may be simple, iterative adjustments to messaging, usability, or target market. On the other hand, they may be far-reaching—switching from a product to a service, targeting a different audience, or dealing with the fallout from a launch gone horribly wrong.

Successful companies adapt better than their competitors. However, it's easy to miss the key point of adaptivity: You can only adjust when you know what's not working. Everyone needs to make mistakes faster, but not everyone can make mistakes properly. Many organizations simply *lack the feedback to know when and how they're messing up.*

The amazing thing about the Web is that your visitors are trying to tell you what you're doing wrong. They do so with every visit cut short, every slow-loading page, every offer spurned, every form wrongly filled out, every mention on a social network, and every movement of a mouse.

You just need to know how to listen to them.

This book is about understanding the health and effectiveness of your web presence, and the relationship you have with your market, so you can adapt. It will show you how to monitor every website that affects your business—including your own—and how to use the insights you glean to iterate faster, adapting better than your competitors. It's about closing the feedback loop between the changes you make and the impact they have on your web audience.

Close this loop and you'll strike the right balance of brand awareness, compelling offers, web traffic, enticing site design, competitive positioning, grassroots support, usability, and site health that leads to success. Ignore this feedback and you risk being out-marketed, out-innovated, and out of touch.

A Fragmented View

At many of the companies we've surveyed, nobody sees the big picture. Different people watch different parts of their web presence: IT tests a site's uptime; marketers count visitors; UI designers worry about how test subjects navigate pages; and market researchers fret over what makes customers tick.

The walls between these disciplines can't stand. Each discipline needs to know how the others affect it. For example, bad design, leads to abandonment and missed revenue. A misunderstanding of visitor motivations makes otherwise compelling offers meaningless. Slow page loads undermine attractive websites. And successful marketing campaigns can overwhelm infrastructure capacity. As a result, these once-separate disciplines are being forced together.

Your online marketing battle isn't just fought on your own website. Right now, someone online is forming a first impression of you. Right now, your brand is being defined. Right now, someone's comparing you to a competitor. The battle is happening out in the open, on Twitter, Facebook, reddit, and hundreds of other blogs, mailing lists, and forums.

While it's essential to watch your own website, it's just as important to keep track of your competitors and the places where your established and potential communities hang out online. If you want to survive and thrive, you need to be aware of all your interactions with your market, not just those that happen within your own front door.

Complete web monitoring isn't just about watching your own website—*it's about watching every website that can affect your business*. Those sites may belong to you, or to someone else in the company, or to a competitor. They may even belong to someone who's not a competitor, but who strongly influences your market's opinions.

Out with the Old, in with the New

Traditional marketing takes time. For consumer packaged goods, months of work and millions of dollars go into the development and launch of a new soft drink or a rebranded dish soap.

Consider a traditional company, circa 1990. Once a year, executives gathered for strategic planning and set the long-term objectives of the firm. Then, every quarter, middle managers reviewed the performance of the organization against those goals.

Quarterly results showed whether sales targets were being met. By analyzing these results, the company could tell whether a particular region, product, or campaign was working. They adjusted spending, hired and fired, and maybe even asked research and development to change something about the product.

The call center also yielded good insights: in the last quarter, what were the most common customer complaints? Which issues took the longest to resolve? Where were the most refunds issued, and what products were returned? All of this data was folded into the quarterly review, which led to short-term fixes such as a product recall, training for call center operators, or even documentation sent to the channel.

At the same time, the company gathered marketing data from researchers and surveys. Market results showed how well campaigns were reaching audiences. Focus group data provided clues to how the target market responded to new messages, branding, and positioning.

After a couple of quarters, if something wasn't going smoothly, the issues were raised to the executive level for next year's strategic planning. If one standout product or service was succeeding beyond expectations, it might have been a candidate for additional development or a broader launch.

In other words, we'd see the results next year.

The Web has accelerated this process dramatically, because feedback is immediate. Customers can self-organize to criticize or celebrate a product. A letter-writing campaign might have taken months, but a thousand angry moms reached Motrin in a day. As a result, product definition and release cycles are compressed, making proper analysis of web traffic and online communities essential.

For web applications, the acceleration is even more pronounced. While a soft drink or a car takes time to build, ship, sell, and consume, web content is scrutinized the second it's created. On the web, next year is too late. Next *week* is too late.

Web users are fickle, and word of mouth travels quickly, with opinions forming overnight. Competitors can emerge unannounced, and most of them will be mining your website for ideas to copy and weaknesses to exploit. If you're not watching your online presence—and iterating faster than your competitors—you're already obsolete.

But it's not all doom and gloom. Watch your online presence and you'll have a better understanding of your business, your reputation, and your target market.

A Note on Privacy: Tracking People

Throughout this book, we'll be looking at user activity on the Web. Integrating analytical data with user identity will help your business. But integrating user data also has important moral and legal implications for privacy.

To make sense of what's happening online, you need to stitch together a huge amount of data: page retrievals, blog mentions, survey results, referring sites, performance metrics, and more. Taken alone, each is just an isolated data point. Taken together, however, they reveal a larger pattern—one you can use to make business decisions. They also show you the activities of individual users.

Collecting data on individuals' surfing habits and browsing activities is risky and fraught with moral pitfalls. In some jurisdictions, you have a legal obligation to tell visitors what you're collecting, and in the event of a security breach, to inform them that their data may have been compromised.

As a result, you need to balance your desire for complete web visibility with the risks of collecting personally identifiable information. This means deciding what you'll collect, how long you'll store it, whether you'll store it in aggregate or down to the individual visit, and who gets access to the data. It also means being open about your collection policies and giving visitors a way to opt out in plain, simple terms, not buried away in an obscure Terms of Service document nobody reads.

Being concerned with privacy is more than just knowing the legislation, because legislation is capricious, varies by legislature and jurisdiction, and changes over time. Marketers are so eager to understand their customers that tracking personal identity will likely become commonplace. Ultimately, we will tie together visitors' activity on our own sites, in communities, on the phone, and in person. Already, retail loyalty cards and call center phone systems can associate a person's behavior with his or her online profile.

As a web professional, you will need to define what's acceptable when it comes to tracking users. You can make great decisions and watch websites without needing to look at individual users. Individual visits can sometimes be a distraction from the "big picture" patterns that drive your business anyway. But as we start to extract more from every user's visit, you'll need to think about the moral and legal implications of watching web users throughout your monitoring strategy.

CHAPTER 2

What Business Are You In?

Every web business tracks key performance indicators (KPIs) to measure how well it's doing. While there are some universally important metrics, like performance and availability, each business has its own definitions of success. It also has its own user communities and competitors, meaning that it needs to watch the Web beyond its own sites in different ways.

Consider, for example, a search engine and an e-commerce site. The search engine wants to show people content, serve them relevant advertising, then send them on their way and make money for doing so. For a search engine, it's good when people leave—as long as they go to the right places when they do. In fact, the sooner someone leaves the search engine for a paying advertiser's site, the better, because that visitor has found what she was looking for.

By contrast, the e-commerce site wants people to arrive (preferably on their own, without clicking on an ad the e-commerce site will have to pay for) and to stay for as long as it takes them to fill their shopping carts with things beyond what they originally intended.

The operators of these two sites not only track different metrics, they also want different results from their visitors: one wants visitors to stay and the other wants them to leave.

That's just for the sites they run themselves. The retailer might care about competitors' pricing on other sites, and the search engine might want to know it has more results, or better ones, than others. Both might like to know what the world thinks of them in public forums, or whether social networks are driving a significant amount of traffic to their sites, or how fast their pages load.

While every site is unique and has distinct metrics, your website probably falls into one of four main categories:

Media property
> These sites offer content that attracts and retains an audience. They make money from that content through sponsorship, advertising, or affiliate referrals. Search engines, AdWords-backed sites, newspapers, and well-known bloggers are media properties.

Transactional site

> A site that wants visitors to complete a transaction—normally a purchase—is *transactional*. There's an "ideal path" through the site that its designers intended, resulting in a goal or outcome of some kind. The goal isn't always a purchase; it can also be enrollment (signing up for email) or lead generation (asking salespeople to contact them), and that goal can be achieved either online or off.

Collaboration site

> On these sites, visitors generate the content themselves. Wikis, news aggregators, user groups, classified ad listings, and other web properties in which the value of the site is largely derived from things created by others are all collaborative.

Software-as-a-service (SaaS) application

> These sites are hosted versions of software someone might buy. SaaS subscribers expect reliability and may pay a monthly per-seat fee for employees to use the service. Revenues come from subscriptions, and a single subscriber may have many user accounts. On some SaaS sites, users are logged in for hours every day.

It's common for parts of a site to fall into different categories. An analyst firm that sells reports is both a media property and a transactional site. A popular blog is as much about collaborative comments its users leave as it is about Google AdWords on the pages it serves. A video upload site is a media property filled with content users provide. And a free-to-try SaaS site that encourages users to subscribe to a premium version has a transactional aspect, as well as embedded ads similar to a media site.

The key is to decide which of these categories fits each part of your site, and then to determine which metrics and tools you should use to understand your web properties, your community, and your competitors. Let's look at each type of site in more detail.

Media Sites

Media sites are high-profile web destinations that provide content to attract visitors. They then display relevant advertising to those visitors in return for pay-per-click or pay-per-view revenues. In a few cases, the sites also sell premium subscriptions, which can be treated as separate transactional sites. Sites like CNET.com, MTV.com, NYTimes.com, and TechCrunch.com fall within this category.

Business Model

Figure 2-1 illustrates the basic elements of a media site's business model.

1. The media site embeds a link to an ad network's content within the media site's pages.
2. A visitor retrieves the media site's page with the embedded ad.

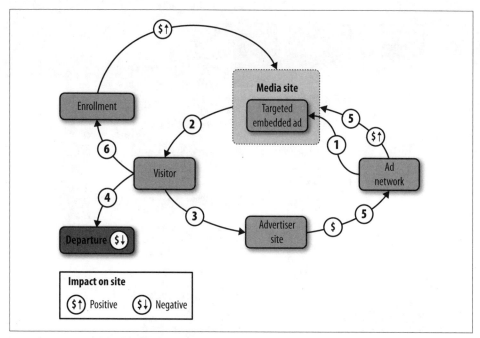

Figure 2-1. Elements of a media site's business model

3. The visitor clicks on the advertisement and visits the advertiser's site.

4. Alternatively, the visitor may leave the site without clicking on an ad, in which case the visit has not helped the media site make money (but it *has* raised the site's traffic numbers, which may attract advertisers).

5. The advertiser pays the ad network for ad traffic, and the ad network shares a portion of the money with the media site.

6. Additionally, the visitor may subscribe to a feed, bookmark the site, invite friends, or do other things that make him more likely to return, boosting the media site's number of loyal visitors.

Transactional Sites

Sergio Zyman, Coca-Cola's first chief marketing officer, describes marketing success as "selling more stuff to more people more often for more money more efficiently."

This is a great summary of what transactional companies care about: completing transactions, increasing visitors, maximizing shopping cart size, and encouraging repeat business. BestBuy.com, ThinkGeek.com, Expedia.com, and Zappos.com fall within this category.

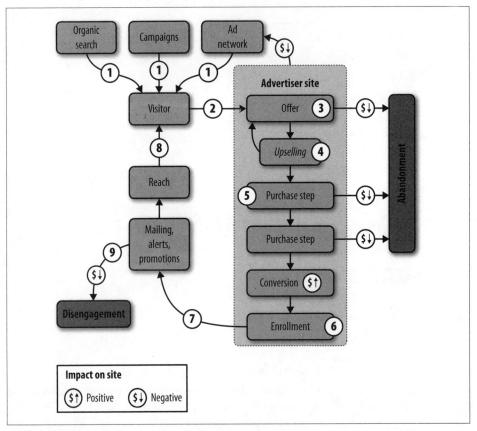

Figure 2-2. Elements of a transactional site's business model

Business Model

Figure 2-2 illustrates the basic elements of a transactional site's business model.

1. A visitor learns about the transactional site through organic (unpaid) search, advertising campaigns, word of mouth, social networks, or paid advertising.

2. The visitor visits the transactional site.

3. The transactional site makes some sort of offer.

4. If the visitor accepts the offer, the site may try to upsell the visitor by suggesting additional merchandise.

5. The visitor then moves through the various purchase steps, hopefully without abandoning the process, until payment.

6. The site tries to enroll the visitor, particularly since it now has the visitor's contact information as part of the purchase process.

7. This allows the site to contact the visitor as a part of ongoing campaigns.

8. The visitor may act on the new offers, increasing her lifetime value to the transactional site.

9. The visitor may also disengage by blocking messages or changing contact preferences.

Most brick-and-mortar companies' websites are a variant of this model, in which the transaction completes offline when a sales representative contacts a customer or a buyer finds the location of a store. While the outcome doesn't happen on the site itself, the organization derives revenue from it and wants to encourage the conversion from visitor to buyer.

An important exception to the goal of conversion is the support site. In the case of support organizations, some outcomes have a negative impact on the business. Visitors who first seek support from a company's website, but who ultimately contact a human for that support, cost the company money. Negative outcomes, like call center calls, are just as important to measure as positive ones.

Collaboration Sites

A collaboration site makes money from advertising in the same way a media site does. But early on, as the site is trying to grow its user community so that it becomes a big enough player to command good advertising, it's more focused on ensuring that the site contains valuable content that is well regarded by search engines, that it's retaining a vibrant user community, and that the site is growing virally.

In the beginning, collaboration sites may need to seed content manually. Wikipedia recruited thousands of librarians to kick off the creation of its content, and Flickr relied on manual rankings of published pictures by its editors until the community was big enough to vote for popular images on its own. Since most online communities are collaboration sites, we'll look at some metrics for monitoring these kinds of sites in the community section of the book.

Wikipedia.com, Facebook.com, and Slideshare.com fall within this category. Many internal sites, such as intranets built on SharePoint or wikis like Socialtext, use similar KPIs to externally facing collaboration sites.

Business Model

Figure 2-3 illustrates the basic elements of a collaboration site's business model.

1. A visitor comes to the site via an invitation, a social network recommendation, organic search, word of mouth or search results.

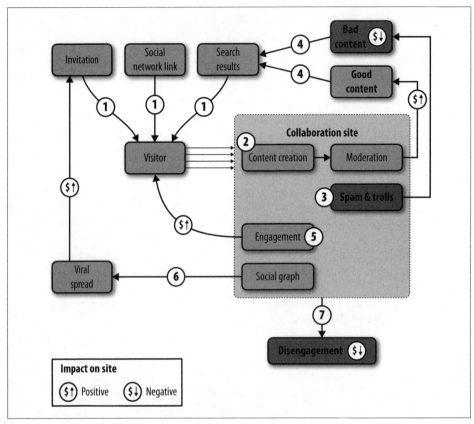

Figure 2-3. Elements of a collaboration site's business model

2. The visitor uses the site, hopefully creating content in the form of comments, uploads, and postings.

3. On the other hand, the visitor could create "bad" content, such as link spam or offensive content.

4. The content on the site is judged by subsequent search indexes, which affect the site's search ranking.

5. The visitor may engage with the site by subscribing to comment threads, befriending other users, voting on content, and so on, which encourages him to return.

6. In time, the visitor may also invite members of his social graph to join the site, leading to a larger user base.

7. On the other hand, he may stop using the site, disengaging and resulting in "stale" content.

Software-as-a-Service Applications

SaaS is a burgeoning segment of the software industry. With ubiquitous broadband and a move toward mobile, always-on Internet access, applications that once lived in an enterprise data center now run on-demand on the Internet, often with less hassle and cost than running them in-house.

SaaS sites have significantly different monitoring requirements from the other three site categories we've looked at. For one thing, they're often targeted at business users rather than consumers. They're also usually competing with in-house desktop alternatives—think Microsoft Office versus Google's SaaS-based office offering, Google Docs.

From a monitoring point of view, SaaS is less about acquiring or converting customers and more about delivering an already-purchased service well enough to convince subscribers to renew their subscriptions. It's also about measuring productivity rather than a goal or an outcome. Salesforce.com, Standoutjobs.com, Basecamphq.com, Fresh books.com, and Wufoo.com fall within the SaaS category.

Business Model

Figure 2-4 illustrates the basic elements of a SaaS site's business model.

1. An enterprise signs up for the site, presumably through a transaction process, and pays a usage fee.
2. This may also involve a per-seat fee for its employees or per-use fee.
3. At this point, the employees use the SaaS site. If performance is unacceptable, the SaaS site may be liable for an SLA refund and may receive helpdesk calls.
4. If the site is hard to use, employees may call the SaaS site's support desk.
5. Numerous problems will lead to high support costs for the SaaS operator.
6. If, as a result of poor performance and usability, employees aren't productive, the enterprise will cancel the subscription, resulting in customer churn, lost revenues, and higher cost of sales.
7. On the other hand, if users are productive, the enterprise will renew or extend its subscriptions and may act as a reference.

In the next chapter, we'll look at some of the metrics you can collect to understand the health of your site. In the Appendix, we'll look at these metrics and show which of them matter for each of the four types of site.

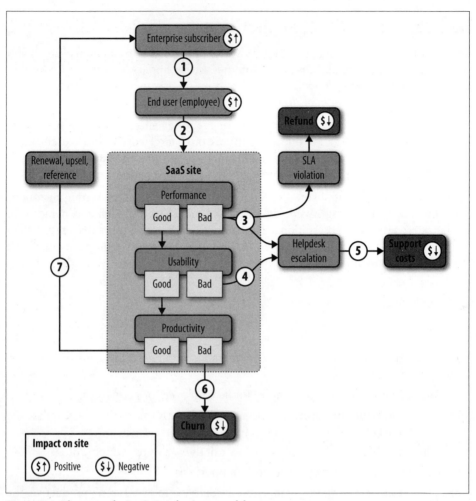

Figure 2-4. Elements of a SaaS site's business model

CHAPTER 3

What Could We Watch?

Regardless of the type of site you're running, there are many things you can track: the actions visitors took, the experiences they had, how well they were able to use the site, what they hoped to accomplish, and most importantly, *whether your business benefited in some way from their visits.*

Here's a quick overview of some of the things you'd like to know, and the tools you'll use to collect that knowledge.

What we'd like to know	Tool set to use
How much did visitors benefit my business?	Internal analytics
Where is my traffic coming from?	External analytics
What's working best (and worst)?	Usability testing, A/B testing
How good is my relationship with my market?	Customer surveys, community monitoring
How healthy is my infrastructure?	Performance monitoring
How am I doing against my competitors?	Search, external testing
Where are my risks?	Search, alerting
What are people saying about me?	Search, community monitoring
How are my site and content being used elsewhere?	Search, external analytics

We're now going to look at many of the individual metrics you can track on your website. If you're unfamiliar with how various web monitoring technologies work, you may want to skip to Chapter 4 and treat this chapter as a reference you can return to as you're defining your web monitoring strategy.

How Much Did Visitors Benefit My Business?

When you first conceived your website, you had a goal in mind for your visitors. Whether that was a purchase, a click on some advertising you showed them, a contribution they made, a successful search result, or a satisfied subscriber, the only thing that really counts now is how well your site helps them accomplish the things you hoped they'd do.

This may sound obvious, but it's overlooked surprisingly often. Beginner web operators focus on traffic *to* the site rather than business outcomes *of* the site.

Conversion and Abandonment

All sites have some kind of goal, and only a percentage of visitors accomplish that goal. The percentage of visitors that your site converts to contributors, buyers, or users is the most important metric you can track. Analyzing traffic by anything other than these goals and outcomes is misleading and dangerous. Visits mean nothing unless your visitors accomplish the things you want them to.

This is so important, we'll say it again: *analyzing web activity by anything other than outcomes leads to terrible mistakes.*

Your site's ability to make visitors do what you wanted is known as conversion. It's usually displayed as a funnel, with visitors arriving at the top and proceeding through the stages of a transaction to the bottom, as shown in Figure 3-1.

By adjusting your site so that more visitors achieve desired goals—and fewer of them leave along the way—you improve conversion. Only once you know that your site can convert visitors should you invest any effort in driving traffic to it.

What to watch: Conversion rates; pages that visitors abandon most.

Click-Throughs

Some sites *expect* people to leave, provided that they're going to a paying advertiser's site. That's how bills get paid. If your site relies on third-party ad injection (such as Google's AdWords or an ad broker) then click-through data is the metric that directly relates to revenue.

A media site's revenue stream is a function of click-through rates and the money advertisers pay for those clicks, measured in cost per mil (CPM), the cost for a thousand visitors. Even if the site is showing sponsored advertising (rather than pay-per-click advertising) for a fixed amount each month, it's important to track click-throughs to prove to sponsors that their money is well spent.

What to watch: The ratio of ads served to ads clicked (click-through ratio); clicks by visitors (to compare to ad network numbers and claims); demographic data and correlation to click-through ratio; and CPM rates from advertisers.

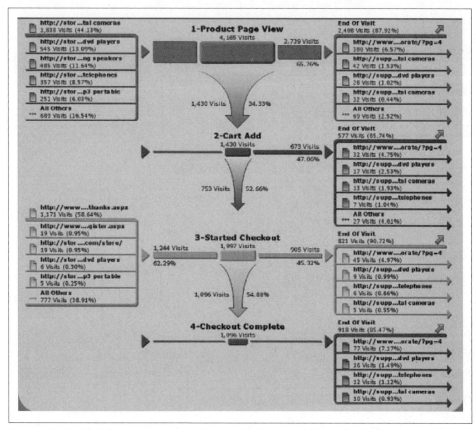

Figure 3-1. A typical e-commerce conversion funnel

Offline Activity

Many actions that start on the Web end elsewhere. These conversions are hard to associate with their ultimate conclusions: the analytics tool can't see the purchase that started online if it ends in a call center, as shown in Figure 3-2.

You can, however, still track conversions that end offline to some degree. Provide a dedicated phone number for calls that begin on the website, then measure call center order volumes alongside requests for contact information. You'll see how much traffic the site is driving to a call center in aggregate, as shown in Figure 3-3.

With a bit more work, you can get a much better idea of offline outcomes. To do this, you'll need an enterprise-class analytics solution that has centralized data warehousing capabilities. First, provide a unique code to visitors that they can then provide to call center operators in return for a discount, as shown in Figure 3-4. Then use this information to associate calls with web visits.

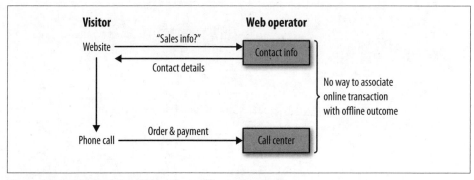

Figure 3-2. A standard web visit with an offline component

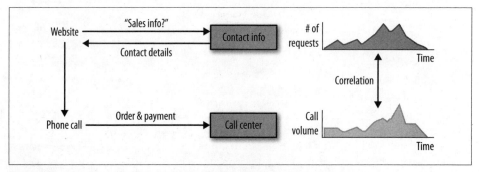

Figure 3-3. Visual correlation of online and offline data sources by time

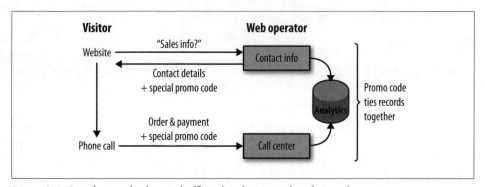

Figure 3-4. Correlation of online and offline data by using identifying information

The need to tie visits to outcomes is one reason many companies are deploying web applets that invite visitors to click to chat with sales or support personnel—it's far easier to track the effectiveness of a website when all the outcomes happen within view of the analytics system.

Figure 3-5. A postcard for FarmsReach.com to be distributed offline includes a custom URL used to tie the marketing message to an online outcome

Real-world beginnings can also lead to online ends, presenting many of the same problems: a tag on a Webkinz toy leads a child to an online portal; an Akoha card prompts someone to create an online account and propagate the card. The postcard pictured in Figure 3-5 contains a unique URL that's tied to an online marketing campaign, allowing an analytics team to compare the effectiveness of this message to others and to optimize offline campaign components.

Standard analytics tools ignore offline components at the beginning or end of a transaction without additional work. If your business has an offline component, you're going to need to get your hands on a system that allows you to integrate your data. Until you can get a system that automates this process, expect to spend a lot of time manually consolidating information.

What to watch: Call center statistics by time; call center data combined with analytics records; on-demand chat usage; lead generation sent to CRM (Customer Relationship Management) applications.

User-Generated Content

If your site thrives on user-generated content (UGC), contribution is key. You need to know how many people are adding to the site, either as editors or commenters, and whether your contributors are creating the content that your audience wants.

UGC contribution can be thought of as the ratio of media consumed to media created. It takes three major forms: new content, editing, and responses. All three are vital to a dynamic site, but too much or too little of any of them can be a bad sign.

- Too many comments are telltale signs of vitriol and infighting that distracts from the main post and discourages more casual visitors.
- Too much new content from a few users suggests spamming or scripting.
- Frequent editing of a single item may signal partisan disagreements.

If you're running a collaborative site, you care about how valuable new content is. Many of the metrics you'll track depend on the platform you're using. For example, if you're running a wiki, you care about incipient links (the links within a wiki entry that point to another, as-yet-unwritten, entry).

If an entry has no incipient links, it's orphaned and not well integrated into the rest of the site. On the other hand, if it has many incipient links that haven't yet been completed, the site isn't generating new content quickly enough. Finally, when many people click on a particular incipient link only to find that the destination page doesn't yet exist that's probably the page you should write next.

We'll return to specific metrics for various kinds of community sites in the community monitoring section of the book.

What to watch: Read-to-post ratio; difference between "average" and "super" users; patterns of down-voting and "burying" new posts; high concentrations of a few frequent contributors; content with high comment or edit rates; incipient link concentration; "sentiment" in the form of brand-specific comments or blog responses.

Subscriptions

Some media sites offer premium subscriptions that give paying customers more storage, downloadable content, better bandwidth, and so on. This can be the main revenue source for analyst firms, writers, and large media content sites such as independent video producers.

Additional bandwidth costs money, so subscriptions need to be monitored for cost to ensure that the premium service contributes to the business as a whole. For example, if users get high-bandwidth game downloads, how much traffic are they consuming? If they buy content through PayPal, what's the charge for payment? In these situations, analytics becomes a form of accounting, and you may need to collect information from networking equipment such as gigabytes of traffic delivered.

What to watch: Subscription enrollment monitored as a transaction goal; subscription resource usage such as bandwidth or storage costs.

Billing and Account Use

If you're running a subscription website—such as a SaaS application—then your subscribers pay a recurring fee to use the application. This is commonly billed per month, and may be paid for by the individual user or as a part of a wider subscription from an employer.

It's essential to track billing and account use, not only because it shows your revenues, but also because it can pinpoint users who are unlikely to renew. You'll need to define what constitutes an "active" user, and watch how many of your users are no longer active. You'll also want to watch the rate of nonrenewal to measure churn.

Small business invoicing outfit Freshbooks lets users invoice only a few customers for free; automated video production service Animoto limits the length of video clips it generates for free; and picture site Flickr constrains the volume of uploads in a month. This "velvet rope" pricing strategy encourages users to upgrade to the paid service. If you take this approach, treat the act of converting free users to paying users as a conversion process in a transactional site.

What to watch: Monthly revenue, number of paying subscribers; "active" versus "idle" users; volume of incidents per subscriber company; churn (nonrenewal) versus new enrollments.

Where Is My Traffic Coming From?

Once a website benefits from visitors in some way, it's time to worry about where those visitors are coming from. Doing so lets you:

- Encourage sites that send you traffic, either by contacting them, sponsoring them, or inviting them to become an affiliate of some sort.
- Advertise on sites that send you visitors who convert, since visitors they send your way are more likely to do what you want once they reach your site.
- Measure affiliate referrals as part of an affiliate compensation program.
- Understand the organic search terms people use to find you and adjust your marketing, positioning, and search engine optimization accordingly.
- Verify that paid search results have a good return on investment.
- Find the places where your customers, your competitors, and the Internet as a whole are talking about you so you can join the conversation.

The science of getting the right visitors to your site is a combination of Affiliate Marketing, Search Engine Marketing, and Search Engine Optimization, which are beyond the scope of this book.

Referring Websites

When a web browser visits a website, it sends a request for a page. If the user linked to that page from elsewhere, the browser includes a referring URL. This lets you know who's sending you traffic.

 The HTTP standard actually calls a referring URL a "referer," which may have been a typo on the part of the standard's authors. We'll use the more common spelling of "referrer" here.

If you know the page that referred visitors, you can track those visits back to the site that sent them and see what's driving them to you. Remember, however, that you need to look not only at who's sending you visitors, but also at who's sending you the ones that *convert*.

Referring URLs, once a mainstay of analytics, are becoming less common in web requests. JavaScript within web pages or from Flash plug-ins may not include it, and desktop clients may not preserve the referring URL. In other words, you don't always know where visitors came from.

What to watch: Traffic volume by referring URL; goal conversion by URL.

Inbound Links from Social Networks

An increasing number of visitors come to you from social networks. If your media site breaks a news story or offers popular content, social communities will often link to it. This includes not only social news aggregators like reddit or Digg, but also bloggers, comment threads, and sites like Twitter.

These traffic sources present their own challenges for monitoring.

- The links that sent you visitors may appear in comment threads or transient conversations that you can't link back to and examine because they've expired.

- The traffic may come from desktop clients (Twitter users, for example, employ Tweetdeck and Twhirl to read messages without a web browser). These clients omit referring URLs, making visit sources hard to track.

- The social network may require you to be a member, or at the very least, to log in to see the referring content.

- The author of the link may not be affiliated with the operator of the social network, so you may have no recourse if you're misrepresented.

Most of the challenges of social network traffic come from difficulties in tracking down the original source of a message. Sometimes, you simply won't see a referrer. But other times, you'll still see the referring URL but will need to interpret it differently.

For example, a referral from *www.reddit.com/new/* means that the link came from the list of new stories submitted to reddit. Over time, that link will move off the New Stories section of reddit, so you won't be able to find the source submission there. But you do know that the referral was the result of a reddit submission. Other social network referrals contain similar clues:

- *Microblogging websites*, such as Twitter, FriendFeed, or Identi.ca may tell you which person mentioned you, but also whether the inbound link came from someone's own Twitter page (*/home*) or a page they were reading that belongs to someone else.
- *URL-shortening services*, such as tinyurl, is.gd, bit.ly, or snipurl may hide some of the referral traffic. These usually rely on an HTTP redirect and won't show up in the analytics data, but some providers such as bit.ly offer their own analytics.
- Referrals from *microblogging search* show that a visitor learned about your site when searching Twitter feeds. Links from other microblog aggregation (such as hash tag sites) are signs that you're a topic online.
- Referrals from *mail clients*, such as Yahoo! Mail, Hotmail, or Gmail mean someone forwarded your URL via email.
- Referrals from *web-based chat* clients, like meebo.com, are a sign that people are discussing your site in instant messages.
- *Homepage portal* referral URLs can be confusing, and you need to look within the URL to understand the traffic source. For example, a URL ending in */ig/* came from an iGoogle home page; */reader/view* came from Google Reader; and */notebook/* came from Google Notebook. Some analytics packages break down these referrals automatically.

In other words, all referring sites aren't equal. It's not enough to differentiate and analyze referring sites by name; you have to determine the nature of the referrer. Different types of referring sites require different forms of analysis.

Since social networks and communities contain UGC, referrals may also show you when people aren't just linking to you, but are instead presenting your content as their own without proper attribution. Tracking social network referrals is an important part of protecting your intellectual property.

People will often mention your content elsewhere but not link to you directly. This will generally result in users searching for the name of your site and the content in question, which will make referral traffic look less significant while overinflating the amount of search and direct traffic, particularly navigational search.

Community managers need to identify the source of the traffic so they can engage the people who brought them the attention and mitigate the inevitable comment battles. And site designers need to make sure that new visitors become returning visitors.

What to watch: Referring sites and tools by group; sudden changes in unique visitors and enrollments from those groups; search results from social aggregators and microblogs.

Visitor Motivation

Knowing how visitors got to you doesn't always tell the whole story. Sometimes the only way to get inside a visitor's head is to ask her, using surveys and questions on the site. Such approaches are generally lumped into the broader category of *voice of the customer* (VOC).

Asking visitors what they think can yield surprising results. In the early days of travel sites, for example, site owners noticed an extremely high rate of abandonment just before paying for hotels. The sites tried many things to improve conversion—new page layouts, special offers, and so on, but it was only when they asked visitors, through pop-up surveys, why they were leaving that they realized the problem: many users were just checking room availability, with no intention of buying.

Use any travel site today and you'll likely see messages encouraging visitors to sign up to be notified of specials or changes in availability. This is a direct result of the insights gained through VOC approaches.

What to watch: What were users trying to accomplish? Did they plan to make a purchase? Where did they first hear about the site? What other products or services are they considering? What demographic do they fit into?

What's Working Best (and Worst)?

You can always do better. Even with large volumes of inbound traffic and a site that guides visitors to the outcomes you want, there's work to be done: filling shopping carts fuller, emphasizing the best campaigns, ensuring users find things quickly and easily, and so on. One of the main uses of web analytics is optimization.

Site Effectiveness

Your site converts visitors, and you have visitors coming in. What could be better than that? For starters, they could buy more each time they check out. A site that convinces visitors to purchase more than what they initially intended is an effective site.

Many e-commerce sites suggest related purchases or offer package deals. A bookstore might try to bundle a book the visitor is buying with another by offering savings, or try to show what else buyers of that book also bought. A hosting company could try to sell a multiyear contract for a discount. And an airline might refer ticket buyers to a partner rental company or try to add in travel insurance.

The total shopping cart value and the acceptance of these upselling attempts are essential metrics for e-commerce sites. You can treat upselling as a second funnel, and you should track upsold goods independently from the initial purchase whenever possible. Because upselling adds to an existing transaction, you should experiment with it.

Effectiveness isn't just for transactional sites, however. For example, on a collaborative site, how many visitors subscribe to a mailing list or an RSS feed? On a static web portal, how many people visit the "About" page or check for contact information?

What to watch: Percentage of upselling attempts that work; total cart value.

Ad and Campaign Effectiveness

Department store magnate John Wanamaker is supposed to have said, "Half the money I spend on advertising is wasted; the trouble is I don't know which half." Yesterday's chain-smoking ad executive, promoter of subjective opinions, espoused the value of hard-to-measure "brand awareness" at three-martini lunches.

Not so online. Every penny spent on advertising can be linked back to how much it benefits the business. Today, the ideal marketer is an analytical tyrant constantly searching for the perfect campaign, more at home with a spreadsheet than a cocktail. The main reason for this shift is the hard, unavoidable truth of campaign analytics.

Referring sites aren't the only method of categorizing inbound traffic. People don't surf the web by randomly entering URLs to see if they exist. With the exception of organic traffic, most visitors arrive because of a campaign. This may be an online campaign—banner ads, sponsorship, or paid content—but it may also be a part of an offline campaign such as a movie trailer or radio spot, or simply good word of mouth and an informal community.

Analytics applications can segment incoming traffic by campaign to measure how much they helped the bottom line. While it's harder to measure the effectiveness of offline advertising, you can still get good results with unique URLs that get press coverage (such as *http://www.watchingwebsites.com/booklink*).

What to watch: Which campaigns are working best, segmented by campaign ID, media ID, or press release.

Findability and Search Effectiveness

Users conduct site searches to find what they're after. Rather than browsing through several hierarchies of a directory, users prefer to type in what they're looking for and choose from the results.

Yet many site owners often overlook internal search metrics in their analyses. You need to know if your users are finding the results they're after quickly so you can better label and index your site.

There are many commercial search engines, as well as robust open source engines like Lucene (*http://lucene.apache.org/java/docs/index.html*), that developers can integrate into a site. Hosted search engines like Google (*www.google.com/sitesearch/*) can also be embedded within a site and configured to only return results from the site itself. Internal and third-party search engines can generate reports (for example, *http://blog.foofactory .fi/2008/08/interactive-query-reporting-with-lucene.html*) on what visitors are searching for; if this search data is tied into analytics, we can measure search effectiveness.

What to watch: How many searches ended with another search? With a return to the home page? With abandonment? What are the most popular search terms whose sessions have a significantly higher abandonment rate? Which search terms lead to a second search term?

Trouble Ticketing and Escalation

An increase in call center activity and support email messages are sure signs of a broken site. Site operators need to track the volume of trouble tickets related to the website, and ideally relate those trouble tickets to the user visits that cause them in order to speed up problem diagnosis.

There are a number of products that can record visits, flagging those that had problems and indexing them by the identities of the visitors or the errors that occurred. We'll look at these tools in depth in Chapter 6, when we consider how visitors interacted with the site, but for now, recognize that capturing a record of what actually happened makes it far easier to fix errors and prove who or what is at fault. You can also use records of visits as scripts for testing later on.

What to watch: Number of errors seen in logs or sent to users by the server; number of calls into the call center; errored visits with navigation path and replay.

Content Popularity

Media sites are about content. The successful ones put popular content on the home page, alongside ad space for which they charge a premium. Knowing what works best is essential, but it's also complex. Popular stories may be one-hit wonders—trashy content that draws visitors who won't stay. Who you attract with your content, and what they do afterward, is an important part of what works best. In other words, content popularity has to tie back to site goals, as well as stickiness metrics, rather than just page views.

But what about the fleeting popularity of transient content such as breaking news? This is a more difficult problem—stories grow stale over time. To balance fresh content with community popularity, social news aggregators like reddit, Slashdot, and Digg count the number of *upvotes* (how many people liked the content) and divide them by the content's age (based on the notion that content gets "stale"). It's not really this simple— other factors, such as the rate of upvoting, make up each site's proprietary ranking algorithms. Upvoting of this kind also shows the voting scores to the community, making it more likely to be seen and voted on.

 While social networking is a relatively recent phenomenon, its origins can be traced back to K. Eric Drexler's concept of "filtered hyperlinks," which Drexler describes as "a system that enables users to automatically display some links and hide others (based on user-selected criteria)," which "implies support for what may be termed social software, including voting and evaluation schemes that provide criteria for later filtering." Drexler's paper was first published at Hypertext 87 (*http:// www.islandone.org/Foresight/WebEnhance/HPEK1.html*).

What to watch: Content popularity by number of visitors; bounce rate; outcomes such as enrollment; ad click-through.

Usability

No site will succeed if it's hard to use. Focus groups and prerelease testing can identify egregious errors before you launch, but there's no substitute for watching a site in production.

There are high-end products to capture and replay every user's visit, but even if you are on a tight budget you can use JavaScript-based tools to monitor click patterns and understand where a user's mouse moved. Some of these are built into free tools like Google Analytics. By combining these with careful testing of different page designs, you can maximize the usability of an application.

Whenever visitors link to help or support pages, track the pages from which they came, which will point you to the source of their problems.

What to watch: Click patterns on key pages, particularly abandonment points and landing pages; referring URLs on support and feedback pages; form abandonment analysis; visitor feedback surveys.

User Productivity

While usability focuses on whether someone could understand how to do something the way it was intended, user productivity looks at whether visitors could accomplish their tasks quickly and without errors. Every website operator should care whether visitors can accomplish goals, but for SaaS sites this is particularly important, as users may spend their entire workday interacting with the application.

With the growth of the Web as a business platform, people are using browsers for tasks such as order entry or account lookups. If someone's employees are using your web application to accomplish tasks, you need to measure the rate at which those tasks are completed. This could be the number of orders entered in an hour or how long it takes to process an account.

The business that pays for a SaaS subscription cares about its employees' productivity. If you release a version of your SaaS website on which employees take twice as long to enter an order, you're sure to hear about it soon from their frustrated employer. On the other hand, if your website lets employees look up twice as many accounts an hour as an in-house alternative, your sales team should use this as a differentiator when talking to future customers.

What to watch: Time to accomplish a unit of work; tasks completed per session; errors made; changes in productivity across releases.

Community Rankings and Rewards

Sometimes, what's working on a website isn't the site itself, it's key contributors. Much of Wikipedia is edited by a small, loyal staff of volunteers; users flag inappropriate content on community sites like Craigslist, and power users promote interesting content on social news aggregators.

What to watch: Top contributors; contributions by user; specific "rewards" for types of contribution.

How Good Is My Relationship with My Visitors?

Once you've got your site in order, traffic is flowing in, and you're making the most of all of your visitors, it's time to be sure your relationship with them is long and fruitful.

In the early days of the Web, the main measure of engagement with your visitors was loyalty—how often they returned to your site. Today's users receive messages from a wide range of social networks, RSS (Really Simple Syndication) feeds, email subscriptions, and newsgroups, all of which push content out to them without them first asking for it.

As a result, visits don't count as much. The definition of loyalty needs to be amended for this two-way relationship. It's not just about how often your visitors return; it's about how willing they are to let you contact them and how frequently they act on your advances and offers.

Loyalty

The best visitors are those who keep coming back. Thanks to browser cookies, most web analytics applications show the ratio of new to returning visitors. Strike a healthy balance here: get new blood so you can grow, but encourage existing visitors to return so they become regular buyers or contributors.

For this, you need to look at two additional metrics. One is the average time between visits, which shows you how much a part of your audience's daily life you are. The other is the number of users who no longer engage with the site. Since users don't usually terminate an account, this is measured by the time since their last login.

What to watch: Ratio of new to returning visitors; average time between visits; time since last login; rate of attrition or disengagement.

Enrollment

Reaching people when they visit isn't enough. Visitors you're allowed to contact are the holy grail of online marketing. Their love is more than loyalty—it's permission.

Enrollment is valuable because consumers are increasingly skeptical of web marketing. Browsers run ad-blocking software. Mail clients hide pictures. Some portals let users hide advertising. An enrolled visitor is reachable despite all of these obstacles.

Enrollment also provides better targeting. You can ask subscribers for demographic information such as gender, interests, and income, then tailor your messages—and those of your advertisers—to your audience.

What to watch: Signups; actual enrollments (email messages sent that didn't bounce); email churn (addresses that are no longer valid); RSS subscription rates.

Reach

It's great to have people enroll, but it's even better to actually be able to reach them. Whether through email subscriptions, alerts, or RSS feeds, *reach* is the measurement of how many enrolled visitors actually see your messages.

In the case of email, this may be the number of people who opened the message. For RSS feeds, it's the number that actually looked at the update you sent them. For video, it could be the number that played the content beyond a certain point. Figure 3-6 shows the FeedBurner report for subscribers (the number enrolled) and reach (the number that actually saw a message) for a blog.

Reach is a far more meaningful measure of subscription, since it discounts "stale" enrollments and shows how well your outbound messages, blogs, and alerts result in action.

What to watch: Reach of email recipients; reach of RSS subscribers.

How Healthy Is My Infrastructure?

Slow page loads or excessive downtime can undermine even the best-designed, most effective, easiest-to-use website. While web analytics shows you *what* people are doing

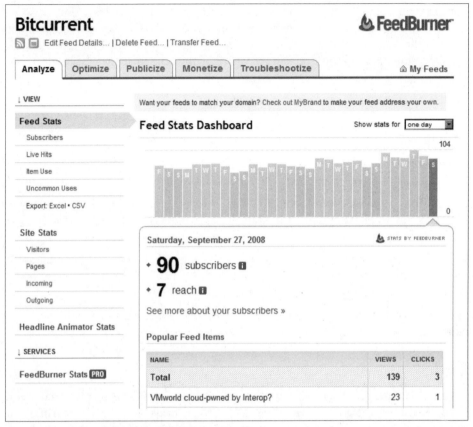

Figure 3-6. Google's FeedBurner Feed Stats

on your site; end user monitoring shows you whether they *could* do it—and how quickly they did it.

Availability and Performance

The most basic metrics for web health are availability (is it working?) and performance (how fast is it?), sometimes referred to collectively as performability. These can be measured on a broad, site-wide basis by running synthetic tests at regular intervals; or they can be measured for every visit to every page with real user monitoring (RUM).

In general, availability (the time a site is usable) is communicated as a percentage of tests that were able to retrieve the page correctly. Performance (how long the user had to wait to interact with the site) is measured in seconds to load a page for a particular segment of visitors.

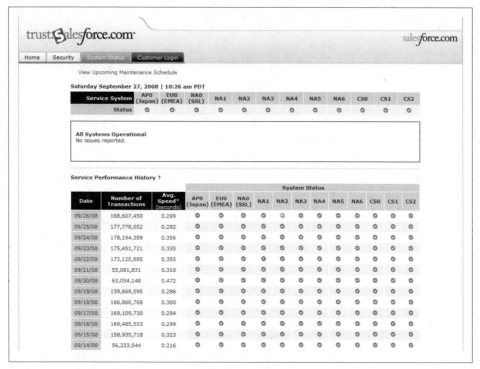

Figure 3-7. Salesforce.com's System Status dashboard

What to watch: Availability from locations where visitors drive revenue; page load time for uncached and cached pages; end-to-end and host time at various traffic volumes; changes in performance and availability over time or across releases.

Service Level Agreement Compliance

If people pay to use your site, you have an implied contract that you'll be available and usable. While this may not be a formal Service Level Agreement (SLA), you should have internal guidelines for how much delay is acceptable. Some SaaS providers, such as Salesforce.com, show current uptime information to subscribers and use this as a marketing tool (see Figure 3-7).

A properly crafted SLA includes not only acceptable performance and availability, but also time windows, stakeholders, and which functions of the website are covered.

Your SLAs may also depend on your partners, so you may have to consider other SLAs when measuring your own. If, for example, your site is part of a web supply chain, you need to measure the other links in that chain to know whether you're at fault when an SLA is missed.

SLAs Are Complex Things

Some web users currently have formal SLAs with their providers, and as more organizations use the Web as the primary channel for business, SLAs will become commonplace. There are many factors to consider when defining an SLA, which is one of the reasons they tend to be either ponderously detailed or uselessly simple.

- From whose perspective are you measuring the SLA?
- Are you measuring the website as a whole or its individual servers?
- Are you watching a single page or an entire workflow or business process?
- Are you measuring from inside your firewall, outside your firewall, or from where your customers are located?
- What clients and operating systems are you using to measure performance?
- Are you watching actual users or simulating their visits?
- Are you measuring the average performance or a percentile (the worst five percent, for example)?
- Does your SLA apply around the clock or only during business hours? Whose business hours?

There's no one correct answer to these questions, but organizations need to know what they're measuring and what they're not. In Figure 3-7, for example, what is Salesforce really measuring? Will they report that a North American instance is not working properly if West Coast users are doing fine but East Coast users are having performance issues?

Measure and report the metrics that comprise an SLA in a regular fashion to both your colleagues and your customers.

What to watch: SLA metrics against agreed-upon targets; customers or subscribers whose SLAs were violated.

Content Delivery

Measuring the delivery of simple, static advertising was once straightforward: if the browser received the page, the user saw the ad. With the advent of interactive advertising and video, however, delivery to the browser no longer means visibility to the user.

Content delivery is important for media companies. A Flash ad may be measured for its delivery to the browser, whether its sites were within the visible area of the browser, and whether the visitor's mouse moved over it. Users may need to interact with the content—by rolling over the ad, clicking a sound button, and so on. Then the ad plays and the user either clicks on the offer or ignores it. This means *each interactive ad has its own abandonment process*.

The provider that served the ad tracks this. The Flash content sends messages back about engagement, abandonment, and conversion. As a result, media site operators don't need to treat this content differently from static advertising.

While rich media often requires custom analytics, there's one kind of embedded media that's quickly becoming mainstream: Web video. David Hogue of Fluid calls it "the new JPEG," a reflection of how commonplace it is on today's sites.

While there are a variety of companies specializing in video analytics (such as Visible Measures, divinity Metrics, Streametrics, TubeMogul, and Traackr), embedded video is quickly becoming a part of more mainstream web analytics packages. It is also becoming a feature of many content delivery network (CDN) offerings that specialize in video.

Most for-pay analytics offerings available today allow a video player to send messages back to the analytics service when key events, such as pausing or rewinding, take place, as shown in Figure 3-8.

Embedded video serves many purposes on websites. Sometimes it's the reason for the site itself—the content for which visitors come in the first place. Sometimes it's a form of advertising, tracked by the ad network that served it. Sometimes it's a lure to draw the visitor deeper into the site. In each case, what you measure will depend on your site.

What to watch: Content engagement; attention; completion of the media; pauses.

Capacity and Flash Traffic: When Digg and Twitter Come to Visit

When a community suddenly discovers content that it likes, the result is a flash crowd. A mention by a popular blogger, breaking news, or upvoting on a social news aggregator can send thousands of visitors to your website in seconds.

For most websites, capacity and bandwidth is finite. When servers get busy and networks get congested, performance drops. The problem with flash crowds is that they last for only a few hours or days—after that, any excess capacity you put in place is wasted. One of the attractions of CDNs and on-demand computing infrastructure is the ability to "burst" to handle sudden traffic without making a long-term investment in bandwidth or hardware.

When you're on the receiving end of a flash crowd, there's a lot to do. Marketing needs to engage the one-time visitors, making them loyal and encouraging them to subscribe or return. IT operators need to ensure that there's enough capacity, working with service providers or provisioning additional resources if applicable. And community managers need to identify the source of the traffic so they can nurture and prolong the attention.

While flash crowds create dramatic bursts of traffic, a gradual, sustained increase in traffic can sneak up on you and consume all available capacity. You need to monitor

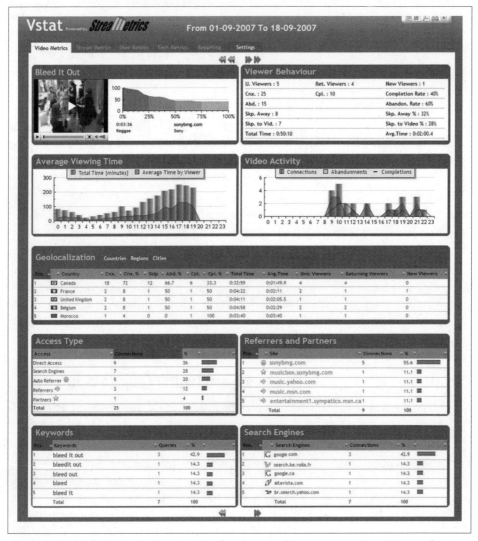

Figure 3-8. Vstat by Streametrics captures information such as average viewing time, geolocation, referrers, and so on

long-term increases in page latency or server processing or decreases in availability that may be linked to increased demand for your website.

Analytics is a good place to start: IT operators should correlate periods of poor performance with periods of high traffic to bandwidth- or processor-intensive parts of the site. If there's a gradual increase in the volume of downloads, you should plan for additional bandwidth. Similarly, if there's a rise in transaction processing, you may need more servers.

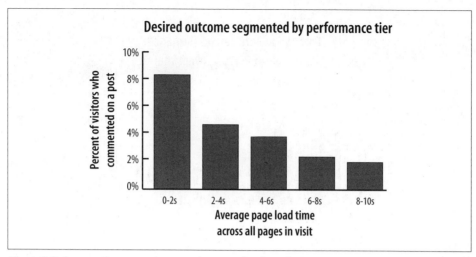

Figure 3-9. Segmenting conversion rates by tiers of web performance

Too often, IT operations and web analytics teams don't talk. The result is last-minute additions to capacity rather than planned, predictable spending.

What to watch: Sudden increases in requests for content; referring URLs; search engine results that reference the subject or the company; infrastructure health metrics; growth in large-sized content or requests for processor-intensive transactions

Impact of Performance on Outcomes

While you can measure the impact of visitors on performance, it's equally important to measure the impact of performance on visitors. Poor performance has a direct impact on outcomes like conversion rate, as well as on user productivity. Google and Amazon both report a strong correlation between increased delay and higher bounce rates or abandonment, and responsive websites encourage users to enter a "flow state" of increased productivity, while slow sites encourage distraction.

The relationship between performance and conversion can be measured on an individual basis, by making performance a metric that's tracked by analytics and by segmenting conversion rates for visitors who had different levels of page latency, as shown in Figure 3-9.

However, this can be hard to do unless you have a way of combining web analytics with the experience of individual end users.

Another way to understand the impact of performance is to compare aggregate page latency with aggregate conversion metrics, as shown in Figure 3-10. To do this properly, you need to eliminate any other factors that may be affecting conversion, such as promotions, daily spikes, or seasonal sales increases.

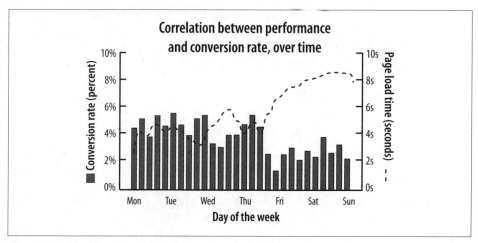

Figure 3-10. Aggregate view of conversion rate alongside site performance

The fundamental question we want to answer is: *does a slow user experience result in a lower conversion rate or a reduced amount of upselling?*

What to watch: Conversion rates segmented by tiers of page latency; daily performance and availability summaries compared with revenue and conversion; revenue loss due to downtime.

Traffic Spikes from Marketing Efforts

Marketing campaigns should drive site traffic. You need to identify the additional volume of visitors to your site not only for marketing reasons, but also to understand the impact that marketing promotions have on your infrastructure and capacity.

Properly constructed campaigns have some unique identifier—a custom URL, a unique referrer ID, or some other part of the initial request to the site—that lets you tie it back to a campaign. This is used to measure ad and campaign effectiveness. You can use the same data to measure traffic volumes in technical terms—number of HTTP sessions, number of requests per second, megabits per second of data delivered, availability, and so on.

Pay particular attention to first-time visitors. They place a greater load on the network (because their browsers have yet to cache large objects) and on applications because of enrollment, email registration, and other functions that occur when a visitor first arrives.

What to watch: Traffic by marketing campaign alongside infrastructure health metrics, such as availability or performance, on as granular a level as possible (ideally per-visit). A summary similar to the one shown in Figure 3-11 is ideal.

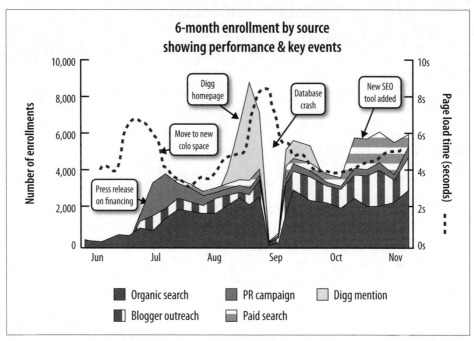

Figure 3-11. A "state of social media" report by month alongside performance information

Seasonal Usage Patterns

If your business is highly seasonal, you need to understand historical usage patterns. The fates of firms like Hallmark and ProFlowers are tied to specific holidays—indeed, at ProFlowers, so much of of the company's transactions happen on Valentine's Day that they refer to it internally as "V-day."

Seasonal usage isn't really a new metric, but it's a requirement for monitoring in general. If you're allowed to, collect at least five quarters of data so you can compare each month to the same month of the previous year. You're doing this for two reasons: to understand usage trends so you can plan for capacity changes, and to confirm that you're meeting long-term SLAs.

You only need to store aggregate data, such as hourly performance and availability, for this long. Indeed, many governments and privacy organizations are looking more closely at the long-term storage of personal information, and some sites have a deletion policy that may limit your ability to capture long-term trends.

What to watch: Page views; performance, availability, and volume of CPU-intensive tasks (like search or checkout) on a daily basis for at least 15 months.

How Am I Doing Against the Competition?

You want visitors. Unfortunately, so do your competitors. If visitors aren't on your website, you want to know where they're going, why they're going there, and how you stack up against the other players in your industry.

In addition to monitoring your own website and the communities that affect your business, you also need to watch your competition.

Site Popularity and Ranking

On the Web, popularity matters. When it comes to valuations, most startups and media outlets are judged by their monthly unique visitor count, which is considered a measure of a site's ability to reach people. Relevance-based search engine rankings reinforce this, because sites with more inbound links are generally considered more authoritative.

Some websites pay marketing firms to funnel traffic to them. Artificial inflation of visits to the site does nothing to improve conversion rates. These paid visitors seldom turn into buyers or contributors, but they do raise the site's profile with ranking services such as comScore, which may eventually get the site noticed by others. That said, raw traffic volumes are a spurious metric for comparing yourself to others.

Several services, such as Alexa and Compete.com, try to estimate site popularity. Use this data with caution. Alexa, for example, collects data on site activity from browser toolbars, then extrapolates it to the population as a whole. Unfortunately, this approach has many limitations, including problems with SSL visibility and concerns that the sample population doesn't match the Web as a whole (see *http://www.techcrunch.com/2007/11/25/alexas-make-believe-internet/*, which points out that according to Alexa, YouTube overtook all of Google at one point).

Compete.com has different methods for determining rankings, and even its estimates don't map cleanly to actual traffic. Figure 3-12 shows a comparison of actual traffic volumes and third-party traffic estimates.

Any mention of accuracy begs the question, "What is actual traffic?" Due to the differences in measurement methodologies across various ranking sites, there's bound to be a difference in traffic estimates. Rough trends should, however, be representative of what's going on, and comparing several companies using the same tools and definitions is a valid comparison.

Traffic estimates work well for broad, competitive analysis across large sites, but fail with low-traffic sites. Alexa, for example, doesn't track estimated reach beyond the top 100,000 websites. This data is still valuable for determining basic growth in an industry—if the large sites with which you compete are growing at 15% a month, a 15% traffic increase on your own site means you're merely holding your own.

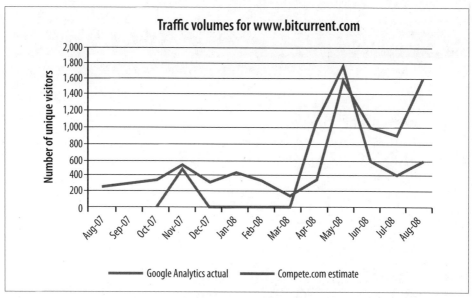

Figure 3-12. Unique visitors compared to Compete.com estimates

There are other ways to measure popularity. You can count how many people type a URL into a search engine—what's known as *navigational search*. This happens a lot. From July to September 2007, Compete.com reported that roughly 17 percent of all searches were navigational searches (*http://blog.compete.com/2007/10/17/navigational -search-google-yaho0-msn/*). We can analyze search terms with tools like Google Trends or Google Insights (*www.google.com/trends* or *http://www.google.com/insights/search/*) and get some idea of relative site popularity. Insights also shows searches by geography, so it can be a useful tool for identifying new markets.

The opposite of navigational search is type-in traffic, where users type a search term like "pizza" into the address bar. This behavior is one of the reasons firms like Marchex buy domains like pizza.com. See John Battelle's article on the subject at *http://battellemedia.com/archives/ 002118.php*.

Technorati and BlogPulse also show the popularity of sites and topics, as shown in Figure 3-13, although their focus is on blogging.

What to watch: Page views; unique daily visitors; new visitors; Google PageRank; Google Trends and Google Insights; incoming links; reach on panel sites like Compete.com or toolbar ranking sites like Alexa. If you're a media site or portal that has to report traffic estimates as part of your business, ComScore and Nielsen dominate traffic measurement, with Quantcast and Hitwise as smaller alternatives.

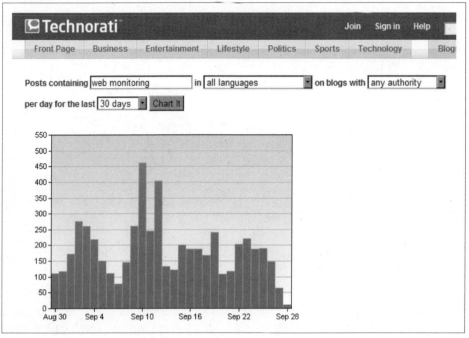

Figure 3-13. Popularity ranking for the term "web monitoring" using Technorati's BlogPulse service

Many "popularity" sites have limited accuracy. They are based on visits from a survey population that has installed a toolbar, and often don't correctly represent actual traffic.

How People Are Finding My Competitors

Your competitors are fighting you for all those visitors. Because they're probably using Google's AdWords, you can find out a good deal about what terms they're using and how much they're spending on searches. Major search engines share data on who's buying keywords and how much they're paying for them—this is a side effect of their keyword bidding model. Services like Spyfu collect this data and show the relative ad spending and keyword popularity of other companies.

Knowing which organic terms are leading visitors to your competitors helps you understand what customers are looking for and how they're thinking about your products or services. On the other hand, using a competitor's web domain, you can find out what search terms the market thinks apply to your product category and change your marketing accordingly.

What to watch: Organic and inorganic search results for key competitors.

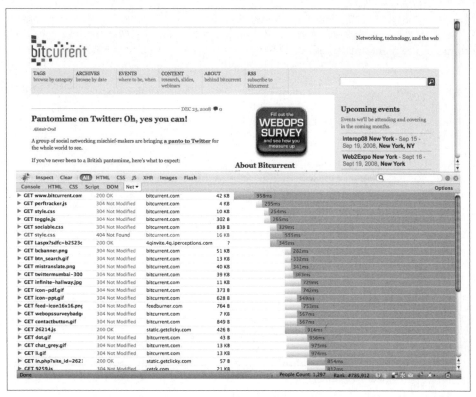

Figure 3-14. Firebug cascading performance analysis diagram of Bitcurrent.com

Relative Site Performance

Now that you have some idea of what traffic your competitors are receiving, how people are finding them, and what they're spending for traffic, you should see how well they're performing. While you can't look at their internal operations, you can compare their performance and availability to yourself and to industry benchmarks.

Synthetic testing companies like Keynote and Gomez publish scorecards of web performance (for example, *http://scorecards.keynote.com* and *http://benchmarks.gomez .com*) at regular intervals that range from simple latency comparisons to detailed multifactor studies.

While these reports give you a good indication of what "normal" is, they're less useful for a narrow market segment. If you're not part of a benchmark index, you can still get a sense of your competitors' performance.

If you simply want to measure a competitor's site, using a browser plug-in like Firebug (*http://getfirebug.com/*) can be enough to analyze page size and load times (see Figure 3-14). This has the added benefit of showing which analytics and monitoring plugins your competitors are using.

If you want to compare your performance to others' over time, you may want to set up your own tests of their site using a synthetic testing service. You may even set up a transactional benchmark that can show the difference in performance across similar workflows.

Don't just consider the latency of individual pages when comparing yourself to others. A competitor may have an enrollment process that happens in only three steps instead of your five-step process, so consider the overall performance of the task.

Be careful when testing competitors' websites, though: some may have terms of service that prohibit you from testing them. You may also want to run your tests from a location or an IP address that isn't linked back to your organization. Not that we'd ever condone such a thing.

What to watch: Industry benchmarks; competitors' page load times; synthetic tests of competitors where feasible. Within these, track site availability, response time, and consistency (whether the site has the same response time from various locations or at various times of the day).

Competitor Activity

No discussion of competitive monitoring would be complete without discussing alerts and search engines. You can use Google Alerts to be notified whenever specific keywords, such as a competitor's brand name, appear on the Web. Additionally, a number of software tools will crawl competitors' websites and flag changes.

What to watch: Alerts for competitor names and key executives online; changes to competitors' pages with business impact such as pricing information, financing, media materials, screenshots, and executive teams.

Where Are My Risks?

Any online presence carries risks. As soon as you engage your market via the Web, aggrieved customers and anonymous detractors can attack you publicly. You may also expose yourself to legal liability and have to monitor your website for abusive content left by others.

Trolling and Spamming

On the Web, everyone's got an opinion—and you probably don't agree with all of them. Any website that offers comment fields, collaboration, and content sharing will become a target for two main groups of mischief-makers: spammers and trolls.

Spammers want to pollute your site with irrelevant content and links to sites. They want to generate inbound links to their sites from as many places as possible in an effort to raise their site's rankings or influence search engines. This is only getting worse.

According to Akismet Wordpress's stats found at *http://www.akismet.com/stats*, SPAM comments have been on a steady rise since they started tracking spammy entries in early 2006.

To combat this, many search engines' web crawlers ignore any links that have a specific *nofollow* tag in them, and blogs that allow commenters to add links routinely mark them with this tag. Nevertheless, blog comment spam is a major source of activity on sites; not only does it need to be blocked, but it must be accounted for in web analytics, since spammers' scripts don't help the business, but they may count as visits.

 In early 2005, Google developed the nofollow tag for the rel attribute of HTML link and anchor elements. Google does not consider links with the nofollow tag for the purposes of PageRank (*http://en.wikipedia.org/wiki/Spamdexing*).

Trolls are different beasts entirely. Wikipedia describes trolling as a "deliberate violation of the implicit rules of Internet social spaces," and defines trolls as people who are "deliberately inflammatory on the Internet in order to provoke a vehement response from other users." Since Wikipedia is one of the largest community sites on the Internet, it has its own special troll problem, and a page devoted to Wikipedia trolls (*http://meta.wikimedia.org/wiki/What_is_a_troll%3F*).

While common wisdom says to ignore the activity of trolls and to block spammers, you should still care about them for several reasons:

- They make the site less appealing for legitimate users.
- They consume resources, such as computing, bandwidth, and storage.
- If your site contains spammy content, search engines may consider it less relevant, and your search rankings will drop.
- You may be liable for harmful, offensive, or copyrighted content others post on your site.

Your antispam software will probably provide reports on the volume of spam it has blocked, as shown in Figure 3-15.

Dealing with spammers and trolls is the job of a community manager. In systems that require visitors to log in before posting, spam is easier to control, since the majority of spam comes from automated scripts run on a hijacked machine rather than from users with validated accounts.

How do you detect spammers and trolls? One way, employed by most antispam tools, is to examine the content they leave, which may contain an excessive number of links or specific keywords. A second approach is to look at their behavior. Spammers and trolls may comment on many posts with similar content, move quickly between topic

Figure 3-15. Spam messages flagged by the WordPress Akismet plug-in

areas, or befriend many community members without having that friendship reciprocated.

A far more effective approach is to harness the power of the community itself. Sites like Craigslist invite visitors to flag inappropriate content so that editors and community managers can intervene. You can capture the rate of flagged content as a metric of how much troll and spammer activity your website is experiencing.

Since every site has slightly different metrics and interaction models, you will likely have to work with your engineering team to build tools that track these unwanted visitors.

What to watch: Number of users that exhibit unwanted behaviors; percent of spammy comments; traffic sources that generate spam; volume of community flags.

Copyright and Legal Liability

If your site lets users post content, you may have to take steps to ensure that this content isn't subject to copyright from other organizations. Best practices today are to ask users to confirm that they are legally permitted to post the content, and to provide links for someone to report illegal content.

It's hard to say what's acceptable in a quickly changing online world. Services like Gracenote and MusicDNS can recognize copies of music, regardless of format, and help license holders detect and enforce copyright. Yet the Center for Social Media (*http://www.centerforsocialmedia.org/files/pdf/CSM_Recut_Reframe_Recycle_report .pdf*) points out that "a substantial amount of user-generated video uses copyrighted material in ways that are eligible for fair use consideration, although no coordinated work has yet been done to understand such practices through the fair use lens."

So as a web operator, you need to track the content users upload and quickly review, and possibly remove, content that has been flagged by the user community.

What to watch: Users who upload significantly more content than others; most popular content; content that has been flagged for review.

Fraud, Privacy, and Account Sharing

If your site contains personally identifiable information, worry about privacy. Safeguarding your visitors' data is more than just good practice—in much of the world, it's a legal obligation.

As with other site-specific questions, you will probably need to work with the development team or your security experts to flag breaches in privacy or cases of fraud. The best thing you can do is to be sure you've got plenty of detailed logfiles on hand that can be searched when problems arise.

One type of fraud that web operators should watch directly, however, is *account sharing*. Many applications—particularly SaaS and paid services—are priced per seat. Subscribers may be tempted to share their accounts, particularly with "utility" services such as online storage, real estate listing services, or analyst firm portals.

We've seen one case of a user who shared his paid account to an analyst's website with his development team, in violation of the site's terms of service. The user carefully coordinated his logins so employees never used the account at the same time. The fraudulent use of the account was only discovered when a web administrator noticed that the user seemed to be traveling from Sunnyvale to Mumbai and back every day.

Pinpointing account fraud can be challenging, but there are some giveaways:

- Accounts that log in while a user is already logged in (concurrent use).
- Accounts that log in from several geographic regions relatively quickly.
- A high variety of browser user agents associated with one account.

- Users who log in despite being terminated from the company that purchased the account.

To track down violators, generate reports that identify these accounts, and provide this data to sales or support teams who can contact offenders, offer to upsell their accounts, or even demand additional payment for truly egregious violations.

What to watch: Number of concurrent-use logins per account; number of states from which a user has logged in; number of different user agents seen for an account.

What Are People Saying About Me?

On an increasingly social web, your marketing has migrated beyond your website into chat rooms, user groups, social networking applications, news aggregators, and blogs. To properly understand what the Internet thinks of you, you need to watch the Web beyond your front door.

Your primary tool for this task is search. Hundreds of automated scripts crawl the Web constantly, indexing what they find. And most community sites offer some form of internal content search these days. A variant on traditional search, known as *persistent search* or *prospective search*, combs the Web for new content that matches search terms, then informs you of it.

Google Alerts dominates the prospective search market, with some other services, such as Rollyo, using competing search engines like Yahoo!. HubSpot offers integrated marketing to small businesses by combining topical searches, lead tracking, and similar functions, as shown in Figure 3-16. There are also many community listening platforms—Radian6, Techrigy, ScoutLabs, Sysomos, Keenkong and so on—that help community managers monitor their online buzz.

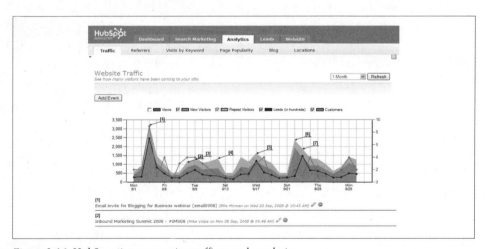

Figure 3-16. HubSpot ties community traffic to web analytics

In all of these models, you subscribe to a keyword across various types of sites (blogs, mailing lists, news aggregators, etc.), then review the results wherever someone is talking about things that matter to your organization.

Site Reputation

In the era of search, nothing matters more than what Google thinks of you. As Figure 3-17 shows, if you're the best result for a particular search term, you don't even pay for your advertising.

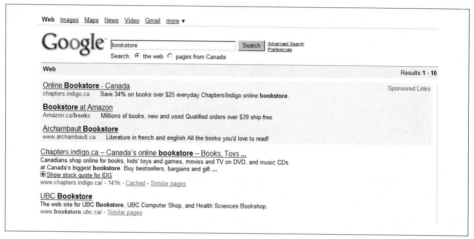

Figure 3-17. Google's ranking of Chapters as a bookstore means the company doesn't have to pay for advertising the way Amazon does in this search

Google's PageRank is a measure of how relevant and significant Google thinks your website is, and encompasses factors such as the number of inbound links to the site and the content on the site itself. The PageRank algorithm Coca-Cola is closely guarded and constantly evolving to stay ahead of unscrupulous site promoters.

Other sites, such as Technorati, use their own approaches, based on similar factors such as the number of inbound links in the past six months.

What to watch: Google PageRank; Technorati ranking; StumbleUpon rating; other Internet ranking tools.

Trends

Google Trends and Yahoo! Buzz show the popularity of search terms, and Google Insights, shown in Figure 3-18, breaks them down over time. If you want to understand the relative popularity of content on the Internet in order to optimize the wording of your site or to downplay aging themes, these tools are useful for more than just competitive analysis.

Figure 3-18. Google Insights, showing spread and growth of searches for "web analytics" worldwide

What to watch: Mentions of yourself and your competitors over time; product names; subject matter you cover.

Social Network Activity

Many social networks have a built-in search you can use to see whether your company, site, or products are being discussed, as shown in Figure 3-19.

What to watch: Search results for your company name, URL, product names, executives, and relevant keywords across social sites like Digg, Summize, and Twitter, as well as any that are relevant to your particular industry or domain.

How Are My Site and Content Being Used Elsewhere?

Other people are using your site, and you don't know it. They may be doing so as part of a mashup. They may be running search engine crawlers to index your content. Or they may be competitors checking up on you. Whatever the case, you need to track and monitor them.

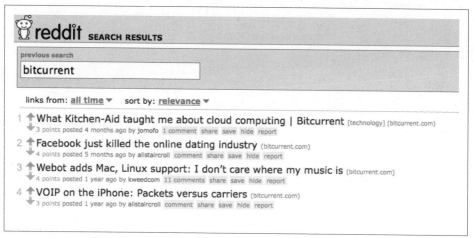

Figure 3-19. A search for a specific topic using reddit's internal search

API Access and Usage

Your site may offer formal web services or Application Programming Interfaces (APIs) to let your users access your application programmatically through automated scripts. Most sites have at least one trivially simple web service—the RSS feed, which is simply an XML file retrieved via HTTP.

Most real web services are more "heavyweight" than syndication feeds, but even the most basic web services need to be monitored. If letting people extend your web application with their own code is important to your business, monitor the APIs to ensure they are reliable and that people are using them in appropriate ways.

Some RSS management tools, like FeedBurner, will showcase "unusual" uses of your feed automatically—for example, someone who is pulling an RSS feed into Yahoo! Pipes for postprocessing, as shown in Figure 3-20.

If your web services include terms of service that limit how much someone can use them, track offending users before a greedy third-party developer breaks your site. Note that many analytics packages that rely on JavaScript can't be used to track API calls because there's no way to reliably embed JavaScript in the content they deliver.

What to watch: Traffic volume and number of requests for each API you offer; number of failed authentications to the API; number of API requests by developer; top URLs by traffic and request volume.

Mashups, Stolen Content, and Illegal Syndication

Your site's data can easily appear online in a mashup. By combining several sites and services, web users can create a new application, often without the original sites

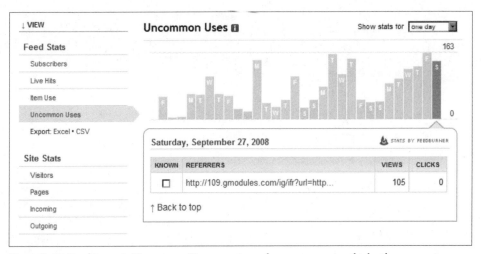

Figure 3-20. FeedBurner's Uncommon Uses report can show you ways in which others are using your RSS feed

knowing it. Tools like Yahoo! Pipes or Microsoft's Popfly make it easy to assemble several services without any programming knowledge.

Some mashups are well intentioned, even encouraged. Google makes it easy for developers to embed and augment their maps in third-party sites. Other repurposing of content, such as republishing blog posts or embedding photos from third-party sites, may not be so innocent. If someone else reposts your content, your search ranking goes down, and if someone else embeds links to media you're hosting, you pay for the bandwidth.

If this is happening to you, you'll see referring URLs belonging to the mashup page, and you can track back to that URL to determine where the traffic is coming from and take action if needed. You may have to look in server logs or on a sniffer, because if the mashup is pulling in a component like a video or an image, you won't see any sign of it in JavaScript-based monitoring.

 Although the term "Sniffer" is a registered trademark of Network General Corporation, it is also used by many networking professionals to refer to packet capture devices in general.

Try to treats mashups as business opportunities, not threats. If you have interesting content, find a way to deliver it that benefits both you and the mashup site.

What to watch: URLs with a high volume of requests for an object on your site without the normal entry path; search results containing your unique content.

Integration with Legacy Systems

Some SaaS applications may connect to their subscribers' enterprise software through dedicated links between the subscriber's data center and the SaaS application in order to exchange customer, employee, and financial data. Relying on these kinds of third-party services can affect the performance and availability of your website. Travel search sites, for example, are highly dependent on airline and hotel booking systems for their search results.

While not directly related to web monitoring, the performance of these third-party connections must be tested and reported as part of an SLA. Excessive use of enterprise APIs or long delays from third-party services may degrade the performance of the website. Keep nonweb activity in mind when monitoring the web-facing side of the business, and find ways to track private API calls to enterprise clients.

What to watch: Volume and performance of API calls between the application and enterprise customers or data partners.

The Tools at Our Disposal

Clearly, there are many metrics and KPIs to gather in order to understand and improve your online presence. Fortunately, there's a wide range of tools available for watching yourself online. The trick is to use the right technologies to collect the metrics that matter the most to your business.

At the broadest level, there are three categories of monitoring technology you can use to understand web activity. You can *collect* what users do from various points in the web connection; you can use *search* engines that crawl and index the web, and may send you alerts for changes or keywords; and you can run scripts that *test* your site directly.

Collection Tools

There are many ways to collect visitor information, depending on how much access you have to your servers, the features of those visitors' browsers, and the kind of data you wish to collect.

Collection can happen on your own machines, on intermediate devices that collect a copy of traffic, through a browser toolbar, or on the visitor's browser. It may also happen with third-party services like FeedBurner (for RSS feeds) or Mashery (for APIs and web services) that proxy your content or manage your APIs.

The volume of data collected in these ways grows proportionally with traffic. Collecting data may slow down your servers, and may also pose privacy risks. You also need to consider what various collection methods can see, since they all have different perspectives. An inline sniffer can't see client-side page load time, for example; similarly, client-side JavaScript can't see server errors.

Search Systems

Another way to monitor your online presence is through the use of search engines that run scripts—called *crawlers*—that visit web pages and follow links on sites to collect and index the data they find.

There are hundreds of these crawlers on the Web. Some of them feed search giants like Google, Yahoo!, and MSN. More specialized crawlers also look for security problems, copyright violations, plagiarism, contact information, archiving, and so on.

Crawlers can't index the entire Web. Many websites are closed to crawlers, either because they require a login, because of their dynamic nature, or because they've blocked crawlers from indexing some of the content. This is common for news sites and blog comment threads. As a result, you need to use site-specific internal search tools alongside global search engines for complete coverage of web activity.

While we may run searches to see what's happening online, a more practical way to manage many search queries is to set up alerts when certain keywords arise or when specific pages change.

Testing Services

In addition to collecting visitor data and setting up searches and alerts, you can run tests against websites to measure their health or check their content. These tests can simulate specific browsers or run from several geographic regions to pinpoint problems with a specific technology or location.

Testing services can also watch your competitors or monitor third-party sites, like payment or mapping, on which your main application depends. There's a wide range of such services available, from open source, roll-your-own scripting to global testing networks that can verify a site's health from almost anywhere and run complex, multistep transactions.

Ultimately, the many metrics we've looked at above, using the collection, search, and testing approaches outlined here, give us complete web visibility. That visibility amounts to four big questions, which we'll look at next.

CHAPTER 4

The Four Big Questions

As you saw in Chapter 3, there's a dauntingly long list of metrics you can track about your website. All of those metrics answer four fundamental questions about your site's visitors:

- What did they do?
- How did they do it?
- Why did they do it?
- Could they do it?

Without the answers to these four questions, you can't tell where to focus your efforts. You don't know what's working well or what's hopelessly broken, and you can't tell how to improve your business. Most companies that can't answer these questions try new strategies randomly, hoping to hit on one that works.

Armed with the answers to the four big questions, you can make informed decisions about where to focus your efforts. Would better design or better infrastructure improve your conversion rates? Are your competitors winning against you before prospective customers ever have a chance to see your offers? Do visitors prefer to search or browse? Are buyers leaving because they didn't like your prices, or because they came to the wrong site altogether? Are your biggest subscriber's users' complaints your fault—or theirs?

These are the kinds of decisions that make or break a business. To make the right decisions, you need to answer the four big questions.

What Did They Do?

The first question concerns what your visitors did, and it's answered through web analytics.

Visitors' actions speak for themselves. Analytics shows you what worked best, but won't tell you *why* something worked. The only way analytics can help improve your site is by showing you which content, messages, designs, and campaigns have the best impact on your business. This is because analytics lacks context—it won't show you what was on a visitor's mind, or whether the site was fast during her visit, or how easy she found it to use.

Analytics was once relatively simple because web transactions were simple. Three things have changed in recent years that complicate matters, however:

Visitor interactions aren't just requests for pages
> Gone are the days of straightforward page-by-page interaction; instead, visitors stay on a single page, but interact with page components through DHTML, JavaScript, or plug-ins.

Visitor impressions start long before users visit your site
> To get a complete picture of a visit, you need to know what people are saying about you elsewhere that led a visitor to your site and set the tone for his visit.

Visits don't follow a set path
> Instead of browsing a predictable sequence of pages to arrive at a result, visitors explore in a haphazard fashion, often relying on searches or recommendations to move through a site.

While the basic building block of analytics is an individual visit, analysts seldom look at web activity with this much granularity. Instead, they look at aggregate analysis—metrics grouped by geography, demographics, campaigns, or other segments. You'll only look at individual visits when you're trying to reproduce a problem or resolve a dispute. Web analytics is more focused on large-scale patterns of interaction.

How Did They Do It?

The second question is all about usability. Most modern websites let users accomplish their goals in a number of ways. Visitors might search for a book or browse by a list of authors. They might click on text hyperlinks or the images next to them. They might take a circuitous route through the application because they didn't see the big button just out of sight on their screens. They might abandon a form halfway through a page. Or they might enter a zip code in the price field, then wonder why they can't complete their transaction.

Usability is a mixture of science and art. Web designers want their work to look fresh and innovative, but need to balance their desire for cutting-edge design with a focus on familiar affordances and button conventions that visitors understand.

Perhaps because of designers' eagerness to introduce fresh sites, they've given users many more interfaces to learn, making the Web less consistent even as its citizens become more savvy.

A profusion of browsers
> Opera, Chrome, Internet Explorer, Firefox, Flock, Camino, Safari, and others all render pages slightly differently. Sometimes the differences can be subtle; other

times, entire ads may not show up, JavaScript may not execute, or functionality may be severely impaired. You may not even be aware of these limitations.

A variety of devices

Notebooks, desktops, mobile phones, and PDAs offer different controls. Some have touchscreens and new gestures, while others have numeric keypads. They all display different resolutions. We use them in many different surroundings—in a car, in bright light, or in a noisy room.

New interaction metaphors

Drag-and-drop, wheelmouse-zoom, modal displays, and so on, are now possible with Flash, Java, and JavaScript.

Usability is most important when it affects goal attainment. It's the domain of Web Interaction Analytics (WIA).

 The term "WIA" was coined by ClickTale, and is also variously referred to as in-page or behavioral analytics.

Why Did They Do It?

The third big question looks at consumer motivation. It's a far softer science than analytics. While analytics might offer tantalizing clues to why visitors tried to do something on your site, you're still left guessing. The only way to be really certain is to ask visitors directly.

On the Web, answering the question of "why" is done through VOC services—surveys that solicit visitor opinions, either within the site itself or through a third-party survey. Respondents are recruited (by mail, for example) or intercepted as they visit the site. They are then asked a series of questions.

With the emergence of online communities, website analysts have new ways of understanding consumers. Web marketing is much more of a two-way conversation than print or broadcast media. As a result, analysts can talk directly with their market and understand its motivations, or analyze its sentiments by mining what communities are discussing.

Could They Do It?

There's no point in asking what users did if they couldn't do it in the first place. Answering the fourth question means measuring the health, performance, and availability of the application. As with web analytics, this task has become more complicated in recent years for several reasons.

An application is no longer made from a single component

Your application may rely on plug-ins and browser capabilities. It may pull some parts of a page from a Content Delivery Network (CDN). It's dependent on stylesheets and JavaScript functioning correctly. It may be delivered to a mobile device. And it may incorporate third-party content such as maps. Even within a single web server there may be various operating systems, services, and virtual machines that make it hard to properly instrument the application.

You're present on sites you don't control

You may have a Facebook group or Twitter profile that drives users to your main site. Your site may incorporate third-party SaaS elements: a helpdesk and FAQ service, a portal for job seekers, and an investor information page with stock prices. Even if your own site works well, your users' experience depends just as much on these external content sources.

Some of your visitors are machines

If you have any kind of API—even an RSS feed—then you need to worry about whether scripts and other websites can correctly access your application. Even if you're just being indexed by search engines, you need to help those engines navigate your site.

Measuring site health from the end user's point of view is the discipline of EUEM, which is the logical complement to web analytics. It's often measured in terms of performance and availability, from both individual and aggregate perspectives.

There are two major approaches to capturing user experience: *synthetic testing*, which simulates user visits to a site, and *real user monitoring* (RUM), which watches actual users on the site itself. Together, they provide an accurate picture of whether visitors could do what they tried to do on the site.

RUM data may be collected through many of the same mechanisms as WIA, and some analytics or RUM products offer WIA functionality, such as replay and click overlays.

Putting It All Together

Answer these four big questions, and you're well on your way to improving your website, because you have a comprehensive understanding of how all its pieces fit together. For example:

- If user experience is suffering, you can increase capacity, reduce page size and complexity, deliver content via CDN, replace components that are too slow, and so on.

- If you're failing to meet agreed-upon service level targets, you can mitigate SLA arguments and reduce subscriber churn by contacting your users to tell them you know they're having problems, and that you're going to improve things.

- If your conversion rates aren't high enough, you can try to better understand what users want or where they get stuck.

- If visitors take too long to complete transactions, or don't scroll down to key information, you can make the site more usable so they can easily find what they're looking for.

- If you're converting the visitors that make it to your site, but aren't getting enough traffic, you can focus on communities and word-of-mouth marketing rather than site design.

Ultimately, by answering the four big questions you close the loop between designing and deploying a website, and adapting quickly. You make mistakes faster and find the right way to connect with your market.

Analyzing Data Properly

Before we look at how to answer these four questions in more detail, we need to talk about data analysis. Many of the monitoring technologies we'll cover rely on statistics to analyze millions of pieces of data about thousands of visitors quickly and easily. If you don't look at that data with an analytical eye, you can easily be misled.

Web analytics, EUEM, VOC, and WIA provide a tremendous amount of raw information about your website. You need to analyze and communicate it properly if it's going to have an impact on your organization. That means comparing metrics to other things, segmenting measurements into useful groups, and using the right math for the job.

Always Compare

Data analysis is all about comparisons. You should always talk about data with words that end in "er": better, faster, stickier, heavier, weaker, later. You can compare groups of visitors, periods of time, or versions of content. You can also compare yourself against your competitors.

To compare, you need a sense of what "normal" is—a baseline or control group. The "always compare" rule means your first job will be to establish baselines for key metrics like conversion rate, visitor traffic, web page performance, visitor satisfaction, email inquiries, and call center volume. Only then can you make useful comparisons in future reports.

Segment Everything

Whether you're trying to judge the effectiveness of a marketing campaign, the cause of a problem, the usability of a page, or the importance of customer feedback, you need to segment your measurements into several groups.

Grouping data into manageable, meaningful clumps is essential. As humans, we naturally try to cluster data into segments. The challenge is in knowing which of the thousands of possible segments is most likely to yield the right data.

Sometimes, the data will segment itself in obvious ways—by geography, browser type, referring site, carrier, and so on. Other times, you'll have to create your own segments along which to analyze the measurements you collect. No matter what you're trying to do, having several segments to compare shows you which is best and which is worst. From there, you can start to fix things.

Segmentation applies everywhere. If you're trying to resolve a performance problem, your first question will be whether there is a particular segment for which the problem is more common: is it offshore visitors? Are all the affected visitors on the same broadband carrier? Is it always the same page, or the same server? Similarly, if users aren't seeing part of the page, is the problem related to age groups? Browser types? Screen resolutions?

Don't Settle for Averages

British Prime Minister Benjamin Disraeli once said, "There are three kinds of lies: lies, damned lies, and statistics" (or perhaps he didn't, though Mark Twain attributed this famous quote to him). The *really* compulsive liars, however, are averages. Averages are misleading, often concealing important information.

Consider, for example, a class of 20 children who are 5 years old. The average age in the room is five. When a 90-year-old grandparent comes to visit, the average age climbs to 9. This is misleading: we'd prefer to know that the average age is five, with one outlier, or that the most common age in the room is five.

For meaningful measurements, insist on histograms (frequency distributions) and percentiles. A histogram is simply a chart of how many times something happened. Figure 4-1 shows the age histogram for the classroom we just visited.

Histograms are particularly useful for analyzing quantitative information (like page load time), as they show us at a glance how many measurements fall outside a certain range or *percentile* of the samples we collected.

When we talk about "the 95th percentile of latency," we mean the delay that 95 percent of visitors experienced. Figure 4-2 shows the distribution of performance across a day's page requests, alongside the average latency and the 95th percentile latency.

Percentiles are good because they let us quantify the impact of a problem. If the 95th percentile page latency suddenly jumps to 16 seconds, as it has in this example, we now know that we have a serious problem affecting a broad swath of visitors (five percent of them, in fact). What do those page requests have in common? Are visitors all asking for the same page? Are they all from the same city? Are requests only from new visitors?

This is where diagnosis begins—by identifying an issue and segmenting to find out what's common across all those who experienced the issue.

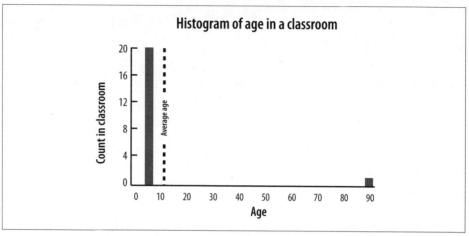

Figure 4-1. A histogram displaying age in a classroom paints a more accurate picture than an average would

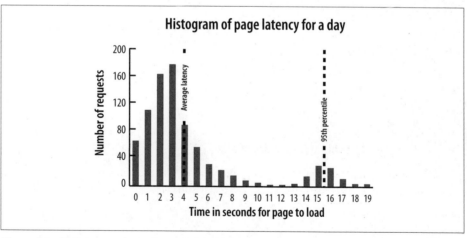

Figure 4-2. A histogram of page latency

A Complete Web Monitoring Maturity Model

As organizations improve their web visibility, they transition through several stages of maturity. You won't achieve the same maturity across all four of the big questions at once. You may already have mature web analytics, for example, but only primitive insights into web performance.

Level 1: Technical Details

At the first level of maturity, monitoring is done for technical reasons. The organization hasn't yet realized that the web application is important to its business strategy, and results are used only to ensure that systems are functioning and to detect egregious errors. If you're at this level, you're probably asking questions like:

- How many hits does my site get each day? Is that number growing?
- What percentage of HTTP GETs to the site receive an HTTP 200 OK response?
- How slowly do my servers respond to the tests I'm running?
- Does my website work on different web browsers?

There's nothing fundamentally wrong with these questions, but on their own, they don't relate to the business because they don't relate to a particular business objective.

Consider, for example, a server that gets a thousand requests. That traffic could all be coming from one energetic user, or it could be a thousand one-page visitors who didn't find what they wanted. Neither is good. Or consider a server that suddenly responds slowly to a test. Unless you know how many users were on your site and what they were doing, you don't know whether the slowdown affected your business.

Level 2: Minding Your Own House

As an organization engages its customers, it first focuses on the things it can control. This leads to questions about the *effectiveness* of the website, such as:

- Does the website convert visitors?
- Is it fast and available when people need it?
- Why do visitors come to the site?
- What are the most common ways people use the site, and which of those lead to good outcomes (goals and conversions)?

These are the key metrics that organizations need to use to make sure the site can do what it needs to: convert visitors. They're a great start.

Because they don't look beyond the borders of the website to the Internet as a whole, however, organizations at this level are still inward-facing and somewhat myopic. They also fail to consider long-term metrics like the lifetime value of a visitor, loyalty, and capacity trends.

Level 3: Engaging the Internet

With your own house in order, you can now turn your attention to the rest of the Web. The main focus here is driving traffic to your site through marketing campaigns, paid search, and organic search. The site engages visitors in other ways, too, forcing you to ask new questions:

- How are people finding the site, and what's working best?
- Are the other components of the Internet, such as online communities, APIs, web services, and CDNs, functioning properly?
- What's the Internet saying about the organization, and how is that affecting traffic rates?
- Are different segments of the Internet showing different levels of usability? Are foreign-language browsers having to scroll down more? Are older visitors with larger font sizes seeing less of certain pages?
- Are there trolling and spamming problems?

To deal with the flood of measurements generated by additional visitors, the organization starts to analyze data differently. For example, it uses software to automatically baseline what "normal" is, pinpoint exceptions, and predict seasonal or growth trends.

Level 4: Building Relationships

The next stage of maturity turns visits into long-term relationships. In the earlier stages of the maturity model, each visit was treated as a unique interaction. As your organization matures, these interactions are stitched together into a user record, and analytics starts to converge with CRM.

The central focus is now a visitor's lifetime relationship with your site. You learn how often he visits the site, what triggers those visits, and what outcomes happen.

Interactions happen beyond the website. Customer loyalty cards show store visits, calls to the call center are tracked by phone number, and so on. The relationship becomes a two-way engagement in which you reach out to visitors who have opted in to mailing lists or who subscribe to RSS feeds.

A relationship-focused organization cares about longevity, loyalty, and capturing the most complete view of its visitors:

- How often do visitors return, and what's their lifetime value to the organization?
- How steep is the learning curve, and how long does it take a new visitor to use 80 percent of the site's features?
- How do word of mouth and the community affect loyalty?
- Which modes of interaction (web, online community, mail, phone, in-person) are most common, and what causes visitors to switch from one to another?
- How many visits does it take before a reader comments?
- How willing are my customers to have me communicate with them?

- What's my reach (ability to get to customers) and how likely are they to act on what I send them?
- What performance or uptime expectations—or implied SLA—do my I have with my power users?

Stitching together individual visits, both offline and online, is a major challenge for any organization. If you're at this stage, you need to handle personal information carefully so you don't run afoul of privacy laws or put your organization at risk.

Level 5: Web Business Strategy

The final stage of maturity occurs when the business makes the Web an integral part of its product and marketing strategies. In the earlier levels of the maturity model, the Web is an "online branch office"—it often has its own profit and loss, and is run like a separate business.

But the Web touches every part of our lives. It's on our phones and in our cars. It's quickly becoming the dominant medium for music, video, and personal messaging. It's how employees work remotely. It's many people's primary source of day-to-day knowledge. The Web's effects reach far beyond a storefront, affecting everything from how companies develop new products to how they support existing ones.

When companies finally embrace the Web as the disruptive force it is, they stop becoming reactive and start becoming opportunistic. Instead of reacting to what web monitoring tells them, they start to wonder whether the Web makes new things possible. When you reach this stage of maturity, you'll be asking questions like:

- Can I develop and roll out products more quickly, and with greater accuracy?
- Can I reduce support costs and make customers self-supporting?
- Can I tie my services to real-world components (for example, linking my CRM portal to a GPS so salespeople can optimize their travel)?
- Can I move into other media or gain presence elsewhere?
- How does the Web remove barriers to entry in my industry? How does it let me erect barriers to new competitors?
- Can I engage target markets in product design and research?

One clear sign that a company views the Web as strategic is that it starts to include Web metrics in employee performance reviews and corporate goal-setting. The Web isn't a side business anymore: it *is* the business.

The Watching Websites Maturity Model

Table 4-1 shows the various levels of maturity. We'll return to the table as we look at how to answer the four big questions.

Table 4-1. *Stages of watching websites*

Component		Maturity Level 1	Level 2	Level 3	Level 4	Level 5
Focus		*Technology: Make sure things are alive.*	*Local site: Make sure people on my site do what I want them to.*	*Visitor acquisition: Make sure the Internet sends people to my site.*	*Systematic engagement: Make sure my relationship with my visitors and the Internet continues to grow.*	*Web strategy: Make sure my business is aligned with the Internet age.*
Who?		Operations	Merchandising manager	Campaign manager/SEO	Product manager	CEO/GM
Web analytics		Technical details: Page view, hits. Focus on operation of the infrastructure, capacity, usage.	Conversion: How many of the visitors complete the goals I intend?	Traffic: How does the Internet learn about me, encourage visits to me?	Relationship: How often do buyers return? What's the lifetime value? Where else do customers interact with me?	Strategy: How can I combine my brick-and-mortar and web businesses? How does the web change my company?
	Synthetic	Availability and performance: Checking to see if the site is available from multiple locations, and reporting on performance.	Transactions and components: Multi-step monitoring of key processes, tests to isolate tiers of infrastructure.	Testing the Internet: Monitoring of third-party components and communities on which the application depends.	Correlation & competition: Using the relationship between load and performance; comparing yourself to the industry and public benchmarks.	Organizational planning: Using performance as the basis for procurement; uptime objectives at the executive level; quantifying outages or slow-downs financially.
EUEM	**RUM**	Errors and performance: Counting hard errors (404, 500, TCP reset) and end user performance grouped by customer or visitor segment.	Analytics integration: Tying user experience to business outcomes within the site to maximize conversions; identifying "soft errors" in transactions.	All components, all functions, automation: Watching content from third-party sites and user actions within a page; automatically forming baselines and diagnosing exceptions.	SLA, CRM: Using RUM information as the basis for SLAs, problem resolution, and release management.	Integrated user engagement: Measuring user experience across call centers, in-person encounters, web transactions, etc., as a single entity; impact of user experience is quantifiable.

Component	Maturity Level 1	Level 2	Level 3	Level 4	Level 5
VOC	"Contact us" buttons and on-site feedback; emphasis on satisfaction.	Surveys within the site via opt-in invitations; emphasis on loyalty.	Engaging the public Internet (chat-rooms, social sites, etc.) and analyzing key topics and discussions; emphasis on word-of-mouth and virality.	Customer collaboration in product and service design; user engagement; emphasis on lifetime value creation, giving the user a sense of ownership.	Consumer feedback tied in to corporate planning through quantitative analysis of VOC and community data; customer as a collaborator in the growth of the company.
WIA	Click diagrams showing "hot" areas on key pages.	Segmentation of user actions (scroll, drag, click) by outcome (purchase, abandonment, enrollment).	Segmentation by traffic source (organic search, campaign) and A/B comparison; visitor replay.	Learning curve analysis; comparison of first-time versus experienced users; automated A/B testing of usability.	Product specialization according to usability and user groups; usability as a component of employee performance.

Web Analytics, Usability, and the Voice of the Customer

Now that we've seen why you need complete web monitoring, it's time to understand what visitors are doing on your website. This is the domain of web analytics, supported by other tools that measure how people interact with the site's interfaces. It's not enough to know what people did on your site, though. You also need to know why they did it, and this is the realm of Voice of the Customer surveys. Part II contains the following chapters:

- Chapter 5, *What Did They Do?: Web Analytics*
- Chapter 6, *How Did They Do It?: Monitoring Web Usability*
- Chapter 7, *Why Did They Do It?: Voice of the Customer*

What Did They Do?: Web Analytics

Think of two web pages you've created in the past. Which of them will work best?

Before you answer, you should know that you're not entitled to an opinion. The only people who *are* entitled to one are your visitors, and they convey their opinions through what they do on your site. If they do the things you want them to, in increasingly large numbers, that's a good sign; if they don't, that's a bad one.

Analytics is that simple. Everything else is just details.

Before you go further in this chapter, we want you to try an exercise. Find a volunteer, and ask him or her to think of something he or she wants. Then draw a picture on a piece of paper, and get him or her to reply "warmer" or "colder" depending on how close the thing you drew was to the thing he or she was thinking. Repeat until you know what they're thinking of.

Go ahead. We'll wait.

How long did it take for you to figure out what he or she wanted? How many iterations did you go through?

Notice that you had plenty of creative input into the process: you came up with the ideas of what might work, and you deduced what your volunteer wanted through small "improvements" to your picture based on their feedback.

The same process takes place as you optimize a website, with some important differences:

- You're listening to hundreds of visitors rather than a single volunteer, so you have to measure things in the aggregate.
- What you define as "warmer" or "colder" will depend on your business model.
- If you know your audience well, or have done your research, your initial picture will be close to what your web audience wants.
- You can ask visitors what they think they want, through surveys, and they may even tell you.

Your visitors are telling you what they want. You just have to know how to listen by watching what they do on your site, and react accordingly. Web analytics is how you listen. Use it wisely, and eventually the website you present to visitors will be the one they were craving.

For a long time, web traffic analysis was the domain of technologists. Marketers were slow to recognize the power of tracking users and tying online activity to business outcomes. Web activity was mainly used for error detection and capacity planning.

The growth of the Web as a mainstream channel for commerce, and the emergence of hosted analytics services—from Omniture, Urchin, WebTrends, CoreMetrics, and others—finally convinced marketing teams to get involved. Today, free services like Google Analytics mean that everyone with a website can get an idea of what visitors did.

There's no excuse not to listen.

Dealing with Popularity and Distance

Imagine that you own a small store in the country. You don't need much market analysis. You can see what people are doing. You know their individual buying preferences, their names, and their browsing habits. You can stock their favorite items, predicting what they'll buy with surprising accuracy.

Now suppose business picks up and the number of customers grows. You slowly lose the ability to keep track of it all. Clients want more merchandise, which in turn requires more floor space. You have to make compromises. The sheer volume of customers makes it impossible to recognize them all. You lose track of their buying habits. Your customers become anonymous, unrecognizable. You have to start dealing in patterns, trends, and segments. *You need to generalize.*

Websites face similar issues. The Web pushes both visitor anonymity and traffic volume to the extreme. You can't know your visitors—in fact, you can't even tell if they're male or female, or how old they are, the way you can in a store. You're also opening the floodgates, letting in millions of complete strangers without any limits on their geographic origins or the times of their visits.

We're going to provide a basic overview of analytics. Analytics is the cornerstone of a complete web monitoring strategy, and many other sources of monitoring—from performance monitoring, to customer surveys, to usability, to community management—all build upon it. Only by consolidating analytics data with these other information sources can you form a truly complete picture. Only through an integrated view can you return to the confidence of someone who's intimately familiar with her market and customers.

There are many excellent books and websites that deal with analytics in much greater depth than we'll attempt—in particular, with the need to focus on outcomes, to experiment, and to focus on data and facts rather than on opinions and intuition. We can recommend Jim Sterne's *Web Metrics: Proven Methods for Measuring Web Site Success* (Wiley), Avinash Kaushik's *Web Analytics: An Hour a Day* (Sybex), and Eric Peterson's *Web Analytics Demystified* (Celilo Group Media). You can find Avinash Kaushik's blog at *www.kaushik.net/avinash*.

The Core of Web Visibility

Web analytics is the piece of a web visibility strategy that's most tightly linked to the business outcome of your website. It captures your users' sessions, segments them in meaningful ways, and shows you how their visits contributed to your business. All of your other monitoring must tie back to it:

- Web performance and availability ensures that visitors can do what they want, when they want.
- Surveys ensure that you understand visitors' needs and hear their voices.
- Usability and interaction analysis measures how easily visitors can achieve the goals you've set for them.
- Community monitoring links what visitors do elsewhere to your site and your brand.

If you're not making decisions about your web presence based on what your analytics tells you, you're making bad decisions. If you're not augmenting analytics with other data, you're making decisions without all the facts.

A Quick History of Analytics

For as long as servers have existed, they've generated logfiles. Early on, these logs were just another source of diagnostic data for someone in IT. Each time a server handled a request, it wrote a single line of text to disk. This line contained only a few details about the request, and it followed the Common Log Format, or CLF. It included information about the user (where she connected from) and about the request (the date and time that it occurred, the request itself, the returned HTTP status code, and the byte length of the document or page transferred).

It was only in the mid-1990s that information such as user agents (the browser) and referrer (where a user came from) was added to logfiles. A slightly more detailed version of HTTP records, known as Extended Log Format (ELF), followed in early 1996. ELF added more client and server information.

ELF gave many companies their first glimpse of what was happening on their websites. Web logs were sparse, and focused on the technical side of the web server: which objects were requested, which clients requested them, when they were retrieved, and the HTTP status codes in response to those requests.

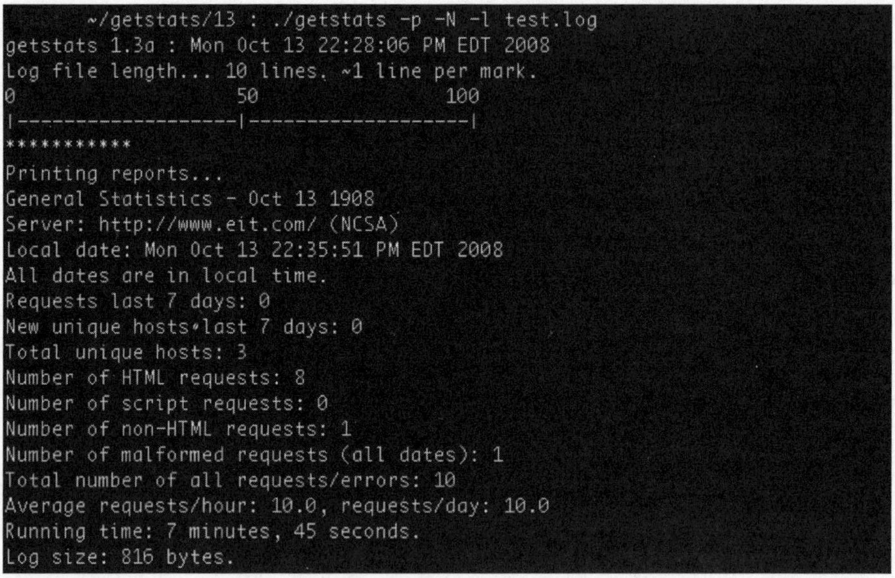

```
      ~/getstats/13 : ./getstats -p -N -l test.log
getstats 1.3a : Mon Oct 13 22:28:06 PM EDT 2008
Log file length... 10 lines, ~1 line per mark.
0               50              100
|-------------------|-------------------|
***********
Printing reports...
General Statistics - Oct 13 1908
Server: http://www.eit.com/ (NCSA)
Local date: Mon Oct 13 22:35:51 PM EDT 2008
All dates are in local time.
Requests last 7 days: 0
New unique hosts last 7 days: 0
Total unique hosts: 3
Number of HTML requests: 8
Number of script requests: 0
Number of non-HTML requests: 1
Number of malformed requests (all dates): 1
Total number of all requests/errors: 10
Average requests/hour: 10.0, requests/day: 10.0
Running time: 7 minutes, 45 seconds.
Log size: 816 bytes.
```

Figure 5-1. GetStats v1.3 output, coded in 1993

At first, web operators parsed these logfiles to find problems, searching for a specific error such as a "404 not found," which indicated a missing file. They quickly realized, however, that they also wanted aggregate data from the logs, such as how many requests the servers had handled that day.

So coders like Kevin Hughes developed applications like GetStats, shown in Figure 5-1, that would "crunch" logs and display the results in a more consumable format.

What Was the First Analytics Tool?

GetStats wasn't the first web server log analysis tool, but it was very influential in terms of the way the data was presented and summarized.

Roy Fielding with wwwstat was the first, as far as I can recall, to present statistics in an easy-to-read paragraph summary form that I think was written in Perl. I also took ideas from Thomas Boutell (wusage) and Eric Katz (WebReport).

—Kevin Hughes,
author of GetStats

Before web analytics became interesting to marketers, however, several things had to happen:

- The Web had to become *mainstream* enough for marketers to care, requiring both a large number of connected consumers and a rich visual experience within web browsers. Clearly, this has happened: there were only 38 million Internet users in 1994, but roughly 1.5 billion by January 2009—a 40-fold increase (*http://www .internetworldstats.com/stats.htm*).

- Analytics had to become *visitor-centric*. To be useful for business, logging had to move from individual requests for pages to user visits so that something a user did on page A could be linked to a purchase on page F. Cookies made this possible and added unprecedented accountability to promotional campaigns.

- Analysts needed ways to *segment* visitors so they could decide which browsers, campaigns, promotions, countries, or referring sites were producing the best business results, and optimize their websites accordingly. Better logging and access to browser data offered good segmentation, which meant that analysts could act on what they saw through experimentation.

By the mid-1990s, established brands were launching their web presence. Secure Sockets Layer (SSL) made it safe to conduct transactions, and companies like Pizza Hut, CDNow, and Amazon were selling real things to real customers through web interfaces. The audiences were there, too: the 1998 World Cup website france98.com served, on average, 180 requests *a second* between April 30 and July 26, for a total of 1.3 *billion* requests.

Web analytics companies like Accrue, NetGenesis, and WebTrends started to process web logs in ways marketing, rather than IT, wanted. This was big-ticket enterprise software, consisting of large up-front licenses and powerful servers on which to run it. This software got its data from logfiles or by sniffing web hits directly from the Internet connection, which meant that IT still had to be involved in the deployment process, and that the company had to maintain servers and storage to collect and analyze all of the data.

From IT to Marketing

Technologists were less and less the audience for the data produced by these tools. Instead, their features focused on marketers eager to embrace their online customers who were clamoring for more insight into what was happening online.

In many companies, something important changes when marketing departments become the customer. Operations is traditionally a cost center—the tools it uses are seen as ways to minimize costs—but the marketing department's tools are about maximizing revenue. Marketing makes it cool.

Three important changes placed analytics firmly in the hands of marketing.

JavaScript collection let marketing bypass IT

As marketers took over web analytics, they wanted ways to monitor their websites without having to ask IT for permission to do so. JavaScript made it possible to collect visitor activity without installing servers or parsing web logs. Marketers were free from the tyranny of IT operators.

In 1996, San Diego-based company Urchin launched a hosted analytics service called Quantified. At the same time, a company in Salt Lake City called Omniture introduced a similar analytics tool, SiteCatalyst, which was focused on large enterprises.

Search engines changed the reports marketers wanted

That year, at Stanford University, two students were working on an algorithm for measuring the importance of pages that would eventually change the way the world found information. Google—and other search engines and directories—spawned an ecosystem of advertisers, buyers, and analysts by selling advertising space alongside search results. The result was a thriving online advertising industry that was cheaper and more accountable than its offline counterpart.

Online ads could be changed at a moment's notice—after all, they were just information. They were also completely trackable. Consequently, marketers could test and adjust campaigns, bidding on the search terms that were most likely to drive buyers to their sites. This changed the kinds of reports analysts wanted. Suddenly, it was all about which campaigns and referring sites were most likely to attract visitors that would complete transactions.

Service models meant pay-as-you-go economics

Third-party analytics services enjoyed better economies of scale than individual enterprises could achieve on their own. An analytics service had enough capacity to handle traffic spikes to one individual site, amortizing the analytics processing across machines shared by all customers. Marketers paid for the traffic they measured, not the up-front cost of servers and software.

Now that analytics belonged to the marketing department, billing was tied to revenue rather than cost. Analytics was paid for as a percentage of web revenues. If traffic increased, a company's monthly analytics bill increased—but so (hopefully) did its revenues.

Around this time, the web analytics industry became much more mature, urged on by people like Jim Sterne (who launched the web analytics conference "eMetrics"), Matt Cutler, Jim Novo, and Bryan Eisenberg (who founded the Web Analytics Association). Other industry thought leaders would later emerge, such as Eric T. Peterson and Avinash Kaushik. We also saw the first steps toward standardization of terms, metrics, and reports.

A Very Short Analytics Glossary

Here's the shortest overview of analytics terms we could come up with. It shows you most of the key terms we'll be looking at in the pages ahead.

A *visitor* arrives at your website, possibly after following a link that *referred* her. She will *land* on a web page, and either *bounce* (leave immediately) or request additional pages.

In time, she may complete a *transaction* that's good for your business, thereby *converting* from a mere visitor into something more—a customer, a user, a member, or a contributor—depending on the kind of site you're running. On the other hand, she may *abandon* that transaction and ultimately exit the website.

Visitors have many external attributes (the browsers they're using, the locations they're surfing from) that let you group them into *segments*. They may also see different *offers* or pages during their visits, which are the basis for further segmentation.

The goal of analytics, then, is to maximize conversions by *optimizing* your website, often by *experimenting* with different content, layout, and campaigns, and analyzing the results of those experiments on various internal and external segments.

From Hits to Pages: Tracking Reach

While early web analytics reports simply counted HTTP requests, or *hits*, marketers quickly learned that hits were misleading for several important reasons:

- The number of hits varies by page. Some pages have dozens of objects, while others have only one object.
- The number of hits varies by visitor. First-time visitors to a site won't have any of the graphical content cached on their browsers, so they may generate more hits.
- It's hard to translate hits to pages. Pages may have JavaScript activity that triggers additional requests. A Google Maps visit, for example, triggers hundreds of hits on a single page as a user drags the map around.

Hits: A Definition

Any server request is considered a hit. For example, when a visitor calls up a web page containing six images, the browser generates seven hits—one for the page and six for the images.

Prior to sophisticated web analytics, this legacy term was used to measure the interaction of the user with the website. Today, hits might be a useful indicator of server load, but are not considered useful for understanding visitor behavior.

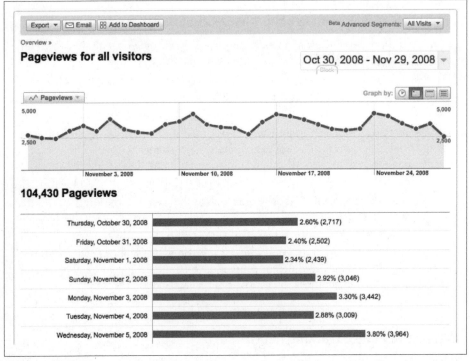

Figure 5-2. A Pageview report in Google Analytics

The first useful web analytics metric is the *page view*, which is a measure of how many pages visitors viewed. A page view report like the one in Figure 5-2 shows you how many times a visitor saw a page of your site.

Page views are still misleading. If your site serves 100 pages, you don't know whether the traffic resulted from a single visitor reading a lot of content or a hundred visitors reading one page each. To deal with this, web analytics started to look at user visits with metrics like unique page views, which ignored repeated page views by the same visitor.

From Pages to Visits: The Rise of the Cookie

One of the main ways in which websites distinguish individual visitors from one another is through the use of cookies—small strings of text that are stored on the browser between visits. Cookies are also a major source of privacy concern because they can be used to identify a visitor across visits, and sometimes across websites, without the visitor's explicit approval.

In the very early days of the Web, each request for content stood alone—two requests from the same visitor weren't related. If a user asked for a page about shoes and then moved on to a page about shirts, those appeared to be completely independent requests.

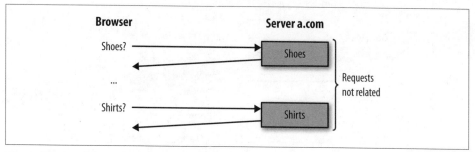

Figure 5-3. A website visit to a server in the early days

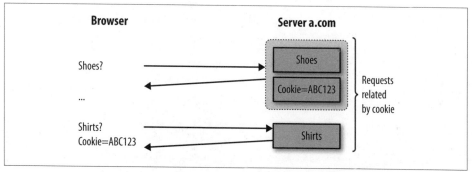

Figure 5-4. A website visit to a server that sends a cookie

This severely limited the usefulness of web applications. You couldn't, for example, put items in a shopping cart; each item looked like an entirely new request, as illustrated in Figure 5-3.

To address this issue, websites started sending cookies to browsers. A *cookie* is a unique text string that a web server sends to the browser when a user first visits. Each time the user returns or requests another page on the site (with the same browser), it passes the cookie to the server, as shown in Figure 5-4. Because of this, *cookies let website operators know which pages they've shown to users in the past.*

A cookie contains long strings of data that aren't supposed to be personally identifiable. (we're using the phrase "supposed to be" because it is technically possible for a website operator to place personally identifiable information inside a cookie—and it happens surprisingly often). In fact, cookies are often encrypted, changed, and updated by the server during a visit. A site cookie also isn't shared: the browser only sends the cookie to the site it got it from in the first place. This means that a cookie can't be used to track a user across several web properties. Figure 5-5 shows the use of a second cookie, this time on site b.com, that is completely unrelated to the cookie sent by site a.com.

Or at least that was the plan.

In Figures 5-4 and 5-5, visitors received two cookies, one from each visit. Consider what happens, however, if those sites serve advertisements from a shared ad network, as

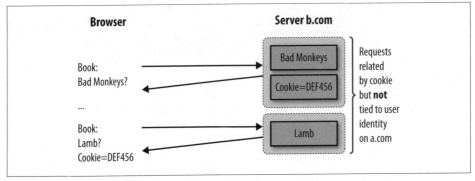

Figure 5-5. The visitor in Figure 5-4 visits a bookstore (b.com) that's unrelated to the previous server, and gets a new—and different—cookie

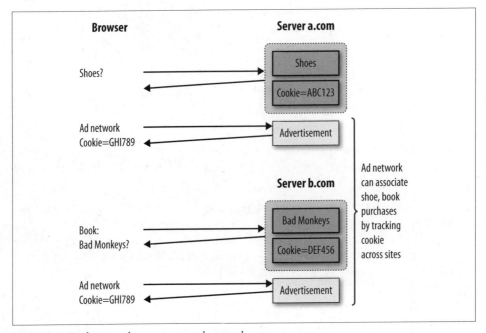

Figure 5-6. Cookies sent by a common ad network

shown in Figure 5-6. When a user retrieves an ad on a.com, then moves to b.com (whose ads are being provided by the same ad network), the ad provider now knows that the two visits are related.

These "tracking cookies" let an advertiser track an individual across many sites, and some consider them a violation of privacy.

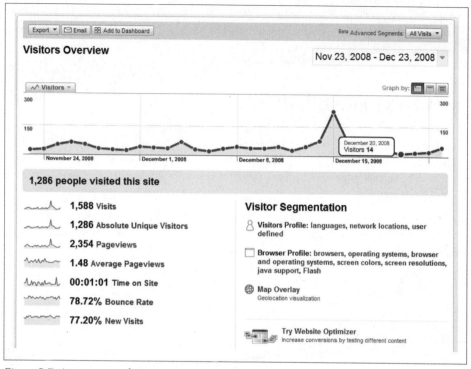

Figure 5-7. An overview of site visitors in Google Analytics

Let's be clear about this: the use of cookies means that if you view an ad on a server with adult content on Monday, then visit a bookstore containing ads from the same ad provider on Tuesday, the ad provider could, in theory, tie the two visits together. When tracking cookies are tied to personal information, such as your bookstore account, the risks are even higher.

In practice, this doesn't happen much—perhaps because of strikingly different ad content—but it's the reason for the public outcry when sites like Facebook try to implement visitor tracking as part of their advertising systems.

On the other hand, cookies make it possible to tie together requests for several pages into a web visit from an individual, making them essential for web analytics.

 There are other ways to tie together page requests, including Uniform Resource Identifier (URI) parameters and hidden form elements. Some sites resort to these when a visitor's browser blocks cookies. These are less likely to cause privacy violations because they're not stored on a browser across visits while still allowing a website to stitch together the pages of a user's visit and keep track of things like shopping carts.

Visitor analysis, such as that shown in Figure 5-7, makes it possible to calculate other data, such as bounce rate (how many people left after seeing only one page), and how many pages a typical visitor looked at. Tracking visits also means we can now see where visitors came from and where they went.

From Visits to Outcomes: Tracking Goals

A visitor's page views are simply steps toward a goal or an outcome. It's these outcomes—not page views—that make your business work. After all, a user who comes to the site and buys something quickly generates far fewer page views but is much better for business than someone who browses around for a while before leaving without buying anything.

Tracking visits by outcome lets you make smarter decisions about your website. For example, if you simply analyze visits to your site based on where visitors came from, you'll see which sites are driving traffic to you, as shown in Figure 5-8.

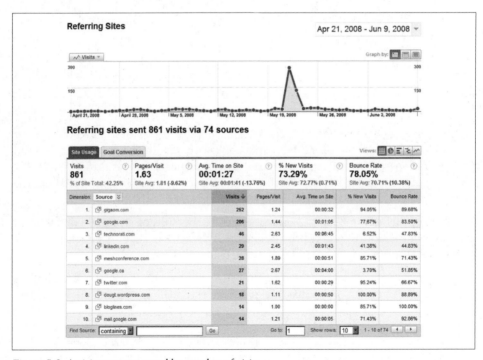

Figure 5-8. A visitor report sorted by number of visits

Now suppose that you have a specific goal, such as having visitors fill out surveys on your website. You can analyze how many people completed that goal over time and measure the success of your business in a report like the one shown in Figure 5-9.

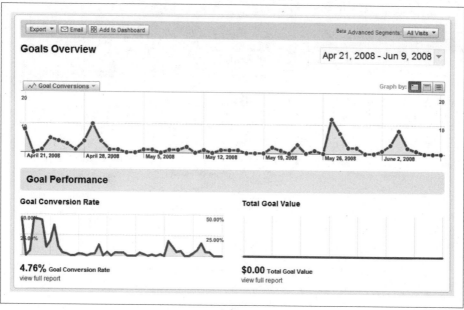

Figure 5-9. Report of how many visitors attained a goal within a website

Goals and outcomes are essential. If you don't have a goal in mind, you can't optimize your site to encourage behaviors that lead to that goal. We typically express the steps a visitor takes towards a goal with a "funnel" like the one shown in Figure 5-10, indicating how many visitors proceeded through several steps and how many abandoned the process.

Funnels are a good way to depict where people are coming from and where they're abandoning a process. However, goals provide more than just accounting for conversions; they let you focus on what's working through segmentation.

Not all traffic is the same. Consider the list of sites that sent you traffic (in Figure 5-8). By segmenting those referring websites by goals, rather than just by total visits, you can see which of them sent you *visitors who mattered*—the ones that achieved the goals you hoped they would.

Referring sites that sent you large amounts of traffic may not have contributed to goals, while those who sent only a small fraction may have contributed significantly, as is the case here. From reports like the one shown in Figure 5-11, you can make the right decisions about where to focus your promotions.

Segmentation goes far beyond just measuring the effectiveness of referring sites. You can segment by anything that your analytics tool collects.

Early analytics tools were limited in what they could capture, however. What if you wanted to understand whether people who bought shoes also bought jackets? Or whether platinum members were more likely to comment on a post? Marketers needed

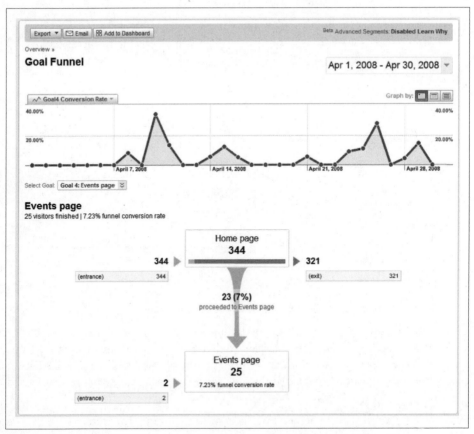

Figure 5-10. A goal funnel showing conversion and abandonment for a simple goal

a way to embed more context into their pages so they could segment in better ways. They needed page *tagging*.

From Technology to Meaning: Tagging Content

In the previous examples, we knew the goal we wanted visitors to accomplish (the "about" page), and we knew the various sites that were sending us traffic. The referring site is part of the default data an analytics tool collects. To add more context to a visit, we need to embed meaning in the page that the analytics system can later extract. This is called page tagging, and to understand it, we need to look briefly at how browser instrumentation works.

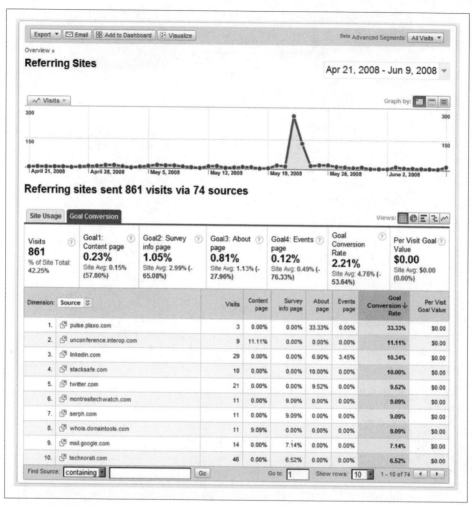

Figure 5-11. The list of referring sites, this time segmented by goal conversion rate, shows which referring sites contributed to your business the most

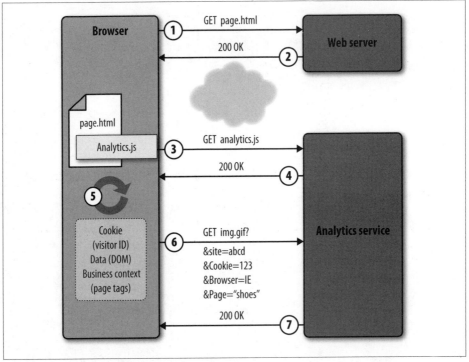

Figure 5-12. How JavaScript-based analytics works

Figure 5-12 illustrates how JavaScript-based analytics systems collect page hits from a visitor.

1. The browser requests a page from the web server.
2. The web server returns the page, which contains a reference to a piece of JavaScript provided by the analytics service.
3. The JavaScript is retrieved from the service. In some cases, it may already be cached on the browser or built into the page itself.
4. The analytics service returns the JavaScript to the browser.
5. The browser executes the JavaScript, which collects information about the page request.
6. The JavaScript puts this information into an HTTP request, often for a tiny image. All of these attributes are sent to the analytics service.
7. The service responds with a small object (which can be ignored).

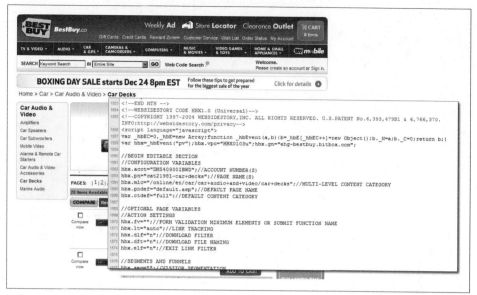

Figure 5-13. Page tags within a page at BestBuy.com

We'll return to this topic later in the chapter when we consider implementation.

The magic of page tagging happens in step 5, in which the JavaScript collects information to send back to the analytics service. This information includes:

- Information about the *technical environment*, such as screen resolution, referring URL, or operating system.

- Information about the *visitor*, such as a cookie that uniquely identifies her and lets the analytics service stitch several pages together into a visit.

- Information within the page itself that provides *business context*. For example, a retailer might tag a page with "shoes" if it offers shoes for sale. Or it might indicate whether the ad the user saw was part of a specific campaign. Figure 5-13 shows an example of page tags on a retail website.

By recording not only technical data on the visit, but also business context, the analyst can then segment outcomes with this context. He can see whether one campaign works better than another, or whether users who buy shoes ultimately buy jackets.

Notice how far we've come from the early days of analytics, in which IT teams looked at hits to measure capacity? Today, web analytics is a marketing discipline used to measure the effectiveness of communications strategies.

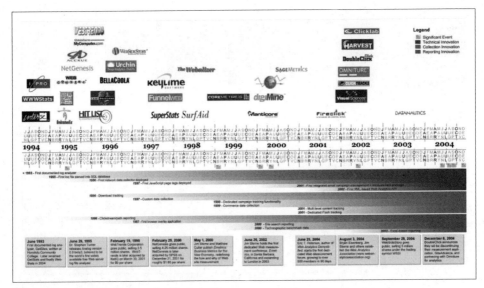

Figure 5-14. A visual history of web analytics

It's also a full-time job, with large organizations often having entire departments devoted to tagging, experimentation, and reporting. A history of web analytics, compiled by the Yahoo! web analytics mailing group as a result of discussions by Eric T. Peterson and John Pestana, is illustrated in Figure 5-14.

Analytics is still changing, particularly around the integration of other data sources and around the move away from page-centric websites.

An Integrated View

Having moved so far from their technical roots, marketers are now realizing that there's more to user experience than just analytics, and that they need to bring other data sources into the fold if they're going to optimize their websites.

- Poor performance undermines user productivity, reduces conversion rates, and discourages visitors from sticking around. Marketers care about the effect performance has on conversion. Figure 5-15 shows an example of a conversion funnel that considers end user performance.

- Visitors have lots on their minds, and marketers ignore it at their own peril. It's easy to survey customers and solicit feedback, so companies are starting to integrate VOC data or page rankings with visits and goals, as shown in Figure 5-16.

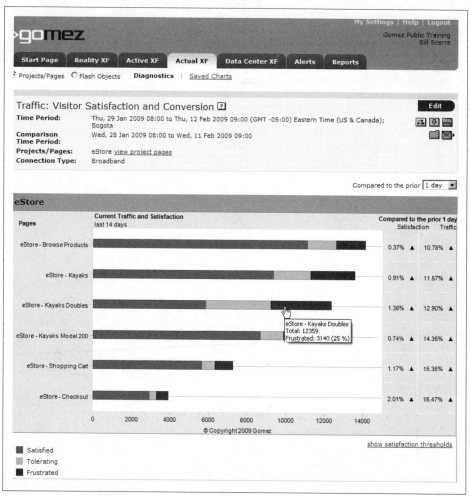

Figure 5-15. End user performance correlated with requests for each page in a goal funnel in Gomez

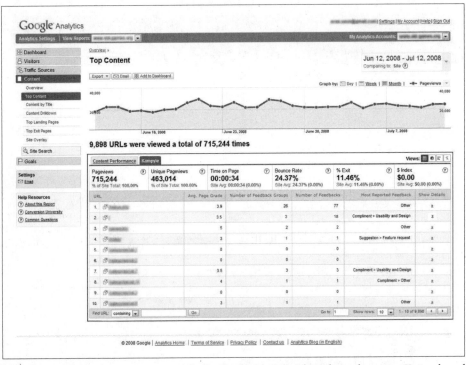

Figure 5-16. Blending visitor survey and feedback information with analytics data using Kampyle and Google Analytics

- Usability has a huge impact on conversion rates, so websites are trying to identify bottlenecks and usability issues at each step of a conversion funnel, like the one shown in Figure 5-17, to optimize their sites and improve user experience.

Making sense of all this information means tying it back to the business context of web analytics. As a result, we're seeing significant advances in web visibility, particularly in the area of data integration for web analytics platforms.

Places and Tasks

At the same time that analysts are integrating performance, visitor opinion, and usability data, the nature of websites is changing. Developers are using programming built into the pages themselves in the form of Ajax-based JavaScript, Flash, Java, and Silverlight, resulting in fewer pages but more actions within a page. Some websites take this to the extreme, consisting of just a rich application embedded in a single web page.

For these sites, the old paradigm of a visitor navigating through several pages toward a goal is no longer accurate. Analytics tools need to adjust what they capture and how

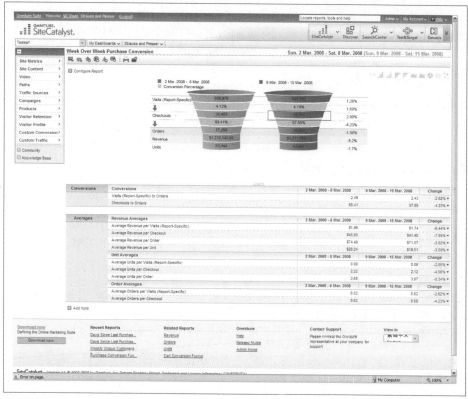

Figure 5-17. A conversion funnel in Omniture's SiteCatalyst linking to Tealeaf in order to analyze and replay visits to understand visitor behavior

they record it accordingly, because not all websites follow the page-centric model on which web analytics was conceived.

Instead of thinking of websites as made up of pages, think of them as a set of "places" and "tasks." Some of the time, users are in a "place"—they're exploring, browsing, and interacting. Occasionally, they'll undertake a "task"—a series of steps toward a goal. They may complete or abandon that task. This is a generalization that works for both page-centric and RIA-based (rich Internet application) websites.

Thinking in terms of places and tasks helps you to understand which outcomes you need to focus on across your entire web business.

Places are where users hang out

On reddit, a list of submissions from others, ranked by popularity, is a place. Visitors can perform a series of small actions like opening linked stories in new tags, or voting things up and down. Similarly, on Wikipedia, a subject entry is a place. In Google Apps, a spreadsheet is a place.

When a user's in a place, you care about his *productivity*—whether the experience was efficient and satisfying. On reddit, are users able to vote submissions up or down smoothly? On Wikipedia, can they scroll through entries easily and find what they're looking for, and do images load? On Google Apps, are they successfully building business projections in that spreadsheet?

You still need to pay attention to abandonment, which happens when a user gets bored doing something and goes elsewhere. But in the case of a place, it's abandonment due to boredom, satisfaction, or disinterest, not due to an inability to accomplish something.

Tasks occur when users have a mission

By contrast, a task is something the user sets out to accomplish. A task comprises several steps, and puts the visitor in a different mental state. Tasks are goal-oriented. On reddit, this is a user creating an account for himself or submitting a new link. On Wikipedia, it's a visitor deciding to edit a page. On Google Apps, it's the employee sharing her spreadsheet with someone.

When a user's trying to accomplish a task, we care about *effectiveness*—whether the task was completed successfully or not. Did the invite result in a new enrollment? Did the user complete the purchase? Was the edit of the article ultimately saved? Was the user able to add the widget to her dashboard? Did the spreadsheet's recipient receive it successfully? When visitors abandon tasks, it's because they ran into a problem. Something was too costly, or confusing, or violated their privacy, or dampened their enthusiasm for the job.

A new way to look at sites

Looking at websites as collections of places and tasks reveals the limitations of page-centric, funnel-minded web analytics.

For one thing, you realize that you need to instrument places and tasks very differently. Places need analysis of actions within the place. How many videos did he watch? How often did he pause them? Did he see the ad? How many entries can she complete an hour? On the other hand, tasks need analysis of progress toward an outcome. Did she send the invite? How far in the form did she get? Which steps took the longest?

To further complicate matters, tasks often involve steps beyond the view of analytics, such as e-mail invitations, instant messages, RSS feeds, and third-party sites. We also want to know other things about tasks. Did the message bounce? Did the invitation's recipient act on it?

Tracking the accomplishment of a task across multiple systems is a challenge, with all manner of tracking cookies, dynamic URLs, and embedded GIFs used to try and follow the task to completion. As outlined in *http://analytics.blogspot.com/2009/02/two-cool -integrations-telephone-leads.html*, some companies coordinate analytics data with offsite goal completion such as phone calls or instant messages. For more information on multichannel monitoring, see *www.kaushik.net/avinash/2008/07/tracking-offline-conversions-hope-seven-best-practices-bonus-tips.html*.

What can you do to get started?

Most web operators have a mental map of their sites. Some even draw it on a wall. You can map out a site, consisting of places and tasks, in this way.

- For each place, make a note of all the events you care about, including timing events ("a video starts playing") and user interactions ("user upvotes and the button's color changes"). Focus on visitor productivity and time spent in the place. Identify the actions that initiate a task (such as "share this spreadsheet"), taking the user out of the place.

- For each task, make a note of all the steps you want to track, including those that aren't on your site. Identify the key metrics you should know (in a mailout, this might be bounce rate, open rate, and click rate.) Focus on conversions, abandonments, and their causes across multiple channels.

The end result is a much more accurate representation of the ebb and flow of your online business. It will probably reveal significant gaps in your web visibility strategy, too, but at least you'll now know where your blind spots are.

The next time you're presenting your web monitoring results, overlay them on the map of places and tasks that you've made. For each place or task, show the analytics (what the users did) and the user experience (whether they could do it). If you have psychographic information (why they did it) such as surveys, or usability metrics (how they did it), include that as well. Also build in any external metrics like mail bounce rates or Facebook group member count.

We'll return to the concept of places and tasks, and web performance considerations for both, when we look at measuring end user experience later in the book.

The Three Stages of Analytics

Any visit can be broken into three distinct stages: finding the website, using the website, and leaving the website. Web analytics is the science of identifying segments of your visitors that go through these three stages in significantly different ways, and then focusing on segments that are the best for your business.

- If a certain marketing campaign helps people find your site, all else being equal, you should put more money into that campaign.

- If a certain kind of site content or layout encourages visitors to use your site in ways that are good for your business, you should focus on that content.

- If certain content, layout, or page load speeds encourage users to stay longer and return more often, you should focus on doing those things more consistently.

- If certain ad locations encourage your visitors to click on those ads, generating revenue for you as they leave the site, you should charge more for those locations and sell more of them.

Finding the Site: The Long Funnel

Getting traffic to your site is often the first thing marketers think about. If all else is equal, more traffic translates into more revenues and a healthy business.

Of course, all else is seldom equal. In addition to search engines, visitors come from a wide range of sources (shown in Figure 5-18), among them hundreds of social networks. Each visit has a different outcome. Some visitors stop by briefly, never to return, while others become lifelong readers or loyal customers.

By optimizing the step *before* users hit the entry page, you can increase traffic volumes, which makes for a larger conversion pool. Let's look at the ways users come to your website.

 It's important to make the distinction between an *entry* page and a *landing* page. The entry page refers to the first page a visitor loaded for a visit to your site. However, a visitor might hit multiple landing pages during a single visit. How is this possible? A visitor may navigate into your site initially (making this both the entry page and the landing page), then navigate away to another site, and finally return to your site in a relatively short time frame while their session cookie is still valid, making it seem like the second landing page was part of the visit. In other words, there can be multiple landing pages but only one entry page.

Visitors can find your site in several ways:

- They can type the URL into the browser directly, known as *direct traffic*.
- They can link to you following an *organic search*.
- They can click on an advertisement or sponsorship link, which counts as banner or *paid search* campaign traffic.
- They can follow a link from another site or from a community such as a social network through *referrals*.
- They can respond to an email invitation as part of an *email marketing campaign*.

Each of these sources is treated differently in web analytics.

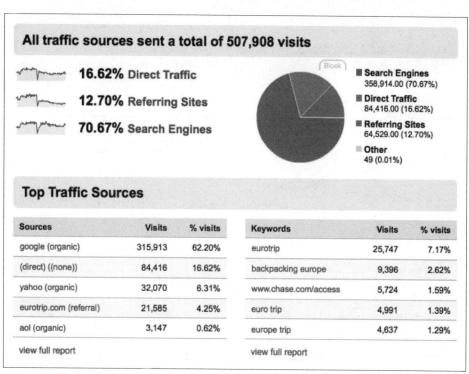

All traffic sources sent a total of 507,908 visits

〰	**16.62%** Direct Traffic
〰	**12.70%** Referring Sites
〰	**70.67%** Search Engines

- ■ **Search Engines** 358,914.00 (70.67%)
- ■ **Direct Traffic** 84,416.00 (16.62%)
- ■ **Referring Sites** 64,529.00 (12.70%)
- ■ **Other** 49 (0.01%)

Top Traffic Sources

Sources	Visits	% visits
google (organic)	315,913	62.20%
(direct) ((none))	84,416	16.62%
yahoo (organic)	32,070	6.31%
eurotrip.com (referral)	21,585	4.25%
aol (organic)	3,147	0.62%
view full report		

Keywords	Visits	% visits
eurotrip	25,747	7.17%
backpacking europe	9,396	2.62%
www.chase.com/access	5,724	1.59%
euro trip	4,991	1.39%
europe trip	4,637	1.29%
view full report		

Figure 5-18. Google Analytics Traffic Sources Overview page

Direct traffic

The simplest way for people to visit your website is by typing it into an address bar. They may have heard your brand name and decided to enter it as a URL, or they may have seen the URL in print advertising.

Direct traffic is the least informative of all traffic sources, because you don't know how the visitor discovered you. As a result, it's hard to optimize. If you need to know more about these visitors, intercept them with a survey and ask them how they found out about your site. We'll look at VOC surveys in Chapter 7.

Direct traffic can also be misleading for several reasons:

Navigational search
> Between 15 and 20 percent of web searches consist of users who type a URL into the search engine (rather than the browser's address bar). A surprising number of Internet users believe that this is how the Internet works, and only ever use searches to find sites, even those they frequent every day. In these cases, the URL of the website—or its name—appears in keyword reports, as shown in Figure 5-19. This is known as *navigational search*, and will usually be shown as organic search traffic rather than direct traffic.

Search sent 80 total visits via 5 keywords

Show: total | paid | non-paid

Visits ⑦		**Pages/Visit** ⑦		**Avg. Time on Site** ⑦
80		**2.84**		**00:02:04**
% of Site Total: 5.63%		Site Avg: 7.91 (-64.14%)		Site Avg: 00:12:26 (-83.35%)

	Dimension: Keyword ⌄	Visits ↓	Pages/Visit
1.	farmsreach	62	2.77
2.	farms reach	14	3.00
3.	farmsreach.com	2	4.00
4.	farmsreach,com	1	1.00
5.	www.farmsreach.com	1	4.00

Figure 5-19. Navigational search results for www.farmsreach.com; high levels of navigational search for the URL can be an indicator of a nontechnical demographic

Type-in traffic

Type-in traffic occurs when a user types a word—such as "pizza"—into a search engine to see what will happen. Depending on the browser or search engine, this will often take the user to a named site such as "*www.pizza.com*" that makes its money from referring visitors to other sites. Type-in traffic is why popular type-in domains are sometimes sold for millions of dollars.

 John Battelle has an excellent write-up on the type-in traffic business at *http://battellemedia.com/archives/002118.php*.

Bookmarking

Users who have visited a site may bookmark it, or the browser may display the most frequently visited sites when it is first launched. Clicking on a bookmark, a "frequently visited" link, or a link in the browser creates a visit with no referring site. The visit is classified as direct traffic by the analytics engine.

Desktop client

> Some desktop tools, such as Twitter and email clients, don't provide any information about a followed link—they simply launch a web browser with a URL that the user received through the desktop tool. As a result, these links are classified as direct traffic, even though they were referred through a community.

JavaScript redirection

> If a user follows a JavaScript-triggered link to your site, the browser might not pass on referring information and the visit will be mislabeled. Careful scripting can fix this, but few web developers make the extra effort.

Browser inconsistency

> Browser behavior can reduce the accuracy of direct visitor count. For example, in certain conditions, Microsoft's Internet Explorer 7 will not pass along referring information if you load a website in another window or tab.

Bots, spiders, and probes

> Scripts and automated visitors, such as search engines indexing your website, may not be properly identified and may be counted as direct visits.

These caveats mean that direct traffic counts can be inaccurate. You can mitigate some of this inaccuracy through the use of custom URLs that make it easier to track the effectiveness of a traffic source. For example, in a print campaign, you can include the name of a specific magazine in the URL that you place in that magazine's ads (i.e., *www.watchingwebsites.com/magazinename*). Then you can track the success of advertising in each magazine by segmenting visits within analytics by the URL.

Organic search

One of the most common ways people find a site is through organic, or unpaid, search. A web user enters a search term and clicks on one of the results, as shown in Figure 5-20. The referring URL includes not only the site that referred the user, but also the search terms that drove the traffic to your site.

Over time, your web analytics package develops a list of the most common keywords driving traffic to your site. Some of the keywords that drive traffic can be quite weird and unexpected, as shown in Figure 5-21, and may be a clue that your visitors' intentions on your site aren't what you think they are, and are worthy of further investigation.

If search engines consider your content more relevant, your site will rank higher than others and appear higher in search results. Many factors influence this, such as the text and metadata on your site and the number of other sites that link to yours.

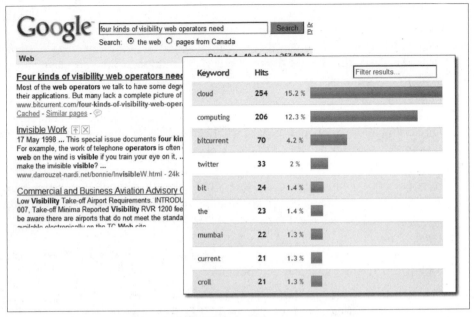

Figure 5-20. Organic search keywords for Bitcurrent.com

Figure 5-21. Keywords that drive traffic to your site can sometimes be unpredictable

As an online marketer, one of your main jobs is to ensure that your site content is ranked highly and that the right organic search terms send traffic to the site. Web analytics helps you to understand whether your site is properly optimized and which keywords are driving traffic, through reports like the one in Figure 5-22.

 The art of search engine optimization (SEO) is interesting, fun, and not covered in this book. We would rather leave you in the hands of the masters. For a great start in understanding SEO ranking, check out *www.seomoz.org/article/search-ranking-factors.*

Either way, the ultimate goal is to get better visibility and more qualified traffic.

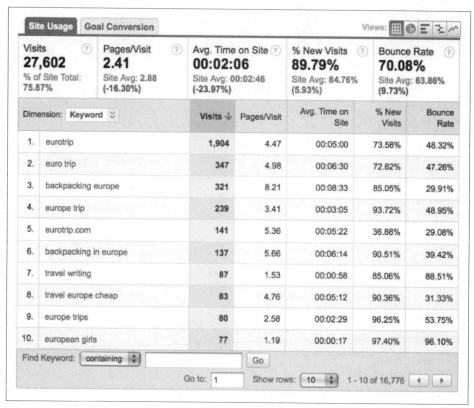

Dimension: Keyword ⌄		Visits ↓	Pages/Visit	Avg. Time on Site	% New Visits	Bounce Rate
1.	eurotrip	1,904	4.47	00:05:00	73.58%	48.32%
2.	euro trip	347	4.98	00:06:30	72.62%	47.26%
3.	backpacking europe	321	8.21	00:08:33	85.05%	29.91%
4.	europe trip	239	3.41	00:03:05	93.72%	48.95%
5.	eurotrip.com	141	5.36	00:05:22	36.88%	29.08%
6.	backpacking in europe	137	5.66	00:06:14	90.51%	39.42%
7.	travel writing	87	1.53	00:00:58	85.06%	88.51%
8.	travel europe cheap	83	4.76	00:05:12	90.36%	31.33%
9.	europe trips	80	2.58	00:02:29	96.25%	53.75%
10.	european girls	77	1.19	00:00:17	97.40%	96.10%

Figure 5-22. A list of keywords visitors used to find eurotrip.com on search engines

Paid search and online advertising

If you're not getting the attention you want organically, you can always pay for it. According to the Interactive Advertising Bureau, online advertising is a $21 billion-dollar industry. Three kinds of ads, shown in Figure 5-23, make up the bulk of online advertising:

- Pay-per-impression, in which advertisers pay media sites for the number of times they show an ad to a visitor.
- Pay-per-click, in which the media site gets paid each time a visitor clicks on the advertiser's message. This is how paid search ads work.
- Pay-per-acquisition, in which the media site is compensated each time a visitor that they send to an advertiser completes a transaction. Affiliate marketing is an example of a pay-per-acquisition model.

The ads may be shown to every visitor or they may be selectively displayed according to keywords and their relevance to search engines.

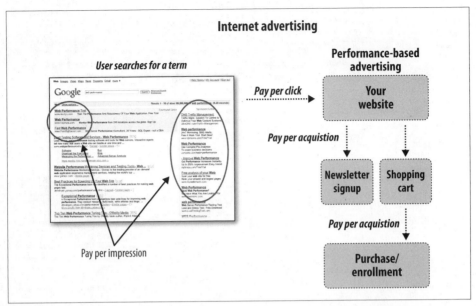

Figure 5-23. The three models of web advertising–pay-per-impression, pay-per-click and pay-per-acquisition

Pay-per-impression advertising is declining, with the possible exception of video advertising. One of the main reasons for this is fraud: marketers prefer to pay for outcomes, and pay-per-click reduces the amount of fraud that they have to deal with. There is also good evidence that picture advertising doesn't encourage users to click on it, so paid search is the most popular form of online advertising today.

Ad payment models vary by advertiser, but are often built around keyword "auctions" and a daily budget for each ad buyer. Ad prices are established by advertisers competing for a keyword.

Google's AdWords, Microsoft's adCenter, and Yahoo!'s Search Marketing display paid ads alongside organic search results. So if you can't optimize your site well enough to qualify for a spot on the first page of their search results, you can still get some real estate on a page of organic results by paying them.

Referrals

A referral occurs when someone mentions your website, which may happen in a blog posting, a mention on a social news aggregator like Digg, a Facebook status message, and so on.

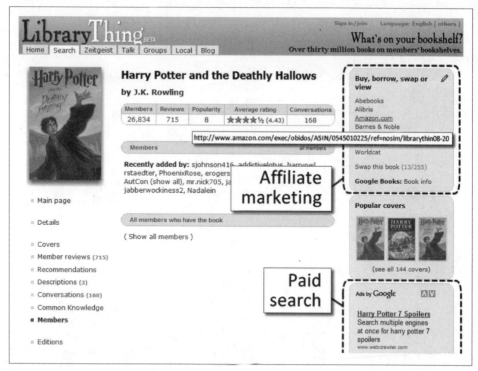

Figure 5-24. Affiliate marketing and paid search on book sharing site LibraryThing showing the unique affiliate URL that identifies the referring site

Many online retailers offer affiliate marketing programs that compensate others for sending them traffic. For example, a blogger might recommend books to her readers, and will be compensated by Amazon.com every time someone follows one of her links and buys a book, as shown in Figure 5-24.

For this to work, the referring link includes an affiliate ID—a code that uniquely identifies the referrer so that she can be credited with the referral.

While referrals do come from a somewhat trusted source, paid referrals have a different URL, and many web users can differentiate between a free referral and a paid affiliate marketing link. Paid referrals work well when the source is trusted and users don't mind that the referrer is being compensated, but they're less genuine than simple, unpaid word of mouth.

Linkbacks

There's a form of referral that is specific to blogging. When one blogger references another, the referencing blog alerts the referenced blog that a linkback (or trackback) has occurred. This is a form of attribution that's built into most blogging systems.

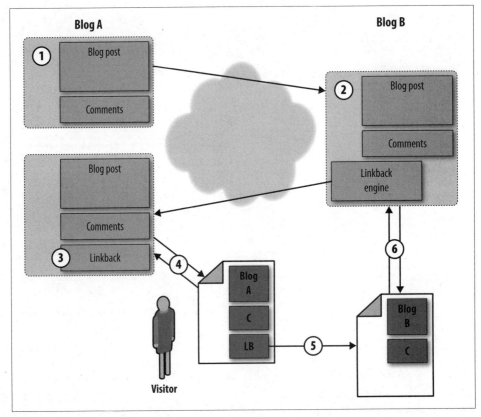

Figure 5-25. How a linkback works

Figure 5-25 shows how a linkback works.

1. The author of Blog A writes a blog post.
2. The author of Blog B mentions A's post in a blog entry he writes, triggering a linkback mechanism on his blogging platform that informs Blog A of the inbound link.
3. Blog A's blogging platform updates the linkback section of the post to include the inbound link from Blog B.
4. A visitor to Blog A reads the post, including the linkback section.
5. The visitor clicks the linkback from Blog B.
6. The visitor sees the post on Blog B that referenced Blog A.

Linkbacks help make blogging dynamic and encourage cross-referencing and attribution of ideas, as shown in Figure 5-26. They allow readers to follow extended conversations between several bloggers.

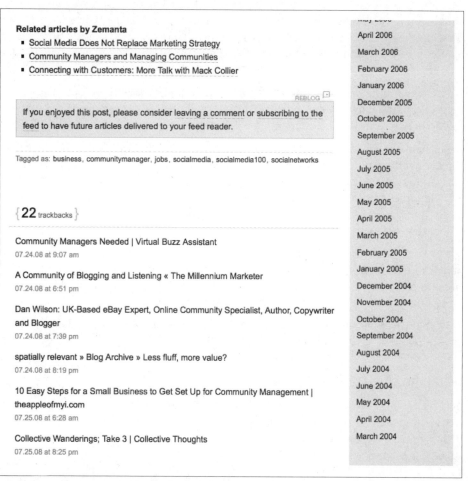

Figure 5-26. 22 linkbacks to an article on chrisbrogan.com

If you're running a blog, linkback analysis is another form of referral analytics you need to track.

The long funnel

Once you've identified visits that came from organic and paid search, advertising campaigns, affiliate referrals, and linkbacks, you're left with direct traffic. Much of this comes from word-of-mouth mentions and social networks—after all, visitors had to hear about you somewhere. However, you can't tell where their first impressions came from, because their visits are the result of several influences you weren't monitoring.

The conversion funnels we've seen so far begin when a visitor first arrives at your site. This isn't an accurate model for the modern Web. Social networks and online communities mean conversion starts long before a visitor reaches your site. Integrating all of these into your understanding of the conversion process is a major challenge for web analysts.

All of these traffic sources represent your "long funnel"—the steps your visitors undergo from initial awareness to conversion—which begins at the conversation prism (by Brian Solis and JESS3, found at *http://theconversationprism.com*).

What happens when users first become aware of your brand on another site affects their conversion once they reach your site. Consider, for example, two visitors from social news aggregator reddit. One visitor may follow a link from the site's home page, while another may read critical comments about your brand within the submission's discussion. Their outcomes on your site are affected by experiences that happened long before they ever saw your home page.

The start of the long funnel is the subject of community monitoring, which we'll look at in Chapters 11 through 14. It's hard to tie community activity to web funnels, in part because of privacy concerns, but long funnel analysis is the holy grail of web analytics and a critical factor to consider in organic traffic acquisition.

Now that we know where users come from, let's figure out what they're doing on your site.

Using the Site: Tracking Your Visitors

You track what your visitors do so that you can identify patterns and improve the site in ways that increase the outcomes you want (such as inviting friends, creating content, or subscription renewals) while reducing those you don't (such as departures or calls to customer support).

Where did they come in?

The first page in a visit is particularly important. Different visits have different first pages, and you can segment outcomes based on the pages at which visitors arrive. You may even have different landing pages for different traffic sources, which lets you tailor your site's message to different segments.

For example, a visitor from reddit might see a message or layout that's designed to appeal to him, while a visitor from Twitter might see a different layout entirely.

Page	Pageviews ↓	Unique Pageviews	Time on Page	Bounce Rate	% Exit
1. /index.php/2008/12/04/myths-entrepreneurs-tell-t...	494	455	00:07:20	91.67%	87.65%
2. /	206	156	00:01:19	58.87%	51.46%
3. /index.php/2008/12/07/testing-and-launching-a-...	150	129	00:05:28	77.88%	77.33%
4. /index.php/2008/05/26/plan-b-five-reasons-comp...	41	38	00:05:52	86.11%	82.93%
5. /index.php/2008/10/08/the-three-kinds-of-ceo/	35	32	00:02:20	85.71%	57.14%
6. /index.php/about/	27	23	00:00:41	66.67%	37.04%
7. /index.php/2008/06/30/dont-use-4by6com-how-...	24	20	00:02:07	73.68%	70.83%
8. /index.php/2008/09/23/the-opposite-of-startup-o...	15	14	00:02:06	85.71%	60.00%
9. /index.php/2008/07/29/startupdrinks-in-montreal/	12	2	00:02:35	0.00%	0.00%
10. /index.php/2008/05/	10	2	00:00:12	0.00%	0.00%

Figure 5-27. Top entry pages in Google Analytics, with corresponding time on page and bounce rate

To understand what visitors first saw, run a Top-N report on entry pages, such as the one shown in Figure 5-27, to see some fundamental facts about your website:

- The total number of page views for the page.
- The number of *unique* page views for the page. Subtracting total page views from unique page views shows you how many visitors viewed a page more than once.
- The time visitors spent on the page before moving on.
- How many visitors arrived at the page and left immediately.
- How many visitors ended their visits on the page.

Armed with just this information, you already know a great deal about which parts of your site are working.

- If your objective is to engage visitors, you want create more pages like those that make visitors linger (lingering indicates visitors found the content useful) and that encourage them to go elsewhere within the site.
- If the page has a high bounce rate, you may need to redesign the page to convince visitors to stay.
- If users exit from this page, you'll need content that draws them back into the site for more than one page by suggesting they enroll or view additional content.

Places and tasks revisited

To understand visitor activity beyond the first page, start thinking in terms of places and tasks. Your visitors have arrived at a place on your site, such as the landing page, and can do various things in that place. They can remain there and interact, they can move to another place, or they can complete some kind of task. Figure 5-28 shows an example of a places and tasks diagram for a social news aggregator.

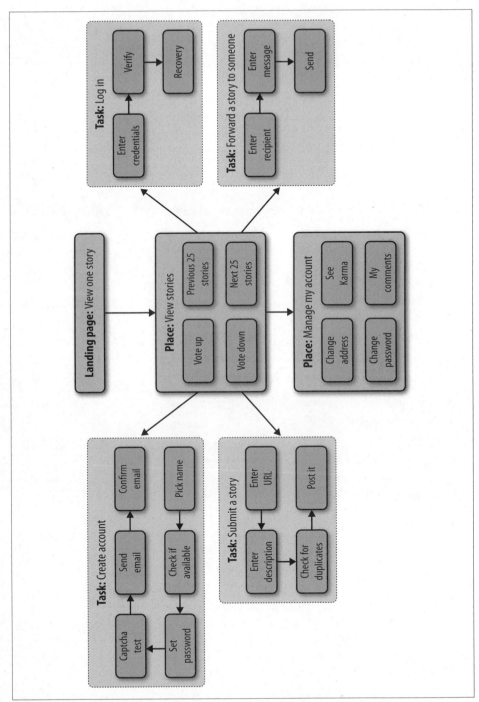

Figure 5-28. An example of places and tasks in a social news aggregator site such as reddit or Digg

The places and tasks model lets you define all the outcomes you want on your site so that you can collect data on them.

For *places*, you care about all-purpose metrics such as exit rate and time spent in the place, as well as where visitors went after each place. You may also have site-specific metrics (for example, number of story votes a visitor placed) that measure productivity, engagement, or value derived from the user's visit.

For *tasks*, you care about progress toward an outcome, and abandonment at each stage. If, for example, a visitor sets out to forward a story to someone, you need to track the many ways in which the task can go awry: the entry of an invalid email address, the visitor abandoning the process, the mail not being delivered, or invitees not acting on mail they've received.

Try to assign each task a dollar value. For e-commerce transactions, this may be the money the visitor spent, but for other kinds of task it may vary. If you have an estimate of the lifetime value of a customer, a completed invitation may have a specific dollar value based on that estimate.

Assigning a dollar value is good for three reasons. First, it makes the somewhat esoteric analytics data concrete: everyone knows what a dollar is. Second, it forces you to quantify how the task outcome helps the business. Third, it makes executives pay attention.

Other examples of tasks you want to track include:

- Adding a contact to a CRM database
- Commenting on a blog post
- Subscribing to an RSS feed or mailing list
- Getting directions or contact information
- Downloading product literature

You need a map of the places and tasks, with key performance indicators (KPIs) for each action. Once you know what these are, it's time to test and segment your website to maximize those KPIs.

Segmentation

All visits are not equal. You need to compare different visitor groups to see which do better or worse than others. Table 5-1 shows some examples of segments by which you may want to analyze your traffic.

Table 5-1. Different ways to segment data

Segment	Example
Demographic segments	The country from which the visitor arrived
Customer segments	First-time versus returning buyers
Technographic segments	Visitors using Macintosh versus Windows operating systems
"Surfographic" segments	Those visitors who surf daily versus those who only do so occasionally
Campaign segments	Those visitors who saw one proposition or offer versus another
Promotion types	Those who saw a banner ad versus a paid search
Referral segments	Those visitors who came from one blog versus those who came from another
Content segments	Those visitors who saw one page layout versus another

Manage Advanced Segments Beta

Segments let you group certain types of visits together.

+ Create new custom segment

Advanced Segment	Actions
Default Segments:	
All Visits	apply to report
New Visitors	copy | apply to report
Returning Visitors	copy | apply to report
Paid Traffic	copy | apply to report
Non-paid Traffic	copy | apply to report
Search Traffic	copy | apply to report
Direct Traffic	copy | apply to report
Referral Traffic	copy | apply to report
Visits with Conversions	copy | apply to report
Custom Segments:	
No custom segments created; click here to create one.	

Figure 5-29. Managing advanced segments in Google Analytics

 Some segmentation data isn't available directly through analytics, and you may need to use customer surveys or enrollment questions to fit visitors into a particular segment—such as those over the age of 40— so that you can analyze a particular KPI or outcome by that segment.

While analytics tools offer many built-in segments, you can create custom segments according to user-defined data or other fields within an analytics tool to slice up traffic in ways that matter to your business, as shown in Figure 5-29.

By comparing KPIs across segments, you can identify what's working and what's not. You might decide that one advertisement works while another doesn't, or that one offer encourages visitors to linger while another makes them leave.

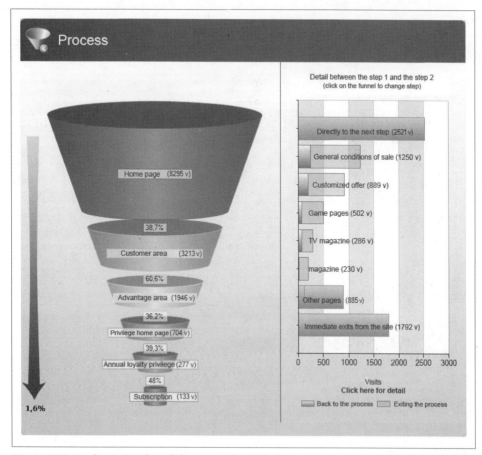

Figure 5-30. A subscription funnel shown in XiTi Analyzer

Goals

The ultimate KPI is the goal, or outcome, of a visit. Within your analytics package, you identify the outcomes you want, such as a purchase confirmation page or an enrollment screen, as well as any steps leading up to that outcome. This is often visualized using a funnel graph (Figure 5-30).

Funnels work well, but they're myopic. They make it hard to identify where users are going, and don't take into account reentry into the process. They also focus on web activity alone, while many websites have goals that include email messages, subscriptions, and other steps that can't easily be captured by simple web requests.

As web analytics tools adapt to today's more distributed conversion processes and usage patterns, we'll likely see places-and-tasks models that track KPIs for each step in a process. The KISSMetrics ProductPlanner (shown in Figure 5-31) is a great resource for determining which metrics to capture for many popular web design patterns.

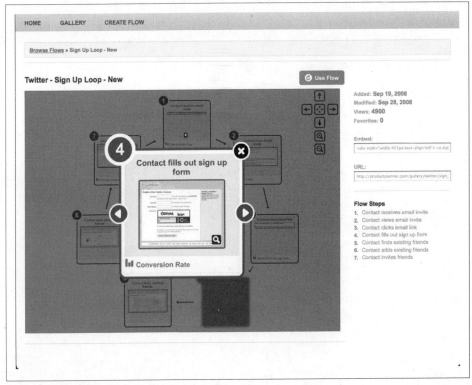

Figure 5-31. Twitter's sign-up process with the associated metric important for this step in the funnel on ProductPlanner

Putting it all together

The core of web analytics is knowing what outcomes you want, describing KPIs that represent those outcomes, scoring various visitor segments against those KPIs, and testing various combinations of content and segments to maximize those KPIs. Despite your best designs and offers, however, visitors often won't do what you wanted—they'll leave. Knowing why they left—and where they went—is equally important.

Leaving the Site: Parting Is Such Sweet Sorrow

While you can't see a visitor's activity once she's left your site, you can tell a lot about her motivations by looking at the last things she saw. You can also intercept parting visitors with a feedback form to try to learn more about their departure, which we'll look at more closely in Chapter 7.

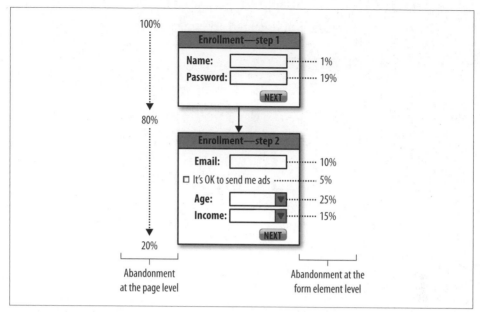

Figure 5-32. While web analytics looks at page-by-page abandonment, visitors usually abandon a process at a specific form element

Abandonment and bounce rate

Abandonment happens when people leave your site before doing something you wanted them to. They may simply leave out of boredom, because of an error or performance issue, or because they changed their minds about completing a particular task.

In recent years, analytics tools have started to look within a single page, to the form elements on that page, in an effort to identify which components of a page drive visitors away.

Consider, for example, a form that asks for personal information such as age. If you analyze the abandoned page at the page level, you won't know that it's a specific form element that's the problem. However, if you use web interaction analytics (WIA) tools to do form-level analytics, you'll have a much better understanding of how visitors are interacting with the site, as shown in Figure 5-32.

We'll look at WIA in more detail in Chapter 6. For now, remember that abandonment is a coarse measurement and there are ways to perform more detailed analyses of where you're losing visitors.

Bounce rate is a close cousin to abandonment. Visitors "bounce" when they enter your site and leave immediately. Bounce rate can mean that visitors didn't find what they expected to find, which may be a sign that search engines aren't properly indexing your content or that you're buying the wrong search keywords. For news sites, blogs, and

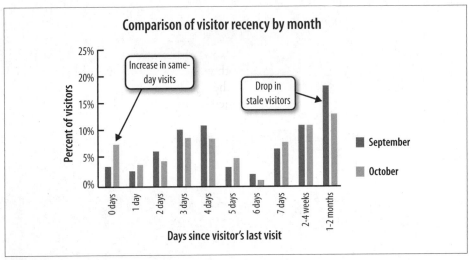

Figure 5-33. A comparison of last visits by user population, showing a reduction in visitor disengagement and attrition from one month to the next

certain content sites, a high bounce rate may be normal—visitors come to read one new article, then leave. Focus on encouraging visitors to bookmark your site, return, share content with others, and subscribe. You should care more about lifetime visitor loyalty than about one-page visits.

Attrition

Unlike bounces and exits, which happen at the end of a visit, attrition happens when your relationship with a visitor grows stale. There are many ways to measure attrition. You might look at the percentage of returning visitors, but if your site is acquiring new visitors, a declining percentage of returning visitors might simply indicate rapid growth.

A better measure of attrition is the number of users that haven't returned in a given period of time. By comparing attrition for two time periods, as shown in Figure 5-33, you can tell whether things are getting better or worse, and whether you're successfully retaining visitor attention.

Many community sites celebrate the number of visitors they have without considering the engagement of those visitors. Any site that depends on user activity and returning visitors must look at attrition carefully and make concerted efforts to reengage users who haven't visited the site in a long time.

Desirable Outcomes

While abandonment is a bad thing, not all departures are bad.

Visitors might leave your site once they've done what they wanted. Perhaps they've successfully completed a purchase, at which point they've finished their transactions.

Your work's not done: you need to learn more about why they bought what they did through surveys, and to encourage them to enroll and return.

Visitors may also have enrolled or subscribed, preferring to hear from you via RSS feed or email subscription. If this is consistent with your business model, it's a good thing. You do, however, still need to monitor whether they're receiving and acting on email messages that you send them or returning when you add new content to your RSS feed to ensure that the enrollment "stuck."

Tracking referrals and ad clicks

You may have sent the visitor elsewhere through a paid referral or an ad, in which case you're making money from the departure. An ad referral is a complex dance between your media website, the ad serving network, and the advertiser to whom a visitor eventually connects. Figure 5-34 shows how this works.

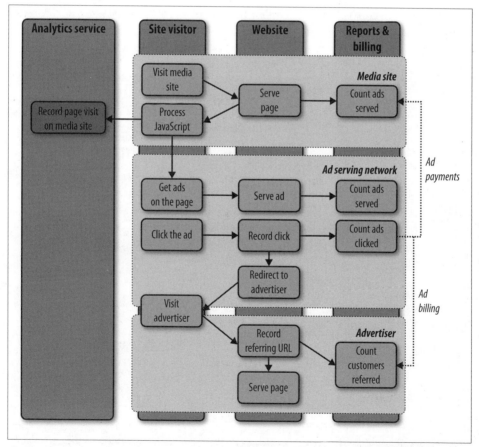

Figure 5-34. How an ad network works with a media site to serve ads and count clicks

Your web page contains a reference to content from an ad inventory provider. This content includes a link to the ad network (such as *http://pagead2.googlesyndication.com/pagead/iclk*) that contains a unique identifier for the ad and the destination site's URL.

The visitor's browser loads the ads that are on the page from an ad serving network, and if the visitor clicks the ad, the network records the click before sending the visitor to the destination (the advertiser's site).

This model doesn't let you count how many ads visitors clicked. As a result, you don't have your own numbers to compare to the ones the ad network tells you—in other words, you're letting the company that's paying the bills say how much it owes you. Not good.

To resolve this issue, you can add JavaScript to your pages to let you know when a visitor clicks an ad, then compare your measurements to the advertiser's numbers in order to resolve any disputes. This additional step is reflected in Figure 5-35 (visit *www.google.com/support/analytics/bin/answer.py?answer=55527&ctx=sibling* for details on using this for the Google Analytics JavaScript).

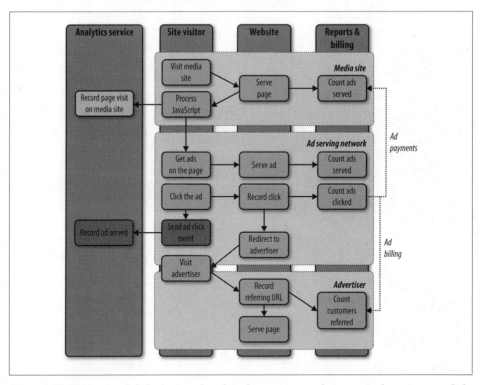

Figure 5-35. Capturing ad click events and sending them to your analytics service lets you count clicks on embedded ads

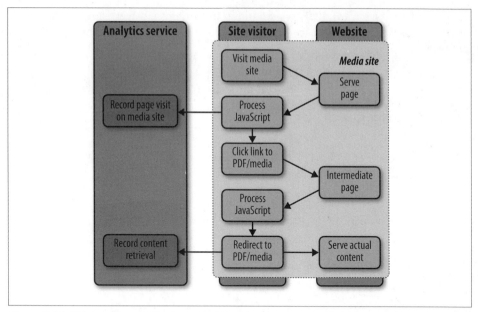

Figure 5-36. Using an intermediate page with JavaScript to capture clicks for non-HTML content

If you're running a sponsored site where advertisers pay for a time period or banner placement, you may be inserting advertising or banners on behalf of your sponsors. To track clicks you can either first redirect the visitor to an intermediate page that records the action or you can use JavaScript to send a message to your analytics system.

You can also do this if you're letting visitors download non-HTML content, such as Adobe PDF documents, since the document you're sending them can't execute Java-Script to let you know it was downloaded. Analytics tools can track requests for this intermediate page, giving you ad-tracking data as shown in Figure 5-36 (visit *http://www.google.com/support/analytics/bin/answer.py?answer=55529&ctx=sibling* for information on using this for Google Analytics).

Implementing Web Analytics

Measuring arrivals, visitor activity, and departures is the lifeblood of a web business. Without this information, you can't make smart decisions about your content, marketing, or business model. You can collect basic analytics data, such as bounce rate and visits, with just a few minutes' work.

More advanced deployment—tracking goals, building custom segments, and tagging content—takes work, however. And if you want to use analytics for accounting data (such as daily orders) as well as for site optimization, you'll probably have to work with the development team to extract additional information from backend databases.

There are free (or cheap) analytics solutions from the big advertising vendors, and Google Analytics has done a tremendous amount to make web operators aware of analytics.

Whatever you're doing, your implementation will have six basic steps:

1. Defining your site's goals
2. Setting up data capture
3. Setting up filtering
4. Identifying segments by which to analyze goals
5. Tagging page content
6. Verifying that everything is working

Let's look at the steps you'll need to take to implement web analytics.

Define Your Site's Goals

Your first step is to understand and map out your web business.

That might sound like a platitude, but it's an essential step in the process. Grab a whiteboard and some markers, and draw out your site using the places-and-tasks model outlined above. In each place, list what makes a visitor "productive." For each task, identify the steps a visitor needs to take to accomplish the task, and the metrics you need to collect in order to track those steps.

Now go through the tasks and see which ones drive your business (this will depend heavily on which of the four kinds of sites you're running). Each of these tasks will become a goal funnel in your analytics system, and you should try to assign a monetary value to each goal, even if it's something as simple as "getting contact information."

Once you've identified the important places and tasks, list the assumptions you're making. Somewhere in your business model, you're assuming that a certain number of visitors perform a given task. Write this number down; if you don't know what it is, guess. Do the same thing for every assumption you're making—how many people will exit from a particular place, how many will be referred by a particular kind of site, and so on.

You'll quickly realize that there are other things you'd like to know. How long does a place take to refresh? What do visitors think of this content? Where in the form do visitors leave? Did the email invitation bounce? You can collect all of these metrics—though sometimes you'll need third-party tools or coding to collect them. In the long term, you should try to integrate these metrics into a single, comprehensive view. For now, however, let's stick with simple web analytics. We'll come back to other kinds of monitoring later in the book.

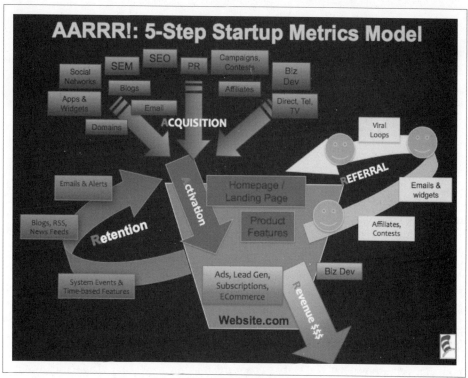

Figure 5-37. Dave McClure's 5-Step Startup Metrics Model

If you need some inspiration for your places and tasks, particularly if you're running a web startup, check out Dave McClure's "Startup Metrics for Pirates – AARRR!" at *www.slideshare.net/Startonomics/startup-metrics-for-pirates-presentation*.

McClure's model, shown in Figure 5-37, shows the many sources from which traffic can arrive (acquisition, retention, and referral), the places within the site (activation), and the tasks visitors can perform, such as enrollment (retention), inviting others (referral), and spending money (revenue). It provides a valuable starting point for thinking about your web business.

Notice that you don't need a web analytics account yet—you're still nailing down the really important question: *what is my site trying to encourage people to do?* Set aside a day for this process—it's a lot of work that often involves a surprising number of people in the company. Mapping your site is a time-consuming and iterative process; if it seems too daunting, pick a subset of the site at first, such as a particular application or a certain domain.

Once you know which metrics you want to collect for each place and task on your site, you need to define them within the analytics package you're using.

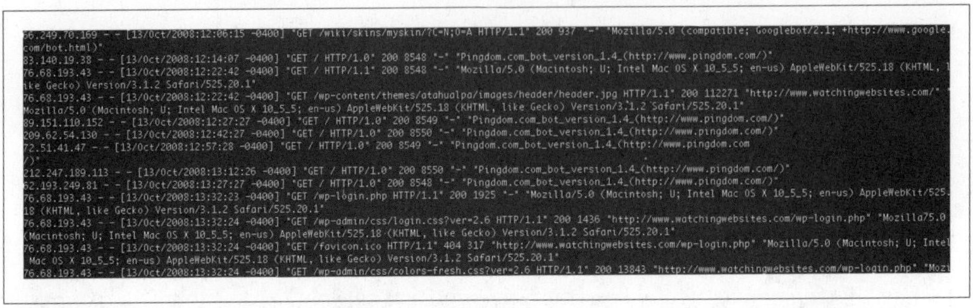

Figure 5-38. A standard httpd logfile

You don't need to configure monitoring of places much, since the default analytics metrics, like bounce rate, time on page, and page views tell you most of what you need to know. If users can perform simple on-page actions, such as upvoting submissions or paginating through content, you may need to record these in-page actions with your analytics tool.

When it comes to monitoring tasks, simply tell the analytics tool what the "goal" page is—for example, a payment confirmation screen—and the pages leading up to that goal.

Set Up Data Capture

Now that you know what you want to capture, it's time to set up the analytics system. There are several ways to deploy web analytics, from server logs to JavaScript-based collection.

Server logs

ELF files that web servers generate contain a limited amount of data, but if you're running a website behind a firewall or have specific privacy restrictions, they may be your best and only source of visitor information.

Popular web servers like Apache httpd and Microsoft IIS generate ELF-formatted logs (shown in Figure 5-38). These logs contain text strings, separated by spaces, that analytics tools import and reassemble into page requests and user visits.

Here's an example of a single hit recorded in an Apache server log:

```
10.100.3.200 - - [28/Mar/2009:11:50:55 -0400] "GET
/ HTTP/1.1" 200 53785 "http://twitter.com/seanpower"
"Mozilla/5.0 (Windows; U; Windows NT 6.0; en-US; rv:1.9.0.7)
Gecko/2009021910 Firefox/3.0.7" 125763
```

Here's a similar hit recorded in a Microsoft IIS server log:

```
2008-08-12 20:05:34 W3SVC216049304 10.0.0.1 GET /
WebUISupportFiles/images/bg_page.gif - 80 - 10.0.0.1 Mozilla/4.0+
(compatible;+MSIE+7.0;+Windows+NT+5.1;+.NET+CLR+2.0.50727;+InfoPath.2) 200 2921
```

Despite some differences in formatting and extra characters, server logs from various web servers are essentially the same. Each string in these logs tells you something about the visitor. For example:

- The *client-ip* field is the IP address of the visitor. You may be able to resolve the address to the organization to which it is registered; you can also look it up using a geolocation database to find out where the IP address is physically located.

- The *cs-referrer* field shows the site that has referred the traffic to you.

- The URI may have special information, such as a campaign identifier (for example, *www.watchingwebsites.com/alnk&campaign_id=130*) or other information against which you can segment visitors.

- The *user agent* field shows information on the browser and operating system the user is running.

- The *HTTP status code* shows errors that may have occurred in the request. However, you can't rely too heavily on HTTP status codes—a "200 OK" message may simply indicate that an apology page was delivered correctly. We'll look at performance and availability monitoring in Chapters 8 through 10.

When using web logs, the analytics tool retrieves logfiles from servers at regular intervals and processes the data they contain. The system usually deletes the logfiles after parsing them. Figure 5-39 shows how logfile-based analytics work.

1. A visitor's browser requests an object from the server. The browser provides a variety of data, such as referring URL and browser type, along with the request.

2. The web server responds with the requested object, as well as information such as object size and compression.

3. The web server writes a line to the logfile describing the request and response.

4. Logfiles from all of the web servers are copied to the analytics tool at regular intervals.

5. The analytics engine stores and aggregates all of the logfile data across all of the servers.

6. At regular intervals—often daily—the analytics engine generates a series of reports.

In more modern logfile analytics, the operator can request reports dynamically from the engine without waiting until a reporting period is complete.

Here are some of the advantages of a server log capture approach:

Better data privacy
> You own the data. You don't need a third party, such as an analytics service, to generate reports. This lets you control where the data goes and who can see it.

Useful for internal websites
> To use a third-party analytics service, you need to be able to send data outside of your organization. If your users' browsers can't reach the Internet, you can't use

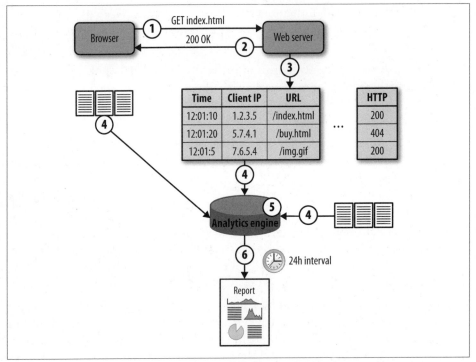

Figure 5-39. Web analytics implementation using logfiles as a data source

these services. This is particularly common for intranets and other in-house environments.

Works when JavaScript won't

Logs may also be useful if your visitor community doesn't use JavaScript, since other analytics collection approaches that rely on client JavaScript won't work. This is common with some mobile devices.

Pages load faster

If you rely on client-side JavaScript to monitor pages, your pages will take longer to load. By analyzing logs instead, you won't add to page delay.

Easier to merge with other data

The server may generate additional logs that contain useful data such as CPU load or custom metrics. This data can be interleaved with HTTP requests according to timestamps. If you don't have web server logfiles, you can't merge these other logs with them.

There are some disadvantages, however, to relying on server logs for capture:

Implementing them will require the efforts of more people
> IT and marketing departments need to work together to deploy log-based analytics. This can mean implementation delays if analytics isn't IT's top priority (and it seldom is!).

Log analysis is a lot of work
> As traffic grows, the amount of processing needed to parse web logs grows alongside it. On large sites, a day's web logs can consume hundreds of megabytes. Logging every request can also consume precious server resources, slowing down the site even further when it's busy. And if you don't plan carefully, it can take more than a day to process a day's web activity, meaning you're constantly playing catch-up while your analytics get increasingly out of date.

It may not be possible for you
> Depending on your hosting environment, you may not be able to get the web server logs in order to process them.

Limited visibility
> Log-based analytics can only report what's in the log files. Browser-side actions and information about the browser itself, such as cookies, screen resolutions, co-ordinates of user clicks, and so on, can't be collected from logs. Furthermore, log-files don't collect POST parameter information unless it's stored within the URI stem as parameters, which can pose a major security risk when users forward URIs to others.

Most of the web analytics industry has moved beyond web logs to client-side collection models unless they have no alternative but to use web log analysis.

Server agents

If you need more data than a web log contains, but absolutely *must* rely on the server itself to collect that data, you can use server agents. These are programs that run on the server and record transaction information, often with additional application context. These agents may also help with logfile collection by pushing logs to a centralized platform. Figure 5-40 shows an example of this.

1. A visitor's browser requests an object from the server. The browser provides a variety of data, including referring URL and browser type, along with the request.
2. The web server responds with the requested object, as well as information such as object size and compression.
3. The agent assembles the request and response data, as well as system health metrics and other data about the environment at the time of the request, and writes a line to the logfile.
4. The agent massages the data it's collected, compresses it, and passes it to a central engine for analysis. The agent may also proactively alert when it detects problems.

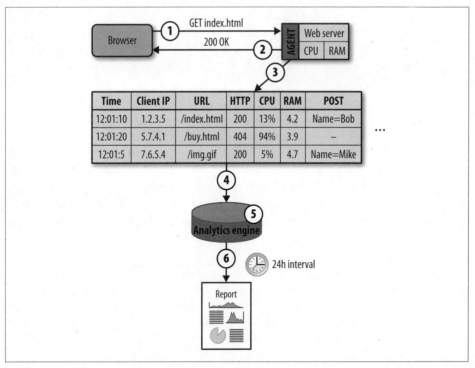

Figure 5-40. Server-agent-enhanced web analytics deployment

5. The analytics engine stores and aggregates all of the logfile data across all of the servers.

6. At regular intervals—often daily—the analytics engine generates a series of reports.

Popular web servers have well-defined interfaces to which an agent process can connect. On Apache servers, the module model (*http://modules.apache.org/*) allows third-party code to see each web request and the corresponding response. On Microsoft's servers, ISAPI (the Internet Server API) provides this connectivity. Companies like Symphoniq also use agent models to capture end user and platform health simultaneously.

Modules and ISAPI extensions may not just be passive listeners in a web conversation between web clients and a server—they may also change it. Many websites use modules for content caching, filtering, and authentication. These modules can get a better understanding of a web visit and can create more verbose logs than what's included in basic ELF.

 See *http://www.port80software.com/products/* for some examples of server agents that provide additional visibility and control over web traffic on the server itself.

Server agents give you more insight into transactions, at the expense of more computing power and additional components you have to manage. If you collect analytics data with server agents, there are some advantages:

Additional information
> Server agents may record additional information about the health of servers (such as CPU, memory, and I/O) and request data (such as POST parameters) that is typically not collected by web servers

Lower data collection overhead
> Server agents can handle some of the aggregation, logfile compression, transfer, and decompression needed to get logs from the servers to the analytics tool. This reduces the amount of work the analytics engine needs to do in crunching log data, as well as the total volume of logfiles that need to be collected from each server.

SSL visibility
> Agents on servers are "inside" the SSL encryption boundary and can see every part of a web transaction without additional work managing SSL keys.

Turnkey solution
> For a small site running on only one machine, a server agent may act as both a collector and a reporting interface.

Data is available, even when the service dies
> Server agents may still collect data when the HTTP service itself isn't functioning properly.

Relying on a server agent to collect data has some downsides, however.

Use of server resources
> Agents consume resources on the servers, which are better used handling requests and serving web pages.

More things to break
> Agents add another point of failure to the web servers. For sites of any size, web servers are stripped-down, single-purpose machines that have been optimized to do a few things quickly and reliably.

No client visibility
> Server agents share many of the limitations of web logs, because they don't have visibility into the client's environment and can't capture important segmentation data about what happens on the client.

Dependent on a working server
> If the server stops working, so does the server agent, leaving you guessing about what went wrong.

Man in the middle: Traffic capture

A third approach to collection is to capture a copy of the data flowing between clients and the web server, either through the mirroring functions of a switch, load balancer, or network tap, and to use this data to re-create the HTTP requests and responses.

To do this, a network device in front of the web server cluster makes a copy of every packet, which is then reassembled into TCP sessions consisting of HTTP requests and responses. The device is usually deployed just in front of or just behind the load balancer, or on the load balancer itself. Figure 5-41 illustrates how this model works.

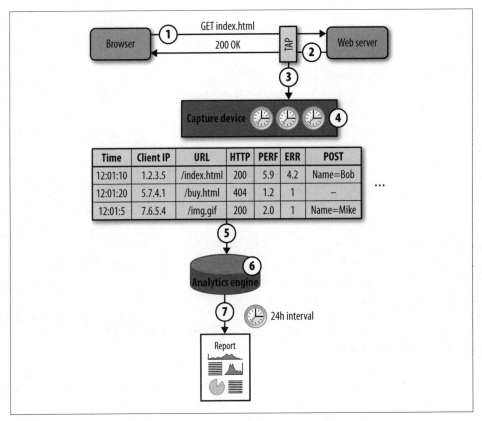

Figure 5-41. Web analytics deployment using passive sniffing technology

1. A visitor's browser requests an object from the server. The browser provides a variety of data, such as referring URL and browser type, along with the request.

2. The web server responds with the requested object, as well as information such as object size and compression.

3. A passive monitoring device sends a copy of the network traffic to a collector device.

4. The collector device reassembles the packets into the HTTP request and response, and may also detect errors, record timing information, and extract snippets of text from the content of the response.

5. The collector stores this data as an augmented ELF containing all the information recorded about the transaction.

6. The analytics engine stores and aggregates all of the logfile data across all of the servers.

7. At regular intervals—often daily—the analytics engine generates a series of reports.

Passive traffic capture was an early favorite for analytics companies like Accrue, who wanted to offload work from web servers, analyze traffic destined for several servers, and get more visibility.

Capturing analytics data through a passive sniffing approach has some advantages:

Aggregates data from many servers
Passive capture can collect many servers' requests at once, provided they cross a single network segment.

Captures network health information
The passive device may record information such as network timings or packet loss.

Works when servers break
If the server dies, the device still records the circumstances surrounding that failure, as well as requests from browsers that went unanswered during an outage.

Custom data collection
Passive capture can collect data from not only default headers, but also from the payload of a transaction (such as the dollar value of a checkout). This can be used as a substitute for some kinds of page tagging.

No network or server delay
This kind of collection introduces no delay to the network and is quick to deploy.

Sees all requests
Passive capture can see requests for content that doesn't execute JavaScript, such as an RSS feed, a media object, or an Acrobat PDF.

Disadvantages of passive capture include:

Requires data center access
A passive capture approach involves network equipment—you need physical access to the data center to deploy this kind of system and you need help from IT to deploy it.

Requires SSL keys
> Passive capture can't sniff encrypted traffic without help. You need to install SSL keys on the sniffing equipment, which may pose a security risk.

Problems with IP addresses
> If the capture device is in front of a load balancer, it won't see the IP addresses of the individual servers. Instead, it sees only the address of the load balancer, making it hard to tell which server handled which request without additional configuration.

Problems with IP addresses, part two
> Conversely, if the passive capture device is installed behind the load balancer, it doesn't see the true IP address of each request, so it can't always determine the source of traffic, making it impossible to segment by region or by domain name.

Lack of client visibility
> A passive capture device can't see what's happening on the browser itself, which is an increasingly important part of any application.

Because of these limitations, most passive capture tools focus on monitoring web performance and end user experience, where their insight into network performance is invaluable. Logfile generation from such devices is a side effect of how they capture data, but web analytics is seldom the main purpose of these tools.

Static image request

Early attempts at hosted analytics were simple: the web operator would embed a third-party object, such as a small banner or one-pixel image, in each page. Each time the page was loaded, the browser would request the small object from the third party, which would keep track of how many web requests it had seen.

Figure 5-42 illustrates this approach to collection.

1. The site operator embeds a reference to an image on the counter's site. The link includes the site operator's unique ID.

2. A visitor's browser requests an object from the server. Along with that request, the browser provides a variety of data such as referring URL and browser type.

3. The web server responds with the page containing the image reference to the site counter.

4. The browser parses the page it receives and sees that it has to ask the counter site (*counter.com*) for an image (*display.gif*).

5. The browser requests *display.gif* from the third-party counter site, and includes the unique ID of the site that was embedded in the page.

6. The counter site looks up the number of hits that the site with that ID had and increments it.

7. The counter site sends a rendered image containing the hit count back to the browser.

8. The browser displays the page containing the hit count.

9. The site engine may also offer reports on traffic volumes, like the one in Figure 5-43.

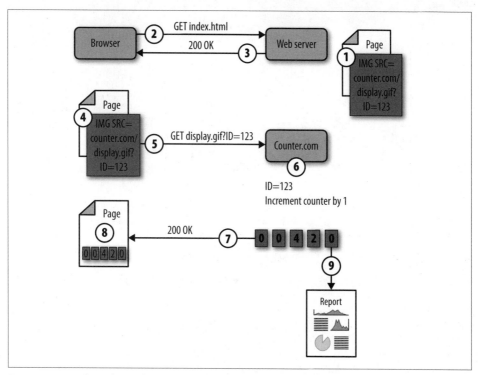

Figure 5-42. Web analytics deployment using an image-based counter

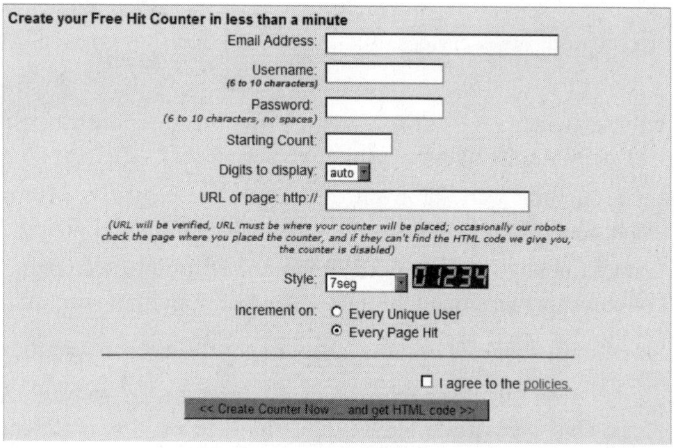

Create your Free Hit Counter in less than a minute

Email Address:
Username: (6 to 10 characters)
Password: (6 to 10 characters, no spaces)
Starting Count:
Digits to display: auto
URL of page: http://

(URL will be verified. URL must be where your counter will be placed; occasionally our robots check the page where you placed the counter, and if they can't find the HTML code we give you, the counter is disabled)

Style: 7seg   01234

Increment on: ○ Every Unique User
                 ● Every Page Hit

☐ I agree to the policies.

<< Create Counter Now ... and get HTML code >>

Figure 5-43. A report from Statcounter.com

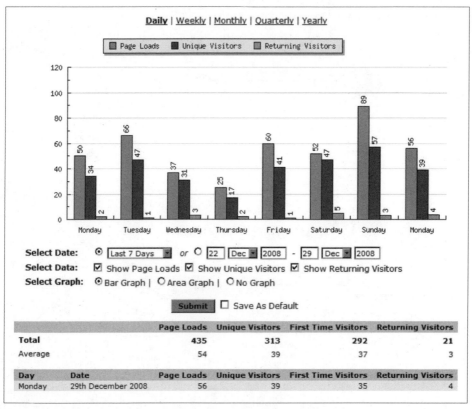

		Page Loads	Unique Visitors	First Time Visitors	Returning Visitors
Total		**435**	**313**	**292**	**21**
Average		54	39	37	3

Day	Date	Page Loads	Unique Visitors	First Time Visitors	Returning Visitors
Monday	29th December 2008	56	39	35	4

Figure 5-44. VisibleCounter hit counter configuration

Initially, services like VisibleCounter, shown in Figure 5-44, simply embedded the number of page hits in an image that was displayed to each visitor.

Companies that created these services quickly realized that they could glean more from each request:

- Several requests from the same browser are probably the same user; this allows the company to estimate the number of visitors.
- Requests that include a cookie from the service are returning visitors who have been to the site before.
- The identity of the page being visited can be found from the referring URL of the retrieved counter image, since that image is a component of the parent page.

As a result, static image requests became a way to get rudimentary traffic statistics.

Static image models aren't really analytics. They only capture page traffic information, and as we've seen, this is misleading data that's hard to act upon. We mention them here because this approach is unfortunately still used by certain media sites that rely on these statistics to share with advertisers to get premium rates.

Static image requests don't carry any custom information about the visitor or the browser environment. Modern JavaScript is superior to this approach. Stat counters are used by casual website creators who aren't interested in understanding their visitors' behaviors or improving their sites. Now that there are free web analytics packages available, there's no excuse to use a static image model for anything other than a lazy way of showing your audience how many visitors you've had.

JavaScript

The first three methods we looked at are all forms of server-side collection. In other words, they're installed near the web server itself. JavaScript (and the static image model) is a client-side collection model. This is by far the dominant model on the Internet today. Browsers are able to collect and send a great deal of information about a user's visit to a third-party system such as an analytics service. Figure 5-45 shows how this works.

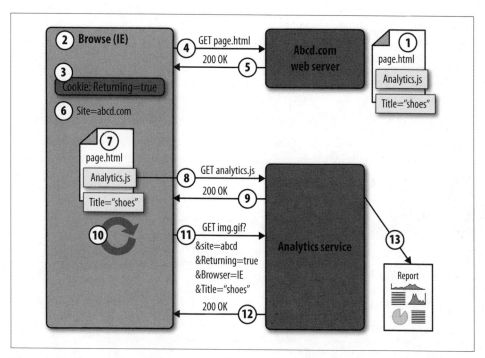

Figure 5-45. JavaScript-based collection of web analytics data

1. The site deploying analytics (Abcd.com) inserts a snippet of JavaScript code into its pages. The code includes a reference to a file, *Analytics.js*, that's stored on the analytics service's servers. The site also includes tags in the page that explain what it's about (in this case, "shoes").

2. A visitor launches his browser. Data about the browser is stored within the Document Object Model (DOM) of the browser.

3. The browser also has site-specific information, such as a cookie associated with Abcd.com. This cookie was created during a previous visit to the site.

4. The browser requests a page from Abcd.com.

5. The web server responds with a page containing the analytics script reference.

6. The browser's DOM also records the fact that the site being visited is abcd.com.

7. As the page is loaded, the browser realizes it has to retrieve *analytics.js* from the analytics service.

8. The browser requests the *analytics.js* code from the service.

9. The analytics service responds with the analytics JavaScript. This may be cached on the browser for future use.

10. When the browser receives the *analytics.js* JavaScript, it executes it. The script collects data from the DOM (browser type, site name), site-specific cookies ("returning=true") and tags within the page ("Title=shoes").

11. The analytics script appends all the data it has collected to a request for a tiny (1×1 pixel) image from the analytics service.

12. The analytics service returns the tiny image, storing the information it received about the visit.

13. The analytics service generates reports based on this data.

For some web platforms, such as blogs, analytics vendors provide plug-ins to make installation even easier. Most of the time, however, implementing JavaScript-based analytics is as easy as entering a few lines of code in a *footer.php* file or changing some file templates.

If you were installing Google Analytics, the JavaScript code would look something like this:

```
<script type="text/javascript">
  var gaJsHost = (("https:" == document.location.protocol) ?
  "https://ssl." : "http://www.");
  document.write(unescape("%3Cscript src='" + gaJsHost +
  "google-analytics.com/ga.js' type='text/javascript'%3E%3C/script%3E"));
</script>
  <script type="text/javascript">
  var pageTracker = _gat._getTracker("UA-xxxxxx-x");
  pageTracker._trackPageview();
</script>
```

Every time a browser loads a page with this script, it will make a request for *http://google-analytics.com/ga.js* and run the script it receives.

JavaScript is a popular collection approach, offering many advantages:

Sees what the visitor sees
> JavaScript sees what the user's environment is like. This includes DOM information such as resolution and activity within the page, or the page's title, or events such as Onload that mark important milestones in a page's delivery.

Works with SSL
> Because it's on the client, it can collect data within SSL connections.

Augments analytics with page context
> It can read tags from the page (identifying things like the products a user is browsing) and send them to the analytics platform, allowing you to more easily segment visits.

Scales with number of users
> In a JavaScript collection model, the browsers do some of the work of collecting user data. They then generate a single hit per page, rather than the hit-per-object rates of server-side collection. This reduces the load on servers somewhat.

Opens the door to hosted services
> Using JavaScript means you can send information to someone else via the visitor's browser. Most companies' IT organizations would be uncomfortable sending logfiles to a third party, but when the visitors' browsers send that data, it's acceptable.

Keeps reporting with RIAs and "long" pages
> JavaScript can manage communications with the analytics service beyond the initial page load. If a page has small subevents, such as a user clicking "play" on an embedded object, JavaScript can capture this client-side activity by sending additional information to the analytics service. JavaScript can also measure things like cursor and mouse movement and even send messages when a user closes a web page.

Reads cookies
> JavaScript can get details on the visitor from a cookie and send this to the service. So, if a user is a "platinum" subscriber to your website, you can store this in a cookie on the user's browser. When the JavaScript executes, it can read values from the `document.cookie` property of the page and turn them into labels for the visitor's session that you can then use for segmentation.

Analyzes mashups
> JavaScript sees requests to third-party sites. Because the JavaScript is executing as part of the container page, it sees all component objects, such as a Google Map or a YouTube video, and can report on them as well. If you're monitoring a mashup, you need JavaScript.

When collecting through JavaScript on a client, you may face the following disadvantages:

Can't see out of the sandbox

JavaScript runs within a browser. Browsers "sandbox" the code they run to prevent malicious websites from gaining control of visitors' computers. The script can't see information such as networking statistics, because it can't communicate directly with the operating system on which it's running.

Requires a loaded page

If the page doesn't load, JavaScript is useless. This is the main reason JavaScript hasn't had the same success for web performance monitoring that it has enjoyed for web analytics. If you want to measure problems, a system that doesn't work when problems occur isn't as useful.

Requires a JavaScript interpreter

Some mobile devices, and some very paranoid visitors, may have disabled JavaScript on their browsers, limiting your ability to collect from them. Users who delete their cookies may also be misreported as new visitors when, in fact, they are returning users.

 According to w3schools (*http://www.w3schools.com/browsers/browsers_stats.asp*), about 5% of users on the Internet today block JavaScript, although this percentage has been in a steady decline since 2005. This study may be misleading, as there is no indication that it takes mobile devices into account.

Page tagging still sucks

JavaScript's real limitation becomes apparent due to the necessity of manually tagging a page to give it meaning. In reports, there is a big difference between seeing a page named "id=3&item=19&size=9&c=5" versus "Big Blue Shoes with Pumps." Consequently, you will need to build analytics into your site from the outset, and it must provide context by individually tagging pages. Many sites have tens of thousands of unique pages. Tagging pages is a web analyst's most hated chore, and a daunting task if done as an afterthought.

Despite the limitations outlined here, JavaScript offers such improved visibility into your visitors that it's by far the leading approach to web analytics collection used online today.

Comparing data capture models

You'll probably use JavaScript for collection unless you have a site that can't take advantage of third-party Internet-connected services. Complementing JavaScript with passive capture for measuring user experience is also increasingly common within large organizations.

Table 5-2 summarizes the trade-offs of each approach.

Table 5-2. A comparison of different methods of deploying analytics

	Weblogs	Server agents	Passive capture	Image request	JavaScript
Deployment					
Requires physical access?			Y		
Requires server administrator?	Y	Y			
Requires changes to page content?				Y	Y
Requires Internet-connected users?				Y	Y
Data stored by third party?				Y	Y
Requires access to SSL keys?			Y		
Requires logfile collection?	Y	Y			
Data visibility					
Sees POST parameters?		Y	Y		Y
Sees network statistics (packet loss)?		Y	Y		
Sees what happened when web service dies?		Y	Y		
Sees what happened when entire server dies?			Y		
Sees failed requests (404s)?	Y	Y	Y		
Sees requests for non-JavaScript objects (PDFs, RSS feeds)?			Y		
Sees browser information (resolution, DOM, last visit)?					Y
Sees third-party content (mashups)?					Y
Sees client-side activity (mouse movement, etc.)?					Y
Performance impact					
Adds to server CPU load?	Y	Y			
Adds to page load time?				Y	Y

Set Up Filters

Now that you've decided how you want to collect analytics data, it's time to decide what you don't want to keep. Much of your web traffic isn't from the visitors you care about: it includes malicious attackers trying to exploit known weaknesses, spammers posting content you're going to block anyway, and crawlers busily indexing the Web. It may include synthetic tests that your web operations team is running to measure the health of the website, or it may consist of visits from internal users and employees whose traffic shouldn't be counted.

No two analytics tools will calculate the same number of hits for a website. This happens for many reasons: different filters, different definitions of a hit or a page, different

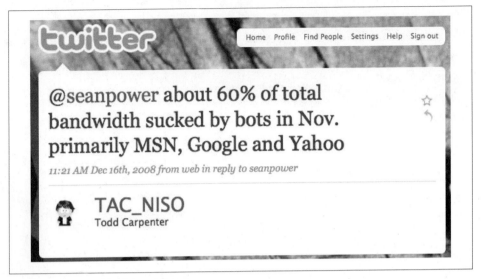

Figure 5-46. Interesting bot stats for niso.org

JavaScript approaches, the impact of bots (which, as Figure 5-46 shows, can be significant), and different ways of inferring pages from objects.

Some of the work of filtering out the noise may be done for you, depending on your chosen approach. Most analytics packages will block visitors with known user agents (those that self-identify as crawlers). If you're using JavaScript collection, crawlers that don't execute JavaScript won't send tracking hits to the analytics providers, but if you're using server-side collection methods you'll need to exclude this traffic.

You may not want to block synthetic traffic—it may be good to know who's crawling your site and how often they're updating their indexes—but you need to identify traffic that shouldn't be included in your estimates of real visitors, or you'll artificially lower your conversion rates.

As your site becomes better known, you'll see an increase in the number of machine-driven visitors that generate hits. We've seen sites where more than half of the total web traffic isn't actual visitors.

Identify Segments You Want to Analyze

Now that you're measuring how well visitors are attaining the goals you've set for them, it's time to segment those visitors up into groups. This way, you can see which groups are performing poorly and help them. You can also try changes to content, promotion, and design and see whether they improve things.

All analytics tools have some amount of built-in segmentation—new versus returning visitors, referring site, browser type, country of origin, and so on. You may want to

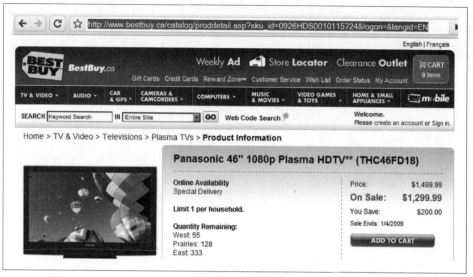

Figure 5-47. A page on BestBuy.ca showing a single URL (/catalog/proddetail.asp) for all product views

create new segments on the fly using page tags, which will allow you to group and analyze visitor performance according to new dimensions.

One quick way to do this is to use unique strings in a URL to create custom segments. For example, if you have pages whose URLs contain the strings "/buyer/" and "/seller/" in them, you can ask most analytics tools to use these to generate a custom segment such as "all URLs that contain /seller/" to see KPIs for just that part of the site.

Tag Your Pages to Give Them Meaning

With collection in place, goals identified, and filtering ready to clean up the noise, it's time to tag your pages. Tagging provides context about the page that's passed from your website to the browser, and from the browser to the analytics provider.

Tagging is where JavaScript-based collection gets complicated. To implement basic JavaScript analytics, all you had to do was include a few lines of text on each page. However, as far as the analytics service knows, those pages are identical. The only thing that makes them different is their URLs.

Consider a retail outlet that has a single page name for all products, which is a common characteristic of many retail sites. In Figure 5-47, that URL is */catalog/proddetail.asp*, and it's the same URL regardless of what product the visitor is looking at. The URL doesn't tell us much about what's on the page. Any human can see that the page is categorized as TV & Video→Televisions→Plasma TVs, but without some help, the JavaScript doesn't know what the page is about.

Fortunately, there's a way to take information (such as "Televisions") and augment the visitor's session record with it. Let's look at the script contained in the page in Figure 5-47 and see how it works.

First, the page includes variables that identify content, such as the user's account, the product name and model, and the multilevel content hierarchy of the product.

```
<!-- BEGIN WEBSIDESTORY CODE HITBOX COMMERCE HBX1.3 (cartadd) -->
<!--COPYRIGHT 1997-2004 WEBSIDESTORY,INC. ALL RIGHTS RESERVED.
 U.S.PATENT No.6,393,479B1 & 6,766,370. INFO:http://websidestory.com/privacy-->
<script language="javascript">
var _hbEC=0,_hbE=new Array;function _hbEvent(a,b){b=_hbE[_hbEC++]
=new Object();b._N=a;b._C=0;return b;}
var hbx=_hbEvent("pv");hbx.vpc="HBX0131.01a";hbx.gn="ehg-bestbuy.hitbox.com";

//BEGIN EDITABLE SECTION
//CONFIGURATION VARIABLES
hbx.acct="DM540930IBWD";//ACCOUNT NUMBER(S)
hbx.pn="0926hds0010115724-panasonic+46+1080p+plasma+hdtv+(thc46fd18)";//PAGE NAME(S)
hbx.mlc="/online/en/tv+and+video/televisions
/plasma+tvs/details";//MULTI-LEVEL CONTENT CATEGORY
hbx.pndef="default.asp";//DEFAULT PAGE NAME
hbx.ctdef="full";//DEFAULT CONTENT CATEGORY
```

The page can also include custom variables (which may be used for experimentation, custom error tracking, or segmentation), as well as details such as the price or whether the visitor was told that quantities were limited. All of this data can be used later to segment buyers and see if pricing affected conversion.

```
//CUSTOM VARIABLES
hbx.hc1="panasonic|panasonic+46+1080p+plasma+hdtv+(thc46fd18)";//BRAND|PRODUCT
hbx.hc2="0926HDS0010115724";//SKU
hbx.hc3="";//CUSTOM 3
hbx.hc4="";//CUSTOM 4
hbx.hrf="";//CUSTOM REFERRER
hbx.pec="";//ERROR CODES

//COMMERCE VARIABLES
hbx.cacct="975410043989";
hbx.pr="panasonic+46+1080p+plasma+hdtv+(thc46fd18)";  //comma delimited products
hbx.bd="panasonic";
hbx.ca="televisions";
hbx.pc="1299.99";     //comma delimited prices
hbx.qn=""; //comma delimited quantities
hbx.sr="1";    //store
hbx.cp="null";  //campaign
hbx.cam="0";  //cart add methodology, 0 = highwatermark, 1 = incremental
hbx.pv=1;  //product view flag, 0 = cart add, 1 = product view

//END EDITABLE SECTION
```

Note that these variables are seldom handcoded into the page. Rather, the application inserts them dynamically each time it renders the page. In other words, implementing tagging means working with developers.

Now the analytics script assembles all of these tags and details, along with browser information such as the type of web browser (`navigator.appversion`), the title of the page (`document.title`), and the referring URL (`document.referrer`):

```
function $ii( a, b,c){ return a.indexOf(b, c?c:0)};
function $is(a,b, c){return b>a.length?
"":a.substring(b,c!=null?c:a.length)};function $a(v){ return
 escape(v) }; var _sv=10, _bn=navigator.
appName,_mn="we74",_bv=parseInt(navigator.appVersion),_rf=$a(document.referrer),
_epg="n&cam="+hbx.cam+"&pv="+(hbx.pv?"1":"0&abd_type
=cart_add")+"&product="+$a(hbx.pr)+
"&quantity="+$a(hbx.qn)+"&brand="+$a(hbx.bd)+"&category="+$a(hbx.ca)+"&
price="+$a(hbx.pc)+
"&store="+$a((hbx.sr=="S"+"TORE")?1:hbx.sr)+"&tz=
PST&aid="+hbx.cacct;if(!$ii(_bn,"Micro"+
"soft"))_bn="MSIE";if(_bn=="MSIE"&&_bv==2)_bv=3;function $l(m,l){return m=="/"?
m:(($ii(m,"/")?"/":"")+(m.lastIndexOf("/")==l?m.substring(0,l):m))};function $n(
a,b){return(a==""||a=="/")?"/":$is(a,hbx.ctdef!="full"?a.lastIndexOf("/",b-2): $ii(
a,"/"),b)};function $o(a,b,c){var d=location.pathname,e=$is(d,d.lastIndexOf("/")+
1,d.length);if(a&&b==c){return(hbx.pndef=="title"&&document.title!=""&&document.
title!=location)?document.title:e?e:hbx.pndef}else{return(b==c)?$n(d,d.lastIndexOf(
"/")):$l(b,b.length-1)}};function $p(a,b,c,d){return ""+(c>-1?$o(b,$is(a,0,c),d
)+";"+$p($is(a,c+1),b,$ii($is(a,c+1),";")):$o(b,a,d))};
hbx.mlc=$p( hbx.mlc,0,$ii(hbx.mlc,
";"),"CONTENT+CAT"+"EGORY");hbx.pn=$p(hbx.pn,1,$ii
(hbx.pn,";"),"PUT+"+"PAGE+NAME+HERE" );
</script><script type="text/javascript" src="/javaScript/hitbox/hbx.js"></script>
<noscript>
```

The result of all this work is a reference to an image embedded in the page. The browser requests the image from the analytics service (in this case, *ehg-bestbuy.hitbox.com*).

```
<img src="http://ehg-bestbuy.hitbox.com?hc=none&
cd=1&hv=6&ce=u&hb=DM540930IBWD&aid=975410043989&n=0926hds0010115724-panasonic
+46+1080p+plasma+hdtv+(thc46fd18)&vcon=/online/en/tv+and+video/
televisions/plasma+tvs/details&seg=&cmp=&gp=&cam=0&pv=1
&abd_type=cart_add&product=panasonic+46+1080p+plasma+hdtv+(thc46fd18)&quantity
=&price=1299.99&store=1&tz=PST&vpc=HBX0131.01an"
 border="0" width="1" height="1"></noscript>
<!-- END WEBSIDESTORY CODE  -->
```

All of the metadata about the page is appended to the request in the form of a set of URI parameters. The publisher of the page never intends to display this image (it's 1 × 1 pixel wide with no border). The image doesn't matter, though: it's the request that matters.

If you have a small site that doesn't have dynamic backends, however, you may have to manually edit page content to provide context about the page to whatever analytics package you're using. Some packages, like Google Analytics, make this fairly simple, while enterprise-grade packages are much more demanding, but provide better segmentation and analytics as a result.

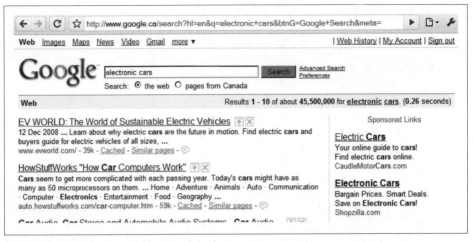

Figure 5-48. A Google search showing the referring URL containing query terms

The tagging process varies by analytics platform. There are a few great analytics integration checklists on the Web. For Google Analytics, check out:

- *http://blog.vkistudios.com/index.cfm/2008/12/5/Google-Analytics-Power-User--Tutorials-and-Screencasts--Part-2--Account-setup*
- *http://analytics.mikesukmanowsky.com/analytics/index.php/2007/08/18/google-analytics-installation-guide/*

For larger platforms, like Omniture, check out:

- *www.wickedsciences.com/blogs/?p=6*
- *www.kpilist.com/2008/10/evar-sitecatalyst-custom-conversion.html*

One final note on JavaScript implementation: page tags can bloat pages and cause them to load slowly, so it makes sense to put this kind of content at the bottom of your pages.

Campaign Integration

By defining your goals, you're able to determine where visitors went, and by tagging your pages, you're able to provide context for what they did. There's one more piece missing, however: you need to capture the things that drove people to your site in the first place.

When a user searches for something, the referring URL contains a list of search terms. A search for "electronic cars," for example, has the URI parameters *q=electronic +cars*, as shown in Figure 5-48. When an analytics script sends the referring URI to an analytics service, the service can parse this URI and see that the keywords "electronic" and "cars" were used.

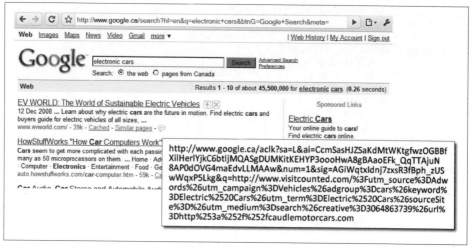

Figure 5-49. The URI for an ad click leading to an intermediate page that records the click-through

If the link is part of a paid campaign, however, the link doesn't go directly to the destination site. Instead, it first requests an ad click object from the search engine (in this case, `www.google.ca/aclk`) with a number of parameters identifying the campaign, the advertiser, and so on, as shown in Figure 5-49.

Your analytics tool needs to extract the campaign information from the subsequent referring URI, which you can then use to segment visitors according to campaigns and goal attainment. Google Analytics does this automatically for Google AdWords, but you can manually create a URL using Google's URL Builder (*www.google.com/support/googleanalytics/bin/answer.py?hl=en&answer=55578*), shown in Figure 5-50.

Campaign information can also be embedded in pages. In the Best Buy example cited earlier, the site operator can embed form, segment, and campaign information within the page according to landing page or elements of the referring URL. Some of these attributes may be passed from page to page during a visit.

```
//OPTIONAL PAGE VARIABLES
//ACTION SETTINGS
hbx.fv="";//FORM VALIDATION MINIMUM ELEMENTS OR SUBMIT FUNCTION NAME
hbx.lt="auto";//LINK TRACKING
hbx.dlf="n";//DOWNLOAD FILTER
hbx.dft="n";//DOWNLOAD FILE NAMING
hbx.elf="n";//EXIT LINK FILTER

//SEGMENTS AND FUNNELS
hbx.seg="";//VISITOR SEGMENTATION
hbx.fnl="";//FUNNELS

//CAMPAIGNS
hbx.cmp="";//CAMPAIGN ID
hbx.cmpn="";//CAMPAIGN ID IN QUERY
hbx.dcmp="";//DYNAMIC CAMPAIGN ID
```

```
hbx.dcmpn="";//DYNAMIC CAMPAIGN ID IN QUERY
hbx.hra="ATT";//RESPONSE ATTRIBUTE
hbx.hqsr="";//RESPONSE ATTRIBUTE IN REFERRAL QUERY
hbx.hqsp="";//RESPONSE ATTRIBUTE IN QUERY
hbx.hlt="";//LEAD TRACKING
hbx.hla="";//LEAD ATTRIBUTE
hbx.gp="";//CAMPAIGN GOAL
hbx.gpn="";//CAMPAIGN GOAL IN QUERY
hbx.hcn="";//CONVERSION ATTRIBUTE
hbx.hcv="";//CONVERSION VALUE
```

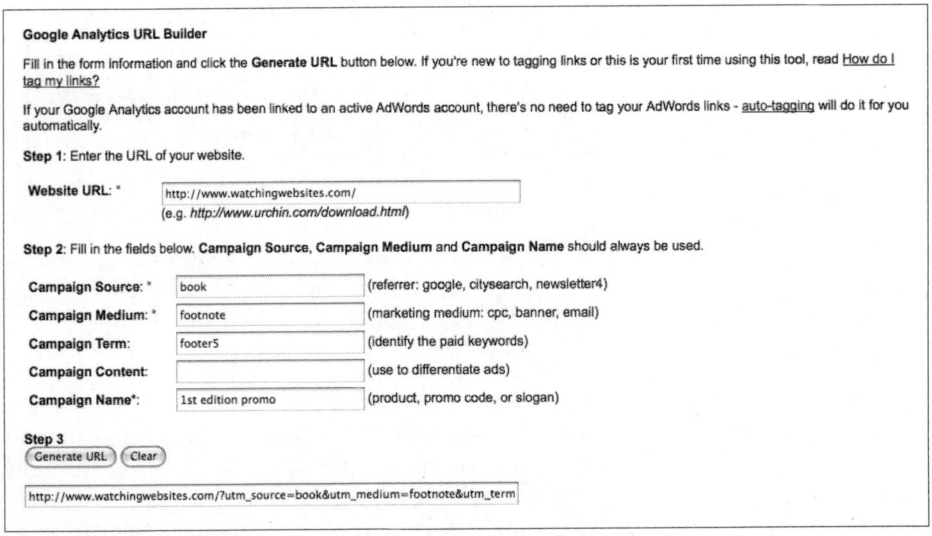

Google Analytics URL Builder

Fill in the form information and click the **Generate URL** button below. If you're new to tagging links or this is your first time using this tool, read How do I tag my links?

If your Google Analytics account has been linked to an active AdWords account, there's no need to tag your AdWords links - auto-tagging will do it for you automatically.

Step 1: Enter the URL of your website.

Website URL: * `http://www.watchingwebsites.com/`
(e.g. *http://www.urchin.com/download.html*)

Step 2: Fill in the fields below. **Campaign Source**, **Campaign Medium** and **Campaign Name** should always be used.

Campaign Source: * `book` (referrer: google, citysearch, newsletter4)

Campaign Medium: * `footnote` (marketing medium: cpc, banner, email)

Campaign Term: `footer5` (identify the paid keywords)

Campaign Content: (use to differentiate ads)

Campaign Name*: `1st edition promo` (product, promo code, or slogan)

Step 3
(Generate URL) (Clear)

`http://www.watchingwebsites.com/?utm_source=book&utm_medium=footnote&utm_term`

Figure 5-50. By entering campaign parameters into Google's URL builder, you create a URL that, if followed, will count toward that campaign's visits

Go Live and Verify Everything

With collection, filtering, goals, tagging, and campaign segmentation in place, it's time to start testing. Here's a quick preflight checklist.

Is performance still acceptable?

Compare pages with JavaScript to those without, and ensure that the time it takes to load the page hasn't increased too much. If it has, you'll negatively affect conversion. For more information on how to measure page performance, see Chapters 8 through 10

Are pages instrumented properly?

Go to each section of the site and view the page source to ensure the scripts appear in the correct places. It can be hard to do this manually, but you can use tools like Stephane Hamel's WASP (*http://webanalyticssolutionprofiler.com/*), shown in

---

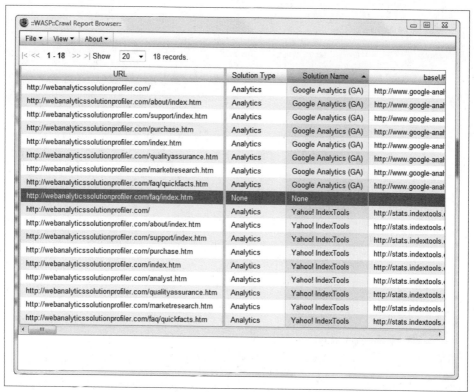

Figure 5-51. A WASP report showing pages with missing tags

Figure 5-51, to crawl a site and identify improperly instrumented pages, or a tool like Firebug to see what's running when you load a site.

Are goals properly configured to measure conversions?

A common problem when implementing analytics is getting goals wrong. If analytics isn't measuring the right steps in a goal, you can't measure success. Fortunately, there are a variety of ways to check goal outcomes and see if you're tracking them properly.

A reverse goal path, like the one shown in Figure 5-52, shows how visitors reached a goal and may identify other paths toward an outcome that you didn't know about. This is a good way to catch missing steps in a conversion process.

You should also compare your financials to your analytics: if the accounting department says you received 20 orders on Monday, but your system only saw 5, something is definitely amiss.

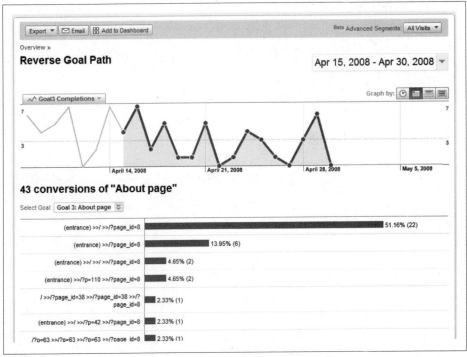

Figure 5-52. Google Analytics' Reverse Goal Path shows how people got to the goal in a funnel

Are tags working correctly?

The easiest way to check page tags is to run reports and see if pages are properly identified. If you want to examine your configuration more closely, change your browser's user agent to something unique, making it easier to identify within your analytics package.

Here's a good way to use Firefox to determine if your tags are working correctly:

1. Open a new browser window.
2. Enter **about:config** in the browser's address bar.
3. Right-click the blank area on the screen and select New from the drop-down list.
4. Select String to create a new browser property as a string.
5. Enter **general.useragent.override** for the preference name.
6. Enter a user agent string of your choosing for the value.
7. To check that it's properly configured, type **useragent** in the Filter field. You should see a screen similar to the one shown in Figure 5-53.

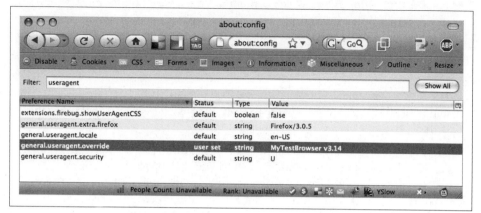

Figure 5-53. Editing the useragent setting in Firefox

Now when you surf with this browser, you'll be identified with the new string instead of your standard browser. You can try out the site, then run an analytics report for this user agent or look for it in logfiles to see the results.

Unfortunately, most analytics tools generate reports only once a day, which means that testing configurations can be time-consuming and iterative, with day-long waits before you can see if your latest changes are working. Some analytics packages, such as Clicky, report data more frequently, making them easier to configure and test.

Sharing Analytics Data

Once your analytics is in place and running correctly, you need to share the data with those who can make use of it.

No organization wants to be overwhelmed with dozens of reports at the outset. Giving your stakeholders open access to the entire system may backfire, as they won't understand the various terms and reports without some explanation. It's far better to pick a few reports that are tailored to each recipient and send regular mailouts. For example:

- For *web designers*, provide information on conversion, abandonment, and click heatmaps.
- For *marketers and advertisers*, show which campaigns are working best and which keywords are most successful.
- For *operators*, show information on technical segments such as bandwidth and browser type, as well as countries and service providers from which visitors are arriving. They can then include those regions in their testing and monitoring.
- For *executives*, provide comparative reports of month-over-month and quarterly growth of KPIs like revenue and visitors.

- For *content creators*, show which content has the lowest bounce rates, which content makes people leave quickly, and what visitors are searching for.

- For *community managers*, show which sites are referring the most visitors and which are leading to the most outcomes, as well as which articles have the most content.

- For *support personnel*, show which search terms are most popular on help pages and which URLs are most often exits from the site, as well as which pages immediately precede a click on the Support button.

Include a few KPIs and targets for improvement in your business plans. Revisit those KPIs and see how you're doing against them. If you really want to impress executives, use a *waterfall report* of KPIs to show whether you're making progress against targets. Waterfall reports consist of a forecast (such as monthly revenue) below which actual values are written. They're a popular reporting format for startups, as they show both performance against a plan and changes to estimates at a glance. See *http://redeye.first round.com/2006/07/one_of_the_toug.html* for some examples of waterfall reports.

Once analytics is gaining acceptance within your organization, find a couple of key assumptions about an upcoming release to the site and instrument them. If the home page is supposed to improve enrollment, set up a test to see if that's the case. If a faster, more lightweight page design is expected to lower bounce rates, see if it's actually having the desired effect. Gradually add KPIs and fine-tune the site. What you choose should be directly related to your business:

- If you're focused on retention, track metrics such as time spent on the site and time between return visits.

- If you're in acquisition mode, measure and segment your viral coefficient (the number of new users who sign up because of an existing user) to discover which groups will get you to critical mass most quickly.

- If you're trying to maximize advertising ROI, compare the cost of advertising keywords to the sales they generate.

- If you're trying to find the optimum price to charge, try different pricing levels for products and determine your price elasticity to set the optimum combination of price and sales volume.

There are many books on KPIs, such as Eric Peterson's *Big Book of Key Performance Indicators* (*www.webanalyticsdemystified.com/*), to get you started.

Repeat Consistently

For your organization to become analytics-driven, web activity needs to be communicated consistently. Companies that don't use analytics as the basis for decision-making become lazy and resort to gut-level decisions. The Web has given us an unprecedented ability to track and improve our businesses, and it shouldn't be squandered.

The best way to do this is to communicate analytics data regularly and to display it prominently. Annotate your analytics reports with key events, such as product launches, marketing campaigns, or online mentions. Build analytics into product requirements. The more your organization understands the direct impact its actions have on web KPIs, the more they will demand analytical feedback for what they do.

If you want to add a new metric to analytics reports you share with others, consider showcasing a particular metric for a month or so, rather than changing the reports constantly. You want people to learn the four or five KPIs on which the business is built.

Start Experimenting

Think of your website as a living organism rather than a finished product. That way, you'll know that constant adjustment is the norm. By providing a few key reports and explaining what changes you'd like to see—lower bounce rates, for example, or fewer abandonments on the checkout page—you can solicit suggestions from your stakeholders, try some of those suggestions, and show them which one(s) worked best. You'll have taught them an important lesson in *experimentation*.

One of the main ways your monitoring data will be used is for experimenting, whether it's simply analytics, or other metrics from performance, usability, and customer feedback. Knowing your site's desired outcomes and the metrics by which you can judge them, you can try new tactics—from design, to campaigns, to content, to pricing—and see which ones improve those metrics. Avoid the temptation to trust your instincts—they're almost certainly wrong. Instead, let your visitors show you what works through the science of A/B testing.

Everything Is an Experiment

You should be implementing A/B testing on some aspect of your site *every day* to maximize its potential.

In fact, any time you find yourself writing one marketing email, laying out one page, preparing one Twitter notice, picking one photo, or drawing one illustration, create two versions instead. Test on a small audience to determine which one works better, then send it to the broader audience.

This simple change will have an immediate, positive impact on all of your online marketing.

A/B testing involves trying two designs, two kinds of content, or two layouts to see which is most effective. Start with a hypothesis. Maybe you think that the current web layout is inferior to a proposed new one. Pit the two designs against one another to see which wins. Of course, "winning" here means improving your KPIs, which is why you've worked so hard to define them. Let's assume that you want to maximize enrollment on the Watching Websites website. Your metric is the number of RSS feed

Figure 5-54. The landing page, in its current form, has an RSS icon in the top-right corner

subscribers. The current layout, shown in Figure 5-54, hides a relatively small icon for RSS subscription in the upper-right of the page.

We might design a new layout with a hard-to-miss enrollment icon (called the "treatment" page) like the one shown in Figure 5-55 and test the treatment against the original or "control" design.

RSS subscriptions are our main metric for success, but you should also monitor other KPIs such as bounce rate and time on site to be sure that you have improved enrollment without jeopardizing other important metrics.

Testing just two designs at once can be inefficient—you'd probably like to try several things at once and improve more quickly—so advanced marketers do *multivariate* testing. This involves setting up several hypotheses, then testing them out in various combinations to see which work best together.

It's hard to do multivariate testing by hand, so more advanced analytics packages can automatically try out content across visitors, taking multiple versions of offers and discovering which offers work best for which segments, essentially redesigning portions of your website without your intervention.

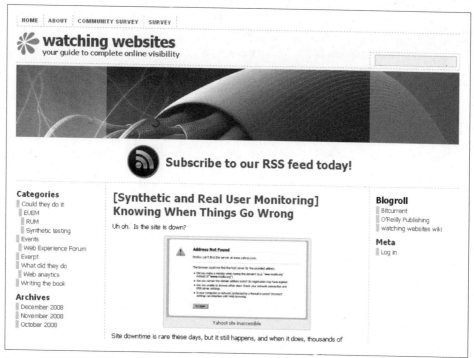

Figure 5-55. The landing page with an alternate RSS feed icon and a call to action

When embarking on testing, keep the following in mind:

Determine your goals, challenge your assumptions

> What you think works well may be horrible for your target audience. You simply don't know. The only thing you know is which KPIs matter to your business and how you'd like to see them change. You're not entitled to an opinion on how you achieve those KPIs. Decide what to track, then test many permutations while ignoring your intuition. Perhaps your designer has a design you hate. Why not try it? Maybe your competitors do something differently. Find out if they have a reason for doing so.

Know what normal is

> You need to establish a baseline before you can measure improvement. Other factors, such as a highly seasonal sales cycle, may muddy your test results. For example, imagine that your site gets more comments on the weekend. You may try an inferior design on a weekend and see the number of comments climb, and think that you've got a winner, but it may just be the weekend bump. Know what normal looks like, and be sure you're testing fairly.

Make a prediction

Before the test, try and guess what will happen to your control (baseline) and treatment (change) version of the site. This will force you to list your assumptions, for example, "The new layout will reduce bounce rate, but also lower time on site." By forcing yourself to make predictions, you're ensuring that you don't engage in what Eric Ries calls "after-the-fact rationalization."

Test against the original

The best way to avoid other factors that can cloud your results is to continuously test the control site against past performance. If you see a sudden improvement in the original site over past performance, you know there's another factor that changed things. We've seen one case where a bandwidth upgrade improved page performance, raising conversion rates in the middle of multivariate testing. It did wonders for sales, but completely invalidated the testing. The analytics team knew something had messed up their tests only because both the baseline and control websites improved.

Run the tests for long enough

Big sites have the luxury of large traffic volumes that let them run tests quickly and get results in hours or days. If you have only a few conversions a day, it may take weeks or months for you to have a statistically significant set of results. Even some of the world's largest site operators told us they generally run tests for a week or more to let things "settle" before drawing any important conclusions.

Make sure both A and B had a fair shake

For the results to be meaningful, all of your segments have to be properly represented. If all of the visitors who saw page A were from Europe, but all of those who saw page B were from North America, you wouldn't know whether it was the page or the continent that made a difference. Compare segments across both the control and the treatment pages to be sure you have a fair comparison.

Repeat and ramp

Once you've completed a few simple A/B tests, you can try changing several factors at once and move on to more complex testing. As an organization, you need to be allergic to sameness and addicted to constant improvement.

For more information on testing, we suggest *Always Be Testing: The Complete Guide to Google Website Optimizer* by Bryan Eisenberg et al. (Sybex) and "Practical Guide to Controlled Experiments on the Web: Listen to Your Customers not to the HiPPO" (*http://exp-platform.com/hippo.aspx*).

Choosing an Analytics Platform

Once you've factored in the collection methods, kinds of reports, and testing capabilities you'll need for your analytics, it's time to choose a platform.

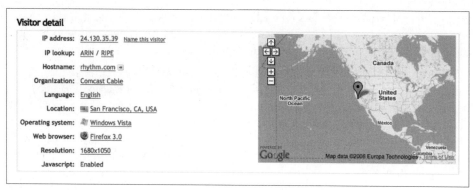

Figure 5-56. Visitor detail in Clicky Analytics

Free Versus Paid

Eric T. Peterson (*http://blog.webanalyticsdemystified.com/weblog/2007/07/the-problem -with-free-analytics.html*) concluded that if you've deployed a free analytics solution, you're probably:

- Only casually making use of web analytics
- Understaffed in the analytics department
- Lacking experience with web analytics in general

Far from criticizing free tools themselves, Peterson makes the point that you're much more likely to take analytics seriously if you've paid for it. What's more, you get what you pay for—free tools provide information, but require much more manual effort on the part of marketers to make and test their changes. In a small organization, analytics often becomes a lower priority than selling or design.

When you pay for something, you'll have access to support, and you may even be able to impact the road map of the product, depending on how common your request is or how much clout you have with the company providing the service.

On the other hand, Brian Eisenberg, CEO at FutureNow, Inc., says:

> My philosophy has always been to "get good at free then pay." There's no sense paying for something until you really operationalize its use. With today's free tools offering 65– 85% of the functionality of high-end tools, I am not sure free is only for the causally involved. About 30% of paid implementations also have Google Analytics or Yahoo! analytics installed.

Real-Time Versus Trending

At the entry level, analytics tools can be divided into two broad groups. Real-time tools, like those from Clicky (Figure 5-56), show you what's going on right now, and are the closest you can get to spying on individual users, giving you their locations, IP addresses, and so on.

Real-time tools concentrate less on goals, outcomes, and conversions. They are useful for root-cause analysis and quick answers (who tweeted that message, who blogged about me first, and so on), but are not designed for long-term decision making.

Goal-oriented analytics tools, on the other hand, have less detailed information. They tend to be tied to backend business models (for example, Yahoo!, Google, and Microsoft's tools automatically integrate with their search and paid search keyword systems). This is consistent with their business models. Services like Google Analytics want you to convert more people through AdWords so that you and they can make more money. Therefore, companies who create these tools want to help you optimize conversions without getting down to individual users.

Some solutions may offer both real-time and long-term perspectives.

Hosted Versus In-House

If you need to keep data to yourself for legal or ethical reasons, or if your visitors' browsers don't connect to the public Internet, you may have no choice but to run your own analytics platform.

If you can use a third-party solution, we suggest you do so—you'll get faster development of features and useful functions, like comparative reports that show how you're faring against others. The only exception to this rule is if you have a business that's tied tightly to some custom analytics.

Data Portability

If you can't get a copy of your data from your analytics provider, you can't leave. Having the ability to bring your data in-house—or better yet, to import it into an alternate analytics provider—means you can negotiate better pricing and keep your service providers honest. This alone is a good reason to care about data portability.

There's an important second reason that's central to the theme of this book. As we'll see in the closing chapters, you'll want to combine analytics data with other information in order to make better decisions about your business. Your analytics solution should either be able to import third-party data for analysis, or export analytics data to a data warehouse you run so that you can analyze it yourself.

The Up-Front Work

Analytics can be hard work, but it pays off. How much time and effort you invest will dictate what you get in return.

What You Get for Free

If you simply add JavaScript for analytics to your website, you'll get basic traffic metrics, some fundamental behavioral data (such as bounce rates) and Top-N reports that show you the most common landing pages, exit pages, and traffic sources.

What You Get with a Bit of Work

By adding goals to your analytics system, you can start to sort traffic according to business outcomes. Your top-N reports will now be about goal attainment, rather than being simply about popularity. You'll not only know who's sending you visitors, but who's sending you the *right* ones. You'll also know the most common paths through your site and where people are leaving.

What You Get with a Bit More Work

By creating custom segments and tags that add context, you'll know more about what's happening on the site. You can segment traffic and understand which products, offers, or users behave in ways you want, and which don't. This focuses your marketing efforts, site design, advertising, and keywords selection.

What You Get with a Lot of Work

Using clients with custom event monitoring, you can see how visitors interact with your site. This isn't for the faint of heart—it requires modifications to JavaScript and a lot more testing. You can also automate multivariate testing (or buy analytics tools that automate it for you), so you're constantly testing competing layouts, offers, campaigns, and content. While this is a lot of work, the result is a site that's constantly and automatically adapting to what its visitors want.

Web Analytics Maturity Model

Throughout this book, we're going to look at a maturity model for web visibility. The first of these models is web analytics, shown in Table 5-3. It borrows heavily from work by Bill Gassman of Gartner and Stephane Hamel of Immeria, and shows how companies progress through various levels of maturity with their web analytics.

Table 5-3. The Web Analytics Maturity Model (adapted from Stephane Hamel and Bill Gassman)

Maturity level	Level 1	Level 2	Level 3	Level 4	Level 5
Focus?	Technology: make sure things are alive	Local site: make sure people on my site do what I want them to	Visitor acquisition: make sure the Internet sends people to my site	Systematic engagement: Make sure my relationship with my visitors and the Internet continues to grow	Web strategy: Make sure my business is aligned with the Internet age
Who?	Operations	Merchandising manager	Campaign manager/SEO	Product manager	CEO/GM
Analytics	Page views, visits, visitors, top ten lists, demographics, technographics	Path analysis, funnel reports, A/B testing, KPIs, dashboards	Merchandising, segmentation, SEO, community referrals, campaign optimization, personas, KPI alerts	Multichannel aggregation, cost-shifting analysis, lifetime visitor value, personalization, dynamic content serving	Multichannel sales reporting, activity-based costing, balanced scorecards, strategic planning, predictive analytics, integrated user experience

Most organizations begin by looking only at traffic. They then turn their efforts inward, trying to get their own sites in order and defining KPIs that should guide their improvements. Once the site is converting its visitors and encouraging them toward the goals you want, organizations focus outward to increase the amount of traffic that's coming in. As the site grows, they incorporate additional data from other sources, such as performance, call center traffic, visitor lifetime value, user feedback, and so on.

Truly mature organizations move beyond even this level of integration, making web analytics a part of their strategic planning process and running their businesses through KPIs. Web metrics become a part of performance reviews and business unit goal-setting, and companies start to use analytical tools to look forward as well as back.

CHAPTER 6

How Did They Do It?: Monitoring Web Usability

Jakob Nielsen once said that in the first 10 years of the Web, doubling conversion rates was easy—you just had to be sure you weren't making silly mistakes with your web design.

Those easy days are over. Now a usable site is the price of admission. Visitors decide whether to leave or remain within seconds of first arriving. Modern sites must surprise and delight visitors by not only giving them what they need, but also by exceeding their expectations. Of course, if your product or service sucks, even the best website in the world won't save you. However, there are plenty of companies with excellent offerings that fail because users can't interact with their websites. We have to change how we think about usability—not just fixing it, but using it as the basis for innovation and differentiation.

Web Design Is a Hypothesis

Websites are invitations to interact. A site's designer intended for it to be used in a certain way, and its usability is a measure of how easily its target audience can interact with it. Counterintuitive sites discourage visitors, while well-designed ones steer visitors toward the desired goals.

User interaction designers base their interfaces on many factors. They strive to use well-understood controls, familiar conventions, consistent structure, and unambiguous terminology. All of these factors only lead to a hypothesis—an educated guess about how visitors will interact—that then needs to be tested and verified. Looking at websites in this way forces us to describe our assumptions about how visitors will use the site, which in turn tells us what we should be monitoring and testing.

Good designers mistrust their designs. They recognize that there's simply no way to know whether a site design will work without first conducting usability testing. They'll be able to tell you why they designed a particular part of the site a certain way, or point to the elements of their designs about which they're least certain. And they're probably your strongest allies when it comes to collecting more data on interactions.

Bad designers, on the other hand, will lobby for their particular layouts or color schemes, regardless of whether they work well in production. They care more about seeing their artwork on the Internet than about whether that artwork helps the business. And they're convinced that they have the answers to usability problems. Bad designers who are eager to push the envelope forget that their visitors aren't like them.

Good design is often boring. Some of the simplest, most plainly designed websites rely on familiar controls like radio buttons and simple black-on-white text. While they work, they frustrate artists seeking a creative outlet.

Web design has a creative aspect, but the process of understanding how visitors interact with that design—and adjusting it accordingly—is unflinchingly scientific. The creative part occurs when inventing visualizations and methods of interacting. Once you have some designs to test out, it's time for science to take over from art. Monitoring how real visitors interact with the website settles all arguments on the matter.

Professional web design teams test usability before they go live. This is an essential step in the design process, but it's not enough. People behave differently on a site that's in production. Their expectations are set by where they've been beforehand and what they're hoping to do. They're constrained by their environments, the kinds of browsers they use, the speed of their network connections, and dozens of other factors you can't anticipate or simulate in prelaunch usability testing.

This is where web interaction monitoring comes in. On any given page, visitors can perform four basic actions:

- They can *consume* what's shown, which may involve scrolling, changing the browser's size, starting and stopping embedded media, or changing font size.
- They can *follow a link* contained in the HTML to continue their navigation, either within your site or elsewhere.
- They can *provide data, typically through a form*, using controls such as radio buttons, checkboxes, text fields, or drop-down lists.
- They can *use their mice and keyboards* to interact with elements of the page, which won't necessarily result in any server actions.

By collecting information about how visitors interact with your website, you see which of these actions visitors perform in response to your designs once the site is live. This is the domain of Web Interaction Analytics, or WIA, a term first coined by Tal Schwartz of ClickTale.

Four Kinds of Interaction

Website visitors misunderstand sites in a surprising number of ways. Lacking the designers' understanding of why the site exists and how data is structured, they'll often fail to notice content or visual cues. They'll click on things that aren't links. They'll navigate through the application in unexpected ways.

We're going to look at four specific kinds of usability problems:

Visitors don't see what you wanted them to
Content is displayed off the visible screen and visitors don't scroll down to view it.

Visitors don't interact as you intended
They don't notice elements designed for interaction, such as buttons or links, or they try to interact with things that aren't part of the design you intended.

Visitors don't enter data correctly or completely
They put the wrong information into a form, linger a long time on a particular form element, or abandon a form halfway through.

You have no idea what's happening on the site
Visitors saw something you didn't intend, or behaved in unusual ways, and you want to know what they saw.

Let's look at these four issues in more detail.

Seeing the Content: Scrolling Behavior

Your site has many different kinds of pages. Some are simple and transactional, presenting visitors with a few options and asking for a decision. Others, such as search result pages or retailer catalogs, contain paginated lists through which a visitor can navigate. Still others, such as blogs and articles, contain content for the visitor to read.

Each page carries its own design and usability constraints.

- *Transactional pages* must offer visitors a clear decision and show possible courses of action.
- *Search result pages* should make it easy to browse through results, with elements such as pagination.
- *Content pages* should provide readable content alongside headings and images.

Any web designer will tell you that this means important content should appear at the top of the page so visitors can see it without having to scroll down. Of the three kinds of pages outlined above, those containing content are usually the longest. When a page extends beyond the visible window of a web browser, visitors need to scroll to read it all.

Scrolling behavior is a measurement of visitor attention. You can tell which content is better by the number of visitors who scroll to the bottom of an article. If all visitors stop scrolling at a certain point in a page, you know they've either found what they were after, or have tuned out.

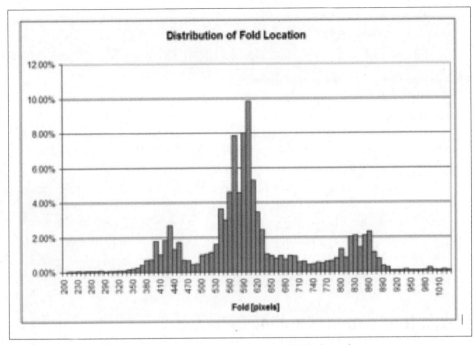

Figure 6-1. Height of the web page fold, in pixels, from the ClickTale study

Scrolling As a Metric of Visibility

In October 2007, WIA vendor ClickTale published a study of 80,000 page views collected over one month in 2007 (*http://blog.clicktale.com/2007/10/05/clicktale-scrolling -research-report-v20-part-1-visibility-and-scroll-reach/*). 91 percent of pages analyzed were long enough to contain a scrollbar—that is, when rendered, their heights in pixels exceeded that of the visitor's browser screen. Data below the visible page is considered "below the fold," and is less likely to be seen.

Scrolling depends a great deal on the typical height of a browser window. The fold varies from visitor to visitor based on the type of browser, any toolbars or plug-ins, and the desktop and window size. As Figure 6-1 shows, there are three clear peaks in the data from the study at 430, 600, and 860 pixels. These represent the most common browser and toolbar settings on the three most common screen heights at the time (800×600, 1024×768, and 1280×1024). Today's variety of monitor sizes and the growing popularity of mobile devices mean these results are changing significantly.

Regardless of the varying heights, a consistent number of visitors—22%—scrolled to the bottom of the page, as shown in Figure 6-2.

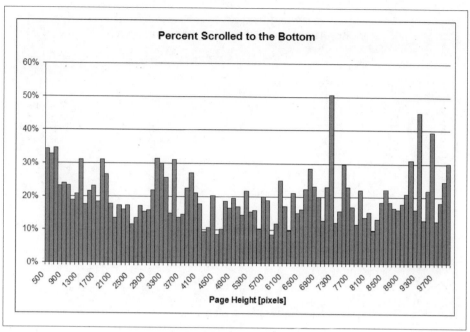

Figure 6-2. Percent of visits analyzed in the ClickTale study that made it to the bottom of the page

As the study says, "the same number of page viewers will tend to scroll halfway or three-quarters through a page, regardless of whether the page size is 5,000 pixels or 10,000 pixels" (*http://blog.clicktale.com/2007/12/04/clicktale-scrolling-research-report-v20-part-2-visitor-attention-and-web-page-exposure/*). Just to be clear: if you have content at the bottom of your page, 71% of visitors won't see it.

Paying attention

Tracking scrolling is only one measure of visitor attention. You also want to know how much time visitors spent on different parts of the page. Did they dwell on the uppermost portions of the page, or spend equal amounts of time throughout it?

To understand this, ClickTale plotted the number of seconds spent on various parts of the page for pages of different heights (shown in Figure 6-3).

The study showed that people spent most of their time in the top 800 pixels of a page regardless of its height, and that they spent slightly longer at the bottom of the page, possibly because they were interacting with navigational buttons or because they simply dragged the slider to the foot of the page.

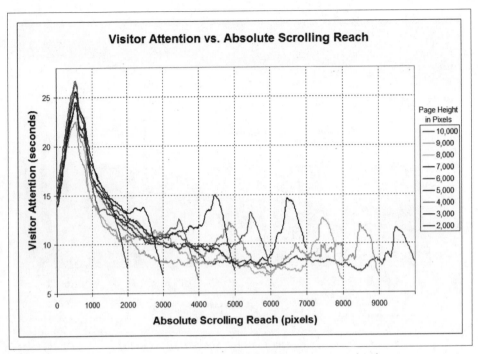

Figure 6-3. Number of seconds spent on parts of the page for various page heights

Your mileage will vary

This study doesn't segment scrolling behavior by page type. We would expect to see significantly different numbers if we were to analyze only content pages where visitors had a reason to review the entire page or only pages with a simple yes/no decision.

You'll also get very different results depending on whether the pages you're analyzing are part of a media site, a SaaS application, a transactional site, or a collaborative portal, and based on what your visitors are trying to accomplish.

Figure 6-4 shows an example of a report for a page, showing how many visitors scrolled how far down within the page.

Using this data, you'll know how far visitors are reading before going elsewhere. For media sites, this kind of WIA report will tell you how likely visitors are to see advertising embedded in the page. For collaborative sites, the report can show you how many visitors are making it to the comment threads below a post.

Proper Interactions: Click Heatmaps

On most websites, visitors have several ways to accomplish their goals. You may let a visitor click on a button or a text link that both link to the same page. Perhaps you have

---

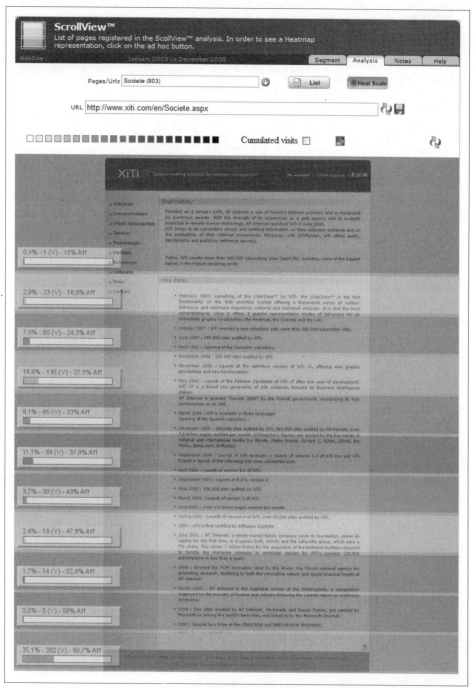

Figure 6-4. A report showing scrolling within a page using AT Internet's ScrollView

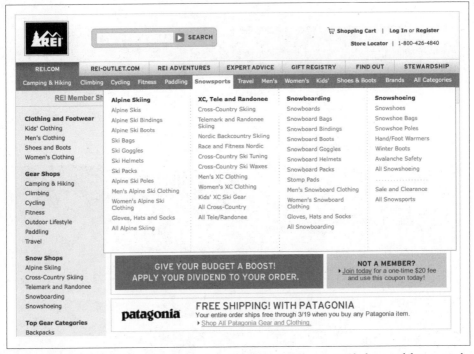

Figure 6-5. Visitors to REI.com have several ways to find a specific item, including a sidebar, specials, a top menu, and a search field

pagination controls at the top and bottom of a page. Or maybe you have top and sidebar menus. There's a tremendous amount of choice for visitors, as Figure 6-5 illustrates.

WIA tools can track the coordinates of each click, showing you what visitors clicked, and what they didn't. By segmenting these clicks, you can get a better understanding of which parts of your interface visitors prefer. You can also streamline the interface: if you discover that the majority of visitors rely on one particular navigational element, such as a top menu, you may be able to reclaim precious above-the-fold real estate by removing redundant or unused controls.

Usability and Affordance

Psychologist J.J. Gibson first coined the term "affordance" in 1977 to refer to things within one's environment that suggest courses of action. A door handle, for example, suggests that the door should be pulled—it "affords" pulling. More recently, the term has been applied to human interface design in general, and computer design in particular, with the fundamental belief that good design makes affordance obvious because it suggests the right actions to the visitor.

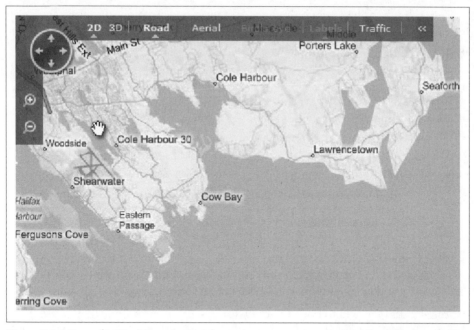

Figure 6-6. Microsoft's Live Maps changes the cursor to encourage certain visitor behaviors, such as dragging

Through years of web use, most consumers have become familiar with the basic building blocks of site interaction, such as text fields, scrollbars, and checkboxes. We recognize that we can expand a drop-down list or toggle radio buttons by clicking other nearby radio buttons. Put another way, a set of radio buttons *affords* us a choice.

On a website, affordances might include graphical elements that have underlying shadows, suggesting you can click them, or text boxes that appear recessed, suggesting they can contain information. More recently, we've added cursors and animations that reinforce a visitor's understanding of how to use the application, as shown in Figure 6-6.

When trying to understand whether a page is usable, we need to ask two questions. The first is whether an affordance exists: do we make it possible for visitors to do what they want to do? If we don't, it may be because the visitor is trying to do something we didn't envision in our design (which is a question better answered through VOC research). It may also be because the visitor doesn't understand the application, which means we need to explain how the site works.

The second question is one of perception: did the visitors notice the affordance we've created for them? Most developers find end visitor testing both frustrating and revealing because testers often can't see buttons that are right in front of their eyes.

Table 6-1 shows the four possible outcomes of these two questions. Problems with web interaction occur either because no affordance exists or because the visitor didn't perceive affordance.

Table 6-1. The four possible outcomes of interaction design

	Visitors perceive affordance	Visitors do not perceive affordance
Affordance exists	*Perceptible affordance* This is good design. The correct functions are present and visitors notice them.	*Hidden affordance* The function is present, but visitors don't notice it. It may be too small or hidden below the fold.
Affordance does not exist	*False affordance* Visitors believe that something is designed for interaction, but it's a red herring, such as an unclickable image that looks like a button.	*Correct rejection* This is good design: visitors are clear about what they can't do, so they're not interacting with the application in unanticipated ways.

In WIA, you will encounter both false affordance (in which your visitors are trying to do things you didn't intend them to) and hidden affordance (in which your visitors can't figure out how to do what you wanted them to).

In other words, we need to track the behavior of visitors, even when they're not using the application as intended, for example, when they click on a background they're not supposed to touch.

Analyzing Mouse Interactions

The most basic WIA report shows where visitors clicked, overlaid on the site, as shown in Figure 6-7.

Even these simple reports are revealing. You'll discover that visitors think graphical elements or backgrounds are actually clickable. Heatmaps like these will show you where visitors are clicking, even if they're not clicking on navigable items.

Segmenting clicks

If you're thinking like an analyst, you're probably eager to segment the data so you can act on it. One kind of segmentation is by click source. Segmented heatmaps color-code clicks according to visitor segments such as referring sites, keywords, browser types, and other information collected by the monitoring script or the analytics service, as shown in Figure 6-8. The segmentation is usually attached to page elements such as `div` tags, rather than just to links. Clicking a color-coded link shows more details about the volume of clicks the area received.

Segmentation of this kind shows you how your visitors' "baggage" affects their interactions. Visitors who come from a Google search, for example, may have different ways of interacting with your site than those who search on Yahoo!. Or users of a particular

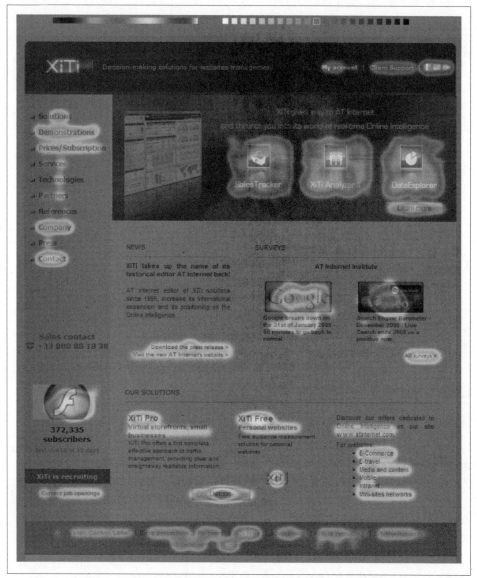

Figure 6-7. A heatmap of clicks overlaid on a web page using XiTi Analyzer

browser may click in one place, while all others click elsewhere—a clue that the page is rendering differently in their browsers.

A good site is not only effective, allowing visitors to achieve goals without making mistakes or getting stuck, it's also efficient, letting them accomplish those tasks quickly. A wizard that helps first-time visitors complete a goal may frustrate "power users" seeking a faster way to achieve the same goals. Segmenting novices from seasoned users

Figure 6-8. Individual clicks segmented by referring URL in Crazy Egg

and examining their clicking behavior can reveal how more experienced visitors use the site differently.

In addition to segmenting by source, you should also segment by the goals you want visitors to achieve. Many analytics tools include click reports that show conversion rates by the elements of a page, as shown in Figure 6-9.

All of these aggregate views help you to understand what's most popular, which links don't lead to desired outcomes, and places where visitors are clicking mistakenly. In other words, they confirm or contest the design assumptions you made when you first built the website.

Data Input and Abandonment: Form Analysis

One of the most valuable, and least analyzed, components of a web application is its forms. A form is where your visitors provide the most detailed information about themselves. It's where they place orders, enter billing information, and leave messages. It's where the rubber meets the road. But companies spend far more time on site design than they do on analyzing form usability and abandonment.

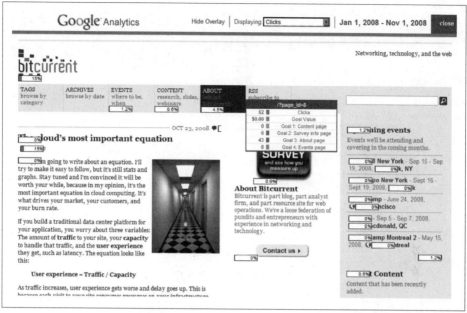

Figure 6-9. Aggregate clicks overlaid on a site, segmented by goal attainment, within Google Analytics

Each form represents a miniature conversion funnel with its own milestones, answering questions like:

- How many visitors saw the form correctly?
- How many visitors started to fill out the form?
- How far down the form did visitors get?
- How long did visitors linger at each field?
- What was the abandonment for each field?
- How many visitors submitted the form?
- How many visitors submitted the form, but missed required fields?
- How many optional fields did visitors skip?
- How many visitors successfully reached the following page?

Given how important form design is to web interaction, it's surprising how little work has been done on getting it right. Figure 6-10 shows an example of a funnel microconversion report.

One excellent resource on form design is Luke Wroblewski's *Web Form Design: Filling in the Blanks* (Rosenfeld Media) and the accompanying blog, "Functioning Form" (*http://www.lukew.com/ff/*).

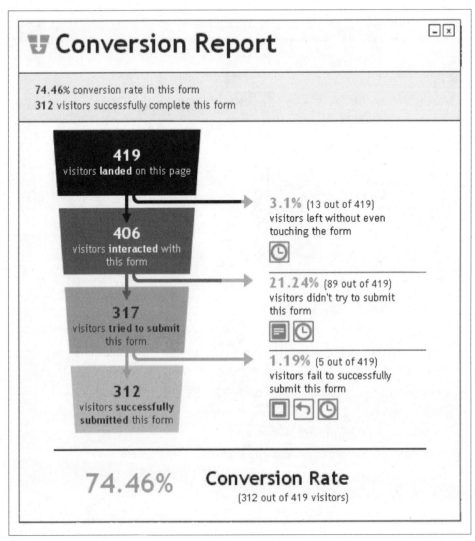

Figure 6-10. Microconversion analysis within a single page

We can peer deeper into the process than just abandonment, however. WIA lets us see how far in a form a visitor progressed, where she dropped out, and which fields she lingered on, as shown in Figure 6-11.

 ClickTale has five reports related to form microconversion, described in the company's forum at *http://forum.clicktale.net/viewtopic.php?f=3 &t=21&p=29&hilit=ssl+https - p29.*

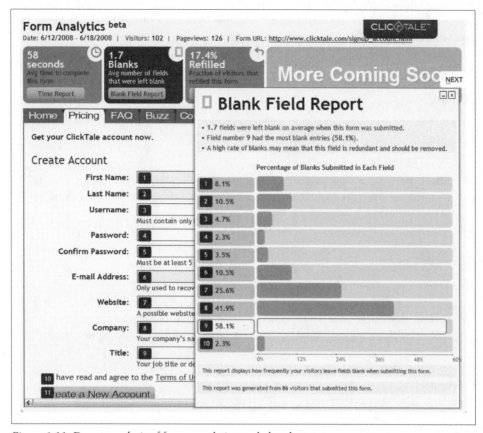

Figure 6-11. Deeper analysis of form completion and abandonment

Aggregation of in-page conversion is a relatively new field. While today's solutions are standalone offerings from ClickTale, FormAlive, and others, we expect to see this kind of analysis become an integral part of web analytics.

Individual Visits: Replay

We've looked at three kinds of analysis that can show whether your design assumptions are valid or wrong in the aggregate. There's another way to understand your visitors better: by watching them individually. Using either JavaScript or an inline device, you can record every page of a visit, then review what happened.

Replaying an individual visit can be informative, but it can also be a huge distraction. Before you look at a visit, you should have a particular question in mind, such as, "Why did the visitor act this way?" or "How did the visitor navigate the UI?"

Without a question in mind, watching visitors is just voyeurism. With the right question, and the ability to segment and search through visits, capturing entire visits and replaying pages as visitors interacted with them can be invaluable.

Stalking Efficiently: What You Replay Depends on the Problem You're Solving

Every replay of a visit should start with a question. Some important ones that replay can answer include:

Are my designs working for real visitors?
> This is perhaps the most basic reason for replaying a visit—to verify or disprove design decisions you've made. It's an extension of the usability testing done before a launch.

Why aren't conversions as good as they should be?
> By watching visitors who abandon a conversion process, you can see what it was about a particular step in the process that created problems.

Why is this visitor having issues?
> If you're trying to support a visitor through a helpdesk or call center, seeing the same screens that he is seeing makes it much easier to lend a hand, particularly when you can look at the pages that he saw before he picked up the phone and called you.

Why is this problem happening?
> Web applications are complex, and an error deep into a visit may have been caused by something the visitor did in a much earlier step. By replaying an entire visit, you can see the steps that caused the problem.

What steps are needed to test this part of the application?
> By capturing a real visit, you can provide test scripts to the testing or QA department. Once you've fixed a problem, the steps that caused it should become part of your regularly scheduled tests to make sure the problem doesn't creep back in.

How do I show visitors the problem is not my fault?
> Capturing and replaying visits makes it clear what happened, and why, so it's a great way to bring the blame game to a grinding halt. By replacing anecdotes with accountability, you prove what happened. Some replay solutions are even used as the basis for legal evidence.

We can answer each of these questions, but each requires a different approach.

Post-launch usability testing: Are my designs working?

It's always important to watch actual visitor behavior, no matter how much in-house usability testing you do. It's hard to find test users whose behavior actually matches

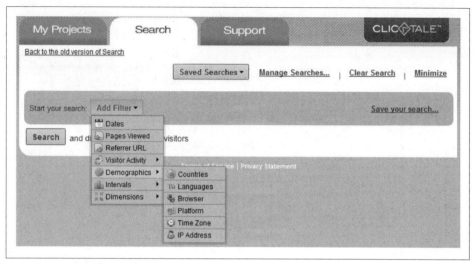

Figure 6-12. Search options in ClickTale

that of your target market, and even if you did, your results wouldn't be accurate simply because people behave differently when they know they're being watched.

For this kind of replay, you should start with the design in question. This may be the new interface you created or the page you've just changed. You'll probably want to query the WIA data for only visitors who visited the page you're interested in. While query setup varies from product to product, you'll likely be using filters like the ones shown in Figure 6-12.

Your query will result in a list of several visits that matched the criteria you provided, such as those shown in Figure 6-13. From this list, you can examine individual visits page by page.

The WIA solution will let you replay individual pages in the order the visitor saw them, as shown in Figure 6-14.

Depending on the solution, you may see additional information as part of the replayed visit:

- The WIA tool may show mouse movements as a part of the replay process.
- The visualization may include highlighting of UI elements with which the visitor interacted, such as buttons.
- The tool may allow you to switch between the rendered view of the page and the underlying page data (HTML, stylesheets, and so on).
- You may be able to interact with embedded elements, such as videos and Java-Script-based applications, or see how the visitor interacted with them.
- You may have access to performance information and errors that occurred as the page was loading.

These capabilities vary widely across WIA products, so you should decide whether you need them and dig into vendor capabilities when choosing the right WIA product.

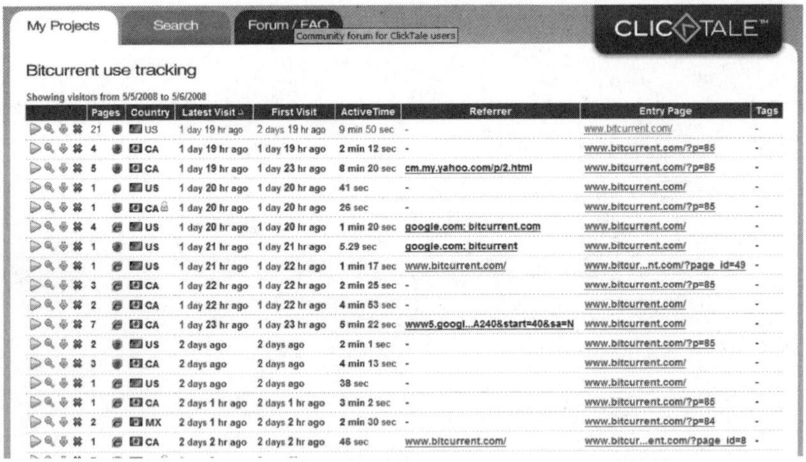

Figure 6-13. Browsing a list of recorded visits in ClickTale

Figure 6-14. A Tealeaf screen replaying a web visit along with visitor actions

Conversion optimization: why aren't conversions as good as they should be?

When you launch a website, you expect visitors to follow certain paths through the site and ultimately to achieve various goals. If they aren't accomplishing those goals in sufficient numbers, you need to find out why.

Lots of things cause poor conversion. Bad offers or lousy messages are the big culprits, and you can address these kinds of issues with web analytics and split testing. Sometimes, however, application design is the problem.

Your conversion funnels will show you where in the conversion process visitors are abandoning things. Then it's a matter of finding visitors who started toward a goal and abandoned the process.

You should already be monitoring key steps in the conversion process to ensure visitors are seeing your content, clicking in the right places, and completing forms, as discussed earlier. To really understand what's going on, though, you should view visits that got stuck.

The Aha Moment: No Foreigners Allowed

A very large U.S.-based computer goods retailer started to receive orders from Canada. While American postal addresses contain a five-number zip code (for example, 01234), Canadian addresses have a six-character postal code (for example, H0H 0H0). The retailer was receiving orders from Canadians, but didn't know it because form field validation on the checkout page was refusing to accept the longer postal codes.

No amount of web analytics, synthetic testing, or real user monitoring (RUM) revealed the issue.

- Web analytics simply showed that visitors were abandoning the process, but since this had been happening for a long time, the company didn't notice an increase. Furthermore, the company didn't anticipate Canadian orders, so it was segmenting by U.S. city and not by country.
- Synthetic tests would have tried a five-number zip code, which would have worked.
- RUM data wouldn't have seen the form rejection on the wire, because form validation was rejecting the contents before they could be submitted via a POST.

By tagging visits that had a rejected form validation, the company identified a pattern of failed transactions that was strongly correlated with Canadian visitors. By replaying the contents of the rejected forms, the company quickly realized what was happening.

Sometimes, problems detected as part of usability monitoring stem from broader concerns—in this case, internationalization and an understanding of target markets.

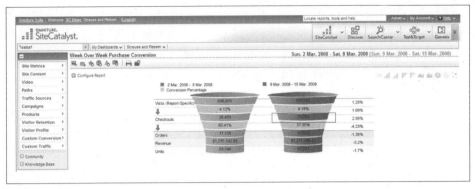

Figure 6-15. Conversion funnel segments in Omniture's SiteCatalyst

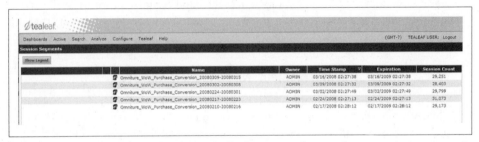

Figure 6-16. A list of captured visits in Tealeaf corresponding to the conversion segments in Figure 6-15; in this case, they're for World of Warcraft purchases

The best place to start is a conversion funnel. For example, in Figure 6-15 a segment of 29,251 visits reached the conversion stage of a sales funnel for an online store.

Let's say you're selling a *World of Warcraft* MMORPG upgrade, and you have a high rate of abandonment on the *lichking.php* page. You need to identify visits that made it to this stage in the process but didn't continue to the following step. There are three basic approaches to this:

The hard way
Find all visits that included *lichking.php*, then go through them one by one until you find a visit that ended there.

The easier way
Use a WIA tool that tags visits with metadata such as "successful conversion" when they're captured. Query for all pages that included *lichking.php* but did not convert.

The easiest way
If your WIA solution is integrated with an analytics tool, you may be able to jump straight from the analytics funnel into the captured visits represented by that segment, as shown in Figure 6-16.

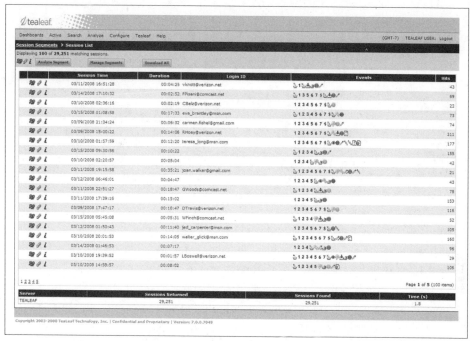

Figure 6-17. A display of captured visits, tagged with specific events or attributes, in Tealeaf

From this segment, you can then select an individual visit to review from a list (such as the one shown in Figure 6-17) and understand what happened.

One final note on conversion optimizations: when you're trying to improve conversions, don't just obsess over problem visits and abandonment. Look for patterns of success to learn what your "power users" do and to try and understand what works.

Helpdesk support: Why is this visitor having issues?

You can't help someone if you don't know what the problem is. When tied into call centers and helpdesks, WIA gives you a much better understanding of what transpired from the visitor's perspective. Visitors aren't necessarily experiencing errors. They may simply not understand the application, and may be typing information into the wrong fields. This is particularly true for consumer applications where the visitor has little training, may be new to the site, and may not be technically sophisticated.

If this is how you plan to use WIA, you'll need to extract information from each visit that helps you to uniquely identify each visitor. For example, you might index visits by the visitor's name, account number, the amount of money spent, the number of comments made, how many reports he generated, and so on.

You'll also want to tie the WIA system into whatever Customer Relationship Management (CRM) solution you're using so that helpdesk personnel have both the visitor's

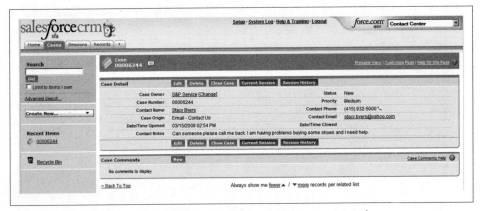

Figure 6-18. Tealeaf session history integrated inside a Salesforce.com record

experience and account history at their fingertips. Figure 6-18 shows an example of Tealeaf's replay integrated into a Salesforce.com CRM application.

The helpdesk staff member receives a trouble ticket that includes a link to her current session. This means she's able to contact the suffering customer with the context of the visit.

Incident diagnosis: Why is this problem happening?

While WIA for helpdesks can assist visitors with the application, sometimes it's not the visitor's fault at all, but rather an error with the application that you weren't able to anticipate or detect through monitoring. When this happens, you need to see the problem firsthand so you can diagnose it.

While you'll hear about many of these problems from the helpdesk, you can also find them yourself if you know where to look.

- When a form is rejected as incomplete or incorrect, make sure your WIA solution marks the page and the visit as one in which there were form problems. You may find that visitors are consistently entering data in the wrong field due to tab order or page layout.

- When a page is slow or has errors, check to see if visitors are unable to interact with the page properly. Your visitors may be trying to use a page that's still missing some of its buttons and formatting, leading to errors and frustration.

- If there's an expected sequence of events in a typical visit, you may be able to flag visits that don't follow that pattern of behavior, so that you can review them. For example, if you expect visitors to log in, then see a welcome page, any visits where that sequence of events didn't happen is worthy of your attention.

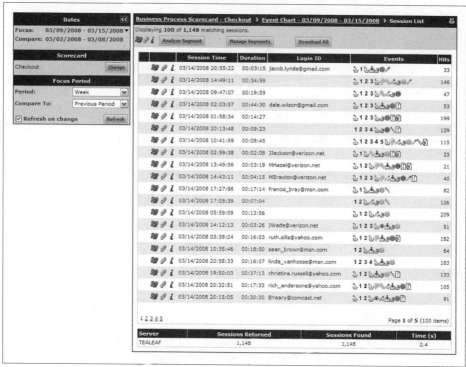

Figure 6-19. A filtered set of visits in Tealeaf

Once you've filtered the visits down to those in which the event occurred, browsing through them becomes much more manageable, as shown in Figure 6-19.

In many WIA tools, the list of visits will be decorated with icons (representing visitors who spent money, enrolled, or departed suddenly, for example) so that operators can quickly identify them. Decorating visits in this way is something we'll cover when we look at implementing WIA.

Test case creation: What steps are needed to test the app?

The teams that test web applications are constantly looking for use cases. Actual visits are a rich source of test case material, and you may be able to export the HTTP transactions that make up a visit so that testers can reuse those scripts for future testing.

This is particularly useful if you have a copy of a visit that broke something. It's a good idea to add such visits to the suite of regression tests that test teams run every time they release a new version of the application, to ensure that old problems haven't crept back in. Remember, however, that any scripts you generate need to be "neutered" so that they don't trigger actual transactions or contain personal information about the visitor whose visit was captured.

Dispute resolution: How do I prove it's not my fault?

There's no substitute for seeing what happened, particularly when you're trying to convince a department that the problem is its responsibility. Replacing anecdotes with facts is the fastest way to resolve a dispute, and WIA gives you the proof you need to back up your claims.

Dispute resolution may be more than just proving you're right. In 2006, the U.S. government expanded the rules for information discovery in civil litigation (*http://www .uscourts.gov/rules/congress0406.html*), making it essential for companies in some industries to maintain historical data, and also to delete it in a timely fashion when allowed to do so. (WIA data that is stored for dispute resolution may carry with it certain storage restrictions, and you may have to digitally sign the data or take other steps to prevent it from being modified by third parties once captured).

Retroactive Segmentation: Answering "What If?"

As a web analyst, you'll often want to ask "what if" questions. You'll think of new segments along which you'd like to analyze conversion and engagement, for example, "Did people who searched the blog before they posted a comment eventually sign up for the newsletter?"

To do this with web analytics, you'd need to reconfigure your analytics tool to track the new segment, then wait for more visits to arrive. Reconfiguration would involve creating a page tag for the search page and for the comment page. Finally, once you had a decent sample of visits, you'd analyze conversions according to the newly generated segments. This would be time-consuming, and with many analytics products only providing daily reports, it would take several days to get it right. The problem, in a nutshell, is that web analytics tools can't look backward in time at things you didn't tell them to segment.

On the other hand, if you've already captured every visit in order to replay it, you may be able to mine the captured visits to do *retroactive segmentation*—to slice up visits based on their attributes and analyze them in ways you didn't anticipate. Because you have all of the HTML, you can ask questions about page content. You can effectively create a segment from the recorded visits. You can mark all visits in which the text "Search Results" appears, and all the visits in which the text "Thanks for your comment!" appears—and now you have two new segments.

This is particularly useful when evaluating the impact of errors on conversions. If visitors get a page saying, "ODBC Error 1234," for example, performing a search for that string across all visits will show you how long the issue has been happening, and perhaps provide clues about what visitors did to cause the problem before seeing that page. You will not only have a better chance of fixing things, you will also know how common the problem is.

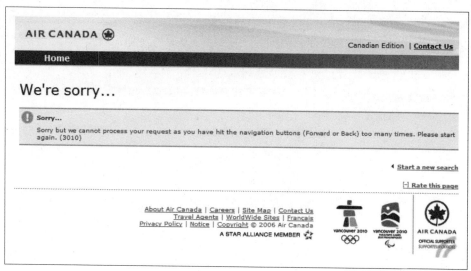

Figure 6-20. Something as simple as clicking the Back button on a travel site can break a visit

Consider the example in Figure 6-20. In this case, the visitor clicked the Back button once, entered new data, and clicked Submit.

The application broke. As an analyst, you want to know how often this happens to visitors, and how it affects conversion. With an analytics tool, you'd create a custom page tag for this message, wait a while to collect some new visits that had this tag in them, then segment conversions along that dimension.

By searching for the string "you have hit the navigation buttons," you're effectively constructing a new segment. You can then compare this segment to normal behavior and see how it differs. Because you're creating that segment from stored visits, there's no need to wait for new visits to come in. Figure 6-21 shows a comparison of two segments of captured visits.

Open text search like this consumes a great deal of storage, and carries with it privacy concerns, but it's valuable because we seldom know beforehand that issues like this are going to happen.

Implementing WIA

Now that you know which questions you want to answer with WIA, it's time to implement it. Which tools you use will depend on what kinds of interaction you want to watch, and how much you need to drill down into individual visits.

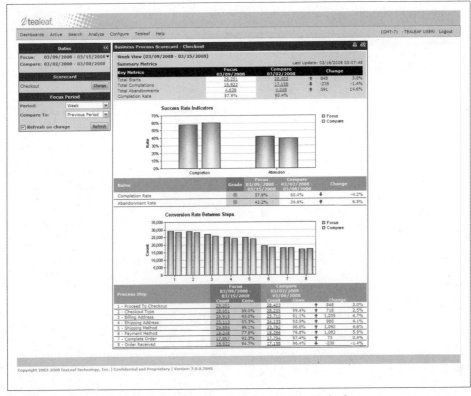

Figure 6-21. Comparing conversion rates across two segments in Tealeaf

Knowing Your Design Assumptions

Your first step is to map out your application and identify all of the assumptions you've made. There will be lots of them, so focus on these three:

Those that drive your business
These may include steps in a funnel, an invitation cycle, publishing content, or making a purchase. This is where you can maximize the effectiveness of your site and optimize conversions.

Those that drive support and helpdesk calls
Which pages make subscribers call you? Which pages give visitors the most ways to do the same things? Where do people usually get stuck? This is how you'll address usability and find problems you haven't identified.

Those that drive productivity

> If there's an action, such as upvoting a topic, adding a comment, updating a record, or browsing through a catalog, in which visitors spend most of their time, analyze it. This is your best chance to improve the efficiency of the application.

Once you've chosen the parts of your site to instrument, start collecting a baseline. Soon, you'll know what "normal" is. You'll have a pattern of clicks, or know which parts of a form make people leave, or know how far visitors scroll, and you can start testing variants on those forms much as you would for page analytics.

Deciding What to Capture

WIA collection happens through client-side JavaScript, server-side instrumentation, or a combination of the two.

If you want to capture things that happen on a browser, like scrolling, partial form completion, and clicks, you'll need to insert JavaScript snippets into the pages you want to monitor. Be sure to baseline website performance before and after inserting these scripts so you know what kind of performance price you're paying for this visibility.

To capture actual visits, you can either use server-side collection (eavesdropping on the conversation between the web server and the browser) or JavaScript. Server-side collection is more complete and provides greater forensic detail, but also costs more. Client-side collection is easier to implement, but is limited in what it can easily record about visits.

If you're planning to review individual visits, be sure you can flip between aggregate and individual views of visitor interactions. The usefulness of a WIA solution depends considerably on your ability to query for specific visits and to analyze across a segment of visits to find patterns within them.

Instrumenting and Collecting Interactions

You capture interactions in much the same way you collect other analytics data: a script on the page records information, such as form completion, click location, or scrolling, and sends it to the service along with something to identify the unique visit. The service does the rest, rendering visualizations and reports. Your work will involve putting the scripts on pages and adjusting those scripts to define what is captured. You'll define the start and end of an "experiment" period to coincide with changes to design and layout.

When it comes to recording entire visits, however, more work is required. Capture and replay of individual visits happens in one of two ways, depending on the WIA solution.

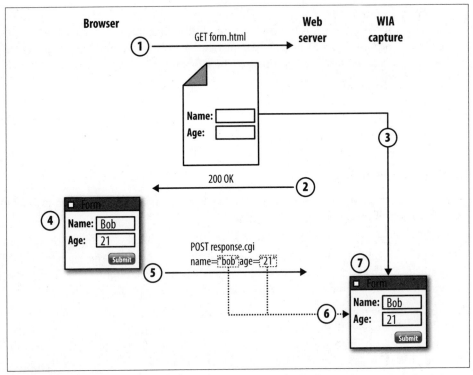

Figure 6-22. How inline capture works in solutions such as Tealeaf

Inline WIA

The first approach to visit capture, employed by WIA products that rely on inline capture of requests and responses, is illustrated in Figure 6-22.

Here's what happens:

1. The browser requests a page from the server.
2. The server responds with the page.
3. The WIA solution captures the HTML of the response, as well as headers and other information such as timing.
4. If the page requires user interaction (such as a form), the visitor carries out the required action.
5. The interaction (in this case, a completed form) is sent back to the server.
6. The WIA solution captures the response (in this case, what the visitor typed into the form).
7. To replay the visit, the system reassembles the original page (*form.html*) and the user input (from *response.cgi*) into a single display.

Capturing user interactions, such as mouse movements, requires additional JavaScript on the client, which is then captured by the WIA solution in the form of small requests similar to those used in an analytics solution. You don't *need* these scripts to see what visitors entered in a form, but if you're concerned about in-page actions or mouse movements, you may need them. If your WIA solution is trying to capture page load time from clients, JavaScript may also be able to augment performance measurements that the inline device collected.

Client-side WIA

The second approach, used by WIA solutions that collect visitor interactions from JavaScript, takes a different approach. The JavaScript on the page tells the service when the visitor has retrieved a page, at which point the service can get its own copy. The JavaScript then records a variety of details about the visit, which may include scrolling behavior, mouse movements, form completions, and clicks (Figure 6-23).

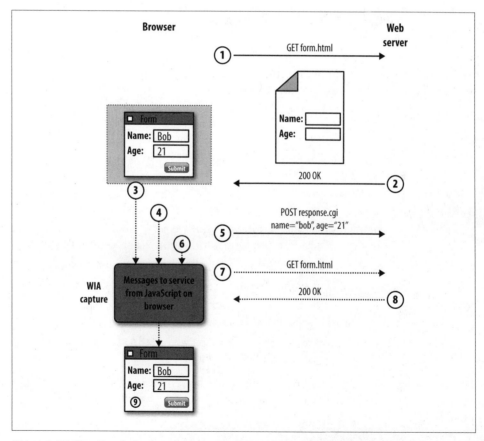

Figure 6-23. How JavaScript-based capture works in solutions such as ClickTale

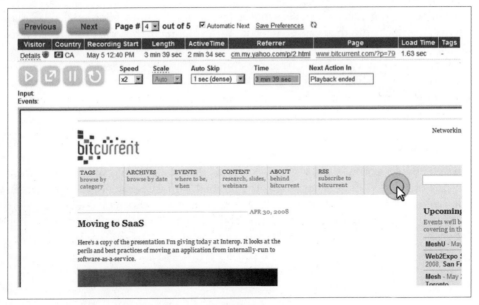

Figure 6-24. Replaying mouse movements in ClickTale

Because the browser is communicating with both the website and the WIA capture service, the steps are somewhat more complex:

1. The browser requests a page from the website.

2. The web server delivers the page.

3. JavaScript on the page tells the WIA capture service that a page (in this case, form.html) has been requested.

4. The visitor moves her mouse, scrolls, and clicks, all of which may be recorded by the JavaScript on the page and sent to the WIA capture service.

5. The visitor completes her interaction with the website, which may include the transmission of data (in this example, form information as part of an HTTP POST).

6. The JavaScript on the page passes this data back to the WIA capture service.

7. During this time, the WIA service retrieves a second copy of the website and stores it.

8. The web server returns a copy of the same page the visitor received.

9. When an analyst wants to review the replayed page, the WIA service shows its copy of the page, mouse movements, and visitor data.

If you're using client-side scripts to capture mouse movements and keystrokes, you can use playback controls to see where the visitor moved and what she typed, as shown in Figure 6-24.

When replaying WIA recordings that include mouse and keystroke data, you'll see the mouse moving around the page, as well as any scrolling the visitor performed. Note, however, that you may not be able to replay actions when the visitor interacted with some site elements. When a browser component had control of the mouse and keyboard, for example, you may not be able to see what the visitor was doing.

Service versus software

Hosted services simply can't give you the complete visibility of inline or server-side equipment, but they're much easier to deploy (and usually cheaper). It's up to you to consider cost, data privacy, and the need for visibility. If you want to use a service, you probably won't be able to capture every component of every visit, so you'll be using the tool more for behavioral analysis and less for diagnostics, retroactive segmentation, and evidence gathering.

Sampling and completeness

Capturing all these visits takes computing power, bandwidth, and storage. Most services limit the number of visits or clicks that they will store. Server-side WIA tools are limited by storage capacity, but all that traffic consumes a lot of storage.

To reduce the amount of data they need to store, these solutions can compress HTML files. They can also store just one copy of frequently requested content, such as stylesheets, JavaScript, and images.

To do this, a tool will look at each HTTP request to see if it has already stored each component on the page. If so, there's no need to store it again. On the other hand, if the requests are for a new object (such as an updated image or a new stylesheet), the system stores a copy. When it's time to assemble the page for replay, the tool simply uses the object that was current at the time the visit occurred.

Filtering and content reduction

Once you deploy a WIA solution, you'll quickly realize that you need to filter some of what's collected, either to reduce the total volume of data or to comply with security and privacy standards. Filters tell the system what to exclude and what to keep.

Often, you'll want to exclude a specific group from collection. You may, for example, only want to record new visitors to the site, or those who come from a Google organic search. Most WIA services allow you to customize collection rules using JavaScript within the pages themselves. For example, you could parse the DOM of the browser and examine the referring URL or check for the presence of a returning visitor cookie. Or you could configure inline server-side tools to check the contents of a page before deciding whether to store it, with similar effect.

Another common form of filtering is to limit the analysis to only a part of the site, such as a checkout process, or only to visits in which an event occurs. Again, using JavaScript,

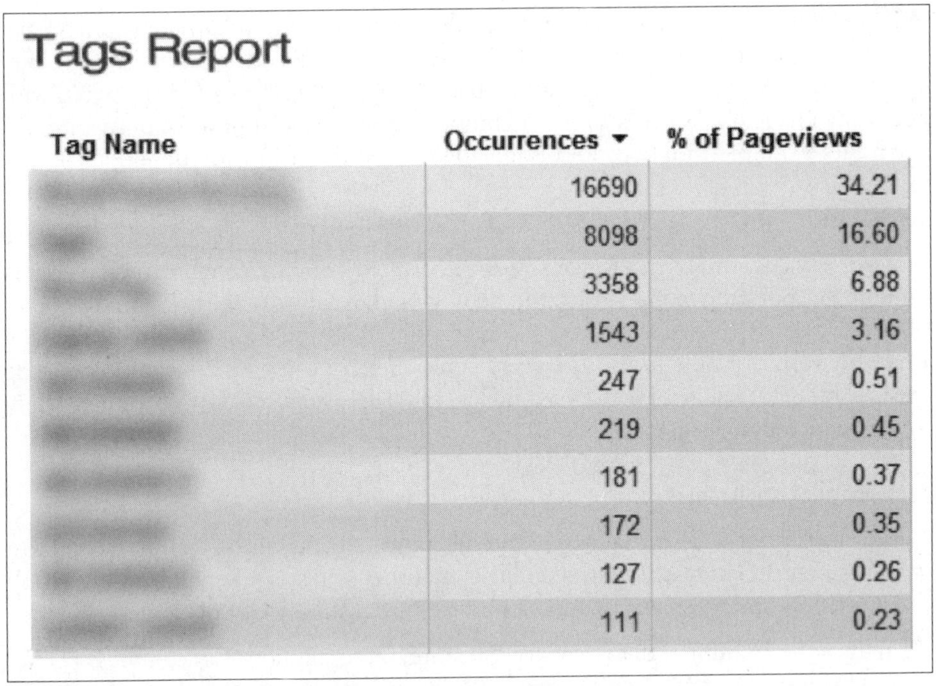

Tags Report

Tag Name	Occurrences ▼	% of Pageviews
	16690	34.21
	8098	16.60
	3358	6.88
	1543	3.16
	247	0.51
	219	0.45
	181	0.37
	172	0.35
	127	0.26
	111	0.23

Figure 6-25. A ClickTale report segmenting occurrences by tags

you can parse for certain conditions on the client. Or you can simply put the collection scripts only on the parts of the site that you want to monitor, which will speed up the time it takes unmonitored pages to load.

Augmenting visits with extraction and tagging

As we've seen, being able to query visits to find the interesting ones is important when replaying visits. Many attributes of a visit, such as browser type, IP address, or time that the visit happened, are automatically collected when the WIA tool records the visit.

Replay tools become far more useful when you add business context to them. Tagging a visit with the visitor's account number means the helpdesk can find it quickly. Flagging visitors who made it to the checkout process or who wound up paying for a product allows you to segment by those dimensions later.

Some client-side WIA tools let you add custom tags to the data they collect. For example, in ClickTale the following HTML marks the page with the `LoginForm Submitted` tag and appends the user's name:

```
<form ... onsubmit="ClickTaleTag("LoginFormSubmitted:" + UserName); ..." ... >
```

Later, you can run an analysis of those tags across captured visits, as shown in Figure 6-25. You can also use the tags as the start of a query for specific visits you want to replay.

You should always tag error conditions, such as a form that the visitor didn't complete correctly or a warning about navigation. That way, when you're analyzing the WIA data you can quickly find visits that encountered those problems and see what caused visitors to get stuck.

For inline, server-side collection, the collecting device identifies information within the HTML (such as the title tag of a page or a visitor's account number) and stores it as metadata about the page. Similarly, you can use certain URLs (such as *thanks.jsp*) to flag a milestone in a visit, such as payment completion. All of this makes captured pages easier to segment and recognize during analysis.

Tying WIA to other sources

If you've found a visit that's interesting, you probably want to share it with others. One way to do this is to forward a link to the WIA system, assuming other members of your team have access. You may also be able to export the data as a video file; a series of screenshots; a test script suitable for QA; a list of page, object, and error records that developers can analyze; or even a raw packet trace that can be examined in a sniffer.

Issues and Concerns

WIA lets us peer deep within the pages of a visit to get a feel for how visitors interacted with components below the page level. It's a technology that must be used judiciously and in the context of the rest of your web monitoring for its results to be meaningful. It also has some important limitations.

What if the Page Changes?

When analyzing visitor interactions, remember that measurements are only valid for a particular page design. You're trying to test a particular page layout to evaluate scrolling, click locations, and form completion. If the page changes in the middle of that test, the results won't make sense.

Some WIA monitoring tools capture the page when you first set up monitoring, and use this as the background when displaying their results. Other tools capture the page at the time the report is generated, and associate results (such as a count of clicks) with specific page elements, so if you delete a button on a page, the click count for that button won't show up in reports.

To avoid this kind of confusion, don't change a page while you're testing it. Treat each WIA analysis as a test with a beginning and end, then make a change, run another analysis, and compare the results.

Visitor Actions WIA Can't See

Client-side WIA captures what happens when a user interacts with the browser, recording such events as changes in dialog box focus, mouse clicks, and keystrokes. Users may do things that have little to do with your site, but that affect how they use the application and your ability to collect data. For example:

- They can copy and paste text, save files, print content, or otherwise cause interactions between the browser and the host operating system.
- They can launch a new window or new browser tab.
- They can close the window and abandon their visits—sometimes by accident.
- They can bring another application to the forefront and interact with it.

If a visitor pastes text into a field from a clipboard, the WIA scripts may not capture that text without considerable additional effort, because there are no "keydown" events for scripts to capture. Similarly, text submitted through autocompletion, in which the browser fills in forms on behalf of the visitor, may not be stored.

When another desktop application has focus, meaning it's "on top," receiving keystrokes and mouse events, mouse and keyboard activity isn't available to the browser and whatever monitoring system you've got installed. This can happen when the visitor is interacting with another window that is covering the current window, when the visitor is browsing through a right-click menu, when the web page is minimized or in the background, or when the visitor interacts with certain Java and RIA plug-in applications or the operating system itself.

In these cases, you'll often know something's happening only because of extended periods of inactivity. Patterns of inactivity that always happen in the same place may indicate a usability issue. For example, visitors may be printing the page, and you could provide them with a Print button at that point in their visit. Or they may be going to fetch a credit card because they didn't know one was needed.

Dynamic Naming and Page Context

If your pages aren't named intuitively, as is the case in Figure 6-26, it may be difficult to understand what visitors were doing on the site.

Page tagging and the extraction of things like title tags can help with this problem, but you'll need to work with developers to ensure URLs are meaningful. Incidentally, this will also help your search engine rankings, since search engines favor sites with descriptive URLs.

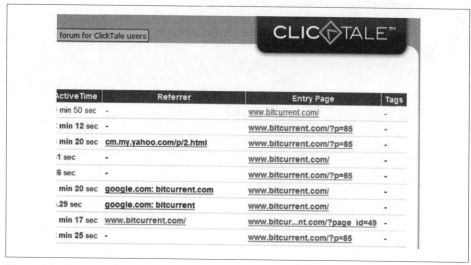

Figure 6-26. Dynamically generated page names don't give you clues about the purpose or content of the page

Browser Rendering Issues

You won't always see what your visitor saw when you replay a visit. Differences in browsers, embedded RIA applications, and other factors mean that for many sites, you'll only have an approximation of the user's visit.

WIA solutions that rely on JavaScript to capture pages are less faithful to the end user experience than inline capture devices, in part because of the way in which they collect data. There are many components of a replayed page, such as embedded components, HTTP POST messages, SSL-encrypted content, and client-triggered JavaScript, that require considerable additional effort on the part of site operators and developers to get working properly in a hosted replay solution.

Different Clicks Have Different Meanings

WIA tools can overcount clicks if they don't understand the context of a page. Consider a report that shows how many mouse clicks each object on the page received in Figure 6-27.

The clear leader in terms of clicks was a `div` tag containing a Flash object. What's not readily apparent from such a report is that the Flash object was a slideshow player. As a result, those clicks were from visitors paginating through the slides in the object, not interacting with the site. Knowing this, you can disregard these outliers. Without application context, however, you might conclude that this object was popular rather than just interactive.

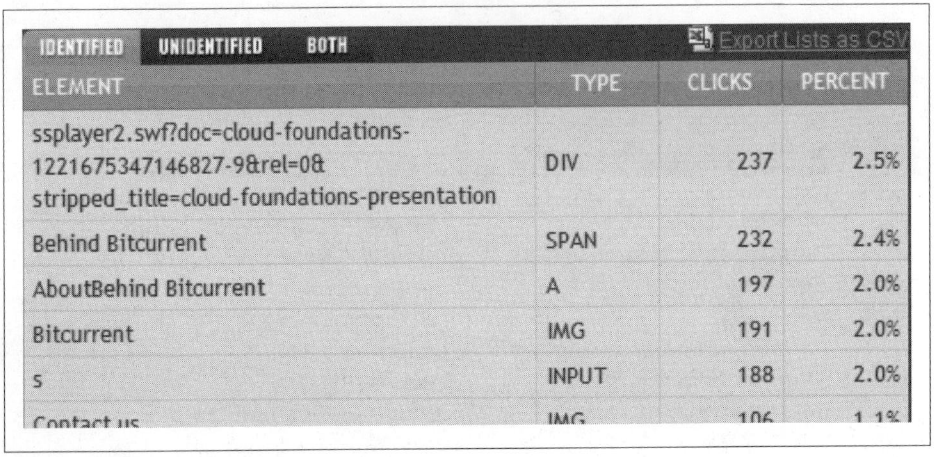

IDENTIFIED | UNIDENTIFIED | BOTH | Export Lists as CSV

ELEMENT	TYPE	CLICKS	PERCENT
ssplayer2.swf?doc=cloud-foundations-1221675347146827-9&rel=0&stripped_title=cloud-foundations-presentation	DIV	237	2.5%
Behind Bitcurrent	SPAN	232	2.4%
AboutBehind Bitcurrent	A	197	2.0%
Bitcurrent	IMG	191	2.0%
s	INPUT	188	2.0%
Contact us	IMG	106	1.1%

Figure 6-27. Embedded RIA objects receive more clicks because they're interactive, not because they are a popular part of the site

The Impact of WIA Capture on Performance

The verbosity of a collection tool depends on the granularity of the data you're capturing. Passive inline collection doesn't introduce delay, but any JavaScript embedded in the browser to capture mouse and keystroke interactions requires an additional GET from the visitor.

For mouse recorders, client-side JavaScript may be transmitting mouse movements at regular intervals as a compressed string. For click-placement recorders, visitors will generate a message to the WIA service each time they click. In fact, if you examine pages that use WIA you'll see the browser's status bar showing that it is communicating with these systems (Figure 6-28).

Vendors claim that their scripts consume no more than 2 Kbps of upstream bandwidth, which is negligible for many broadband visitors. For hosted replay services, there's also the impact of the secondary hit on the server to consider.

Playback, Page Neutering, and Plug-in Components

Replayed pages must be *neutered* so that the act of viewing them doesn't inadvertently trigger real actions. We want to sandbox the replay of the visit from any real-world outcomes so that we don't buy a pair of shoes or send an email message every time we replay the visit.

Since pages are made of many components, you may also have problems replaying visits that included third-party content, such as a map, or a third-party stock quote, many of which require a developer API key.

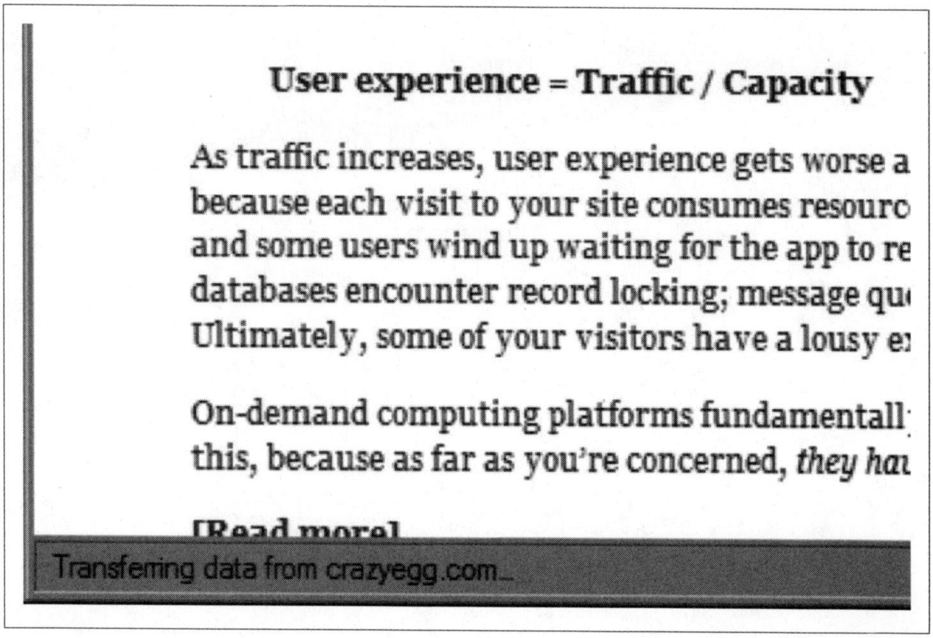

Figure 6-28. Ongoing communication between the browser and the WIA service after the page has loaded

Here's why: when a visitor to your site loads a page that includes a customized Google Map, your browser makes a request to Google. That request includes your URL (as the referring page) and the API key you've received from Google. Google issues these keys for several reasons: to prevent excessive use, to authenticate requests, and so on. When you replay the page from a third party service, the requesting URL (of the service) and the API key (from your site) won't match, so embedded third-party components may behave strangely or not work at all.

Privacy

Of all the technologies we've looked at so far, WIA is the most invasive and risky. Analytics only looks at visits in the aggregate, and RUM is primarily concerned with performance and errors. With WIA's visit capture, you're seeing what visitors saw and typed, usually without them expecting their visits to be seen by someone else.

Depending on the solution you're using, you may be decrypting and capturing a copy of a visitor's entire page, complete with personally identifiable information and credit card numbers. You're also able to uniquely identify individual visitors by the data they enter, further raising the risk of fraud and leakage.

The consequences of privacy violation may be subtler, though. Consider a visitor looking at pages about cancer treatment. Your record of his visit could have a material effect

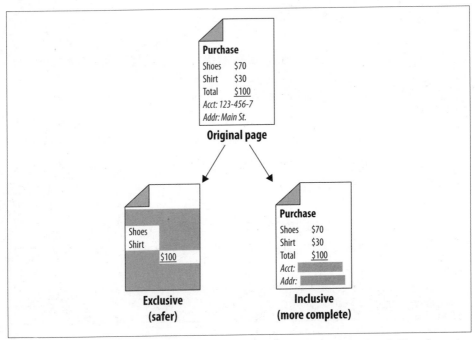

Figure 6-29. Two capture strategies collect different amounts of personally identifiable information from visitors

on his insurance premiums simply because of the pages' URLs. You need to take steps to protect visitor data, control who can access it, and delete it in a timely fashion when it's no longer needed.

Data collection: Inclusive and exclusive

Configure WIA products to capture only what you need. Unfortunately, since you're trying to understand a visitor's behavior, what you need is often the entire visit. You can't just record certain strings within the page—you need the whole page, including images and stylesheets, to see what they saw.

Because of this, WIA tools are usually configured to delete certain strings and even look for common patterns such as the four-sets-of-four-digits of credit cards, and to remove them automatically before displaying them.

A capture strategy that includes all data by default is more promiscuous—it will collect everything it isn't told to ignore, ensuring that you have maximum data for analyzing a visitor, but exposing you to the risk of privacy violations. By contrast, a capture strategy that excludes everything except for what it's told to collect won't capture the full visit, and will limit your ability to segment and review what happened. Figure 6-29 shows these two approaches.

You may want to avoid recording certain form fields, such as credit card numbers. Most WIA services won't record form data that's hidden in the first place (such as passwords), but you can also tag form elements with special strings to tell the collection system to ignore them. With ClickTale, for example, the following HTML tells the system that the form field "Acct" should not be stored by marking it as sensitive.

```
<input id="Acct" type="text" class="ClickTaleSensitive">
```

Inline devices let you define which POST parameters to collect or discard, or to configure regular expressions to extract or remove only certain portions of a page.

Remember, however, that one popular use for WIA is dispute resolution. You may be required to protect historical data without modifying it. In this case, you have to store all of the data collected in a digitally signed, tamper-evident format; you can't scrub the data when it's captured the way you could for RUM data, so your WIA tool needs to hide sensitive data when it's viewed by an operator and delete it when its storage period has expired.

Keystroke capture is an ethical hot potato

What about something a visitor types before she abandons a form?

A visitor might enter data such as her email address into a form on your website, then change her mind about submitting it. Traditional web analysis would just show that the visitor abandoned the visit on that page. On the other hand, a WIA script that's sending back keystrokes and mouse movements might capture a partially completed form. To whom does that data belong? Can you add that visitor to your mailing list? Put another way, *does data collection begin at the keystroke or the Submit button?*

Depending on the vendor, your collection scripts may trap every keystroke or every form field change. If you collect this data and the visitor leaves the site without clicking Submit, you now know things the visitor didn't necessarily want you to.

ClickTale's CEO explains their collection policy as follows:

> While ClickTale does record all keystrokes, we protect the anonymity of visitors who start to fill out online forms and do not submit them. We intentionally mask these visitors' keystrokes since they have not implicitly agreed to share their information with the website.

As consumers become more aware of the ways in which we monitor them, we expect a backlash against this sort of data collection, but given the importance of understanding conversion and abandonment, microconversion analysis will become increasingly popular, and web operators will find themselves employing and defending collection methods that may ultimately pose a privacy risk to the companies for which they work.

Web Interaction Analytics Maturity Model

As with web analytics and EUEM, companies undergo several phases of maturity as they start to embrace WIA technologies. What starts out as aggregate "where did they click" visualization quickly becomes segmentation by internal outcome, external source, usability, and learning curve analysis.

Maturity level	Level 1	Level 2	Level 3	Level 4	Level 5
Focus	Technology: Make sure things are alive	Local site: Make sure people on my site do what I want them to	Visitor acquisition: Make sure the Internet sends people to my site	Systematic engagement: Make sure my relationship with my visitors and the Internet continues to grow	Web strategy: Make sure my business is aligned with the Internet age
Who?	Operations	Merchandising manager	Campaign manager/SEO	Product manager	CEO/GM
WIA	Click diagrams showing "hot" areas on key pages	Segmentation of visitor actions (scroll, drag, click) by outcome (purchase, abandonment, enrollment)	Segmentation by traffic source (organic search, campaign) and A/B comparison; visitor replay	Learning curve analysis; comparison of first-time versus experienced users; automated A/B testing of usability	Product specialization according to usability and user groups; usability as a component of employee performance

CHAPTER 7

Why Did They Do It?: Voice of the Customer

So far, we've looked at how to link what visitors did on your site with the goals you're hoping they'll accomplish. We've seen how to segment visits so you can understand what's working and what isn't, and how to experiment iteratively in order to optimize the effectiveness of your website.

But no amount of optimization will save you if you don't understand what visitors wanted to accomplish in the first place.

Understanding your market's real motivations is difficult at best. According to Lou Carbone of Experience Engineering, Inc., 95 percent of people's motivations are subconscious (*www.smeal.psu.edu/isbm/documents/0206msum.pdf*). People seldom say what they mean. Sometimes they're trying to conform to the interviewer's expectations. Sometimes they want to hide socially unacceptable biases—racism; ethical beliefs, and so on—that nevertheless affect how they behave (the *Bradley effect* is one example of respondents giving more socially acceptable responses that differed from their actual behavior: *http://en.wikipedia.org/wiki/Bradley_effect*).

Often, people simply don't know why they do what they do. Their unconscious mind has decided how to act long before their conscious mind knows about it, leaving people to rationalize their behavior after the fact (*www.mpg.de/english/illustrationsDocumen tation/documentation/pressReleases/2008/pressRelease20080414/index.html*).

That's a frightening prospect, particularly for analytically minded web operators who just want to crunch numbers and measure impacts. Companies invest heavily in consumer research and focus groups. Using eye tracking, recording, surveys, and interviews, they try to get an idea of what's inside their customers' minds. Despite all this, delving into visitor motivations is at best a soft science.

The best way we have to discover our visitors' intentions and concerns is to ask them. Online, we use Voice of the Customer (VOC) technologies. VOC collects customer feedback for analysis and review. Though it affords us only a glimpse of what's on our visitors' minds, that glimpse can be invaluable.

The Travel Industry's Dilemma

In the early days of e-commerce, the best businesses were the ones that had a tremendous amount of information to search and sold big-ticket items: real estate, used cars, and travel. In each of these, the Web fundamentally changed an entrenched sales model where the realtor, salesperson, or agent knew more about the market than the buyer did. Once individual consumers could learn as much about the neighborhood as the realtor, the premium that realtor could charge diminished. Consumers booked their trips online and searched for used cars on eBay.

There's one thing websites can't do as well as humans, though: understand what people are thinking.

Consider, for example, the large travel sites. Companies like Expedia, Travelocity, and Priceline once faced a huge problem with abandonment. Visitors would search for a hotel, find one they liked, check rates and availability—and then leave. This high abandonment rate frustrated all attempts to fix it. The sites tried offering discounts, changing layouts, modifying the text, and more. Nothing would stop potential buyers from abandoning their bookings.

Then one of the sites decided to survey its visitors with open-ended questions like, "Why did you come to the site?" The site's operators quickly discovered that many visitors weren't planning on booking a room at that time; instead, they were simply checking availability. And when they *did* want to book, they were comparing brands for those that matched their hotel loyalty programs.

In other words, the reason they thought visitors were coming to their sites—to book hotel rooms—was often wrong. Instead, visitors were checking whether rooms were free below a certain rate, or trying to find a deal on their favorite hotel chains. Both behaviors were bad for the sites' businesses, since they reduced conversion and undermined the free market on which the discount travel sites relied. The site's operators had a different set of goals in mind than visitors did, and the symptom of this disconnect was the late abandonment.

With this newfound understanding of visitor motivations, travel sites took two important steps. First, they changed the pages of their sites, offering to watch particular searches for the customers and tell them when a deal came along, as shown in Figure 7-1.

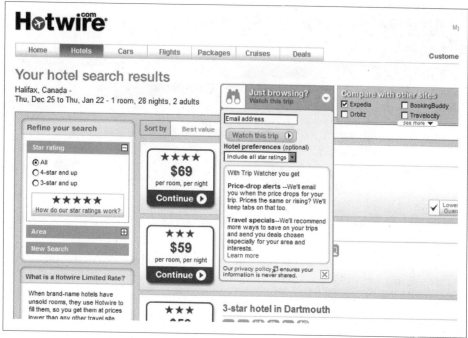

Figure 7-1. Hotwire's "Just browsing?" prompt

Second, they moved the purchasing or bidding to the front of the process, forcing the buyers to commit to payment or to name a price before they found out which hotel they'd booked. This prevented window-shopping for a brand, while allowing the site operators to charge discounted rates.

The results were tremendous, and changed how online hotel bookings work. Today, most travel sites let users watch specific bookings, and many offer deeper discounts than the hotel chains themselves as long as customers are willing to commit to a purchase before finding out the brand of the hotel.

They Aren't Doing What You Think They Are

The lesson here is that your visitors might not be doing what you think they are. While sometimes—as in the travel agency model—they're still doing something related to your business, there are other times when their reasons for visiting are entirely alien.

Consider, for example, online games like PMOG and WebWars. In these games, players install browser plug-ins that let them view websites in different ways than those intended by the site operator. In PMOG, a user can plant traps on your website that other players might trigger, or leave caches of game inventory for teammates to collect.

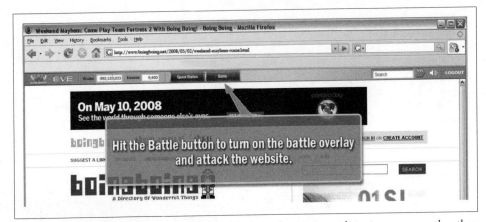

Figure 7-2. A WebWars player competing for the Boingboing.com website in a game, rather than interacting with it

In Webwars, players compete for dominance of popular websites based on the stature of that site in a web-wide version of the board game *Risk*, as shown in Figure 7-2.

Other "overlays" to the web let people comment on sites using plug-ins like firef.ly—shown in Figure 7-3—or use site content for address books and phone directories, as Skype does.

VOC may not show you why these people are visiting the site—indeed, there's no easy way to tell they're playing a game on your site, other than joining their community and playing along with them, which we'll look at in Chapter 11 when we turn to community monitoring.

These plug-ins and overlays might be extreme examples, but they underscore just how disconnected we often are from our visitors' intentions and motivations.

What VOC Is

At its simplest, VOC is just a fancy term for surveys that solicit feedback about your site or your organization. The invitation may come from a pop-up message when visitors first arrive at your site, or from a feedback button on a page. It may even come from an email message that you send to customers.

The surveys use a variety of questions and formats to gauge how respondents feel about things. They also collect data on the respondents so that analysts can correlate the responses to specific groups.

There are four main reasons for companies to conduct VOC studies.

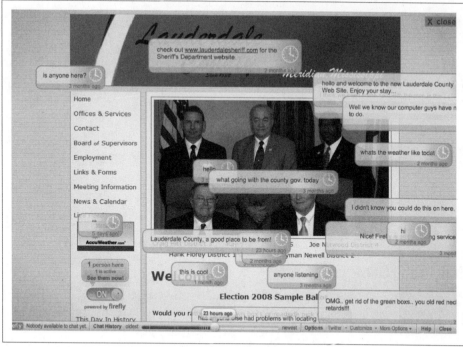

Figure 7-3. The Fort Lauderdale Sheriff's Office website, showing Firefly chats happening atop the site

To get new ideas

Your customers may have motivations or concerns you're not aware of, and asking them can yield new ideas. Once you have an idea, you need to then find out whether it is valid and applies to a broader audience, or is limited to just a few respondents.

To evaluate things you can't find out elsewhere

This can be particularly helpful in evaluating your competitive environment. For example, if you're running a media site, you may want to identify direct competitors (other media sites, for example) and indirect competitors (television or movie theatres) your visitors are aware of.

To see whether improvements worked

This may be a simple evaluation—asking for a user's impression of a new feature —or it may involve comparing satisfaction scores before and after an upgrade to see whether users prefer the new approach.

To collect demographic data (such as age and income) that you can't get elsewhere

This information provides new dimensions along which to segment visitors and learn for whom your site is working best or worst. If you're running a media site, you'll also need independently verified demographic data to attract advertisers.

Let's look at some of the things that VOC studies deliver.

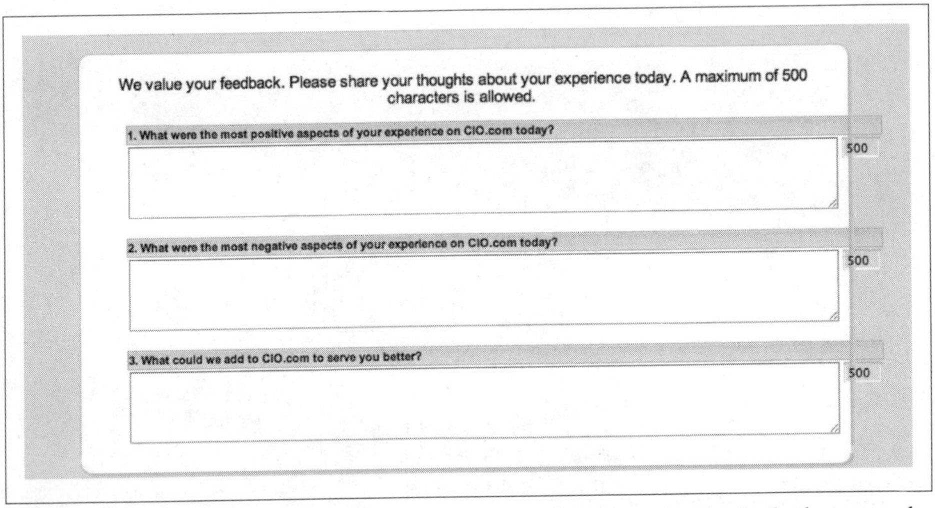

We value your feedback. Please share your thoughts about your experience today. A maximum of 500 characters is allowed.

1. What were the most positive aspects of your experience on CIO.com today?

500

2. What were the most negative aspects of your experience on CIO.com today?

500

3. What could we add to CIO.com to serve you better?

500

Figure 7-4. Open-ended questions can generate insights and unexpected topics for further research

Insight and Clues

First and foremost, VOC gives you clues. It's impossible to get inside your customers' heads and understand their motivations. Many of their motivations are unconscious, so they couldn't tell you why they did something even if you asked them directly.

Mohan Sawhney (*www.smeal.psu.edu/isbm/documents/0206msum.pdf*) of the Kellogg School of Management says:

> Customer insights do not come from quantitative market research. You cannot generate insight out of numbers. Numbers help you to validate insights. A customer insight is a fresh and not-yet obvious understanding of customer beliefs, values, habits, desires, motives, emotions or needs that can become the basis for a competitive advantage. You have to go deeper than what customers themselves say. Insights are not immediately apparent. Anomalies are an excellent starting point for generating insights.

However, you *can* ask open-ended questions like the ones in Figure 7-4 and review the feedback for clues. It's often open-ended responses like these that yield the most insight.

 Throughout this chapter, we're going to use examples from several online surveys we've seen over the last year, administered by multiple VOC services. We don't mean to pick on them—or endorse them—but we *should* thank them. They're trying to learn more about their visitors, and even when they do so clumsily, they're still well ahead of the majority of sites on the Web.

Sifting through hundreds of responses isn't always easy. Fortunately, there are tools and visualizations, such as tag clouds or concordances, that can tease common themes from this kind of unstructured data. By correlating such visualizations with scorecard responses, you can make statements such as, "People who scored their visits badly mentioned these keywords most often in their feedback."

Subjective Scoring

VOC is excellent for collecting subjective information. Suppose you're trying to accomplish the goal of fostering a sense of community. A question like the one in Figure 7-5 can tell you whether you're achieving that goal in the minds of your visitors.

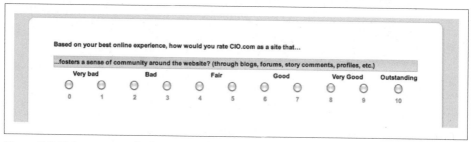

Figure 7-5. Using a range of subjective rankings (known as a Likert scale) is a good way to capture visitor impressions

Unfortunately, many companies first make changes to their sites, then survey to see if the changes worked. This doesn't provide the comparative before/after data that you need to determine whether your change actually had an impact. It's better to launch a VOC survey before a change to develop a baseline against which future adjustments are made.

Demographics

Every website has a target audience in mind. When you launched the site, you intended it to be used by a certain class of visitor. Any marketing message works best when it's tailored to a specific audience. If your actual visitors aren't who you think they are, you may need to adjust your messages, or even your business, accordingly.

Finding out whether your intended visitors match your actual visitors can be hard to do with technology, so sometimes you just have to ask them.

If you want to segment your user population, it's a good idea to ask for demographic data, like gender and age (Figure 7-6). Ensure you are not violating any legislation on the storage of personally identifiable information, however. Some regions require that you let respondents opt out of specific questions (*www.casro.org/codeofstandards.cfm*).

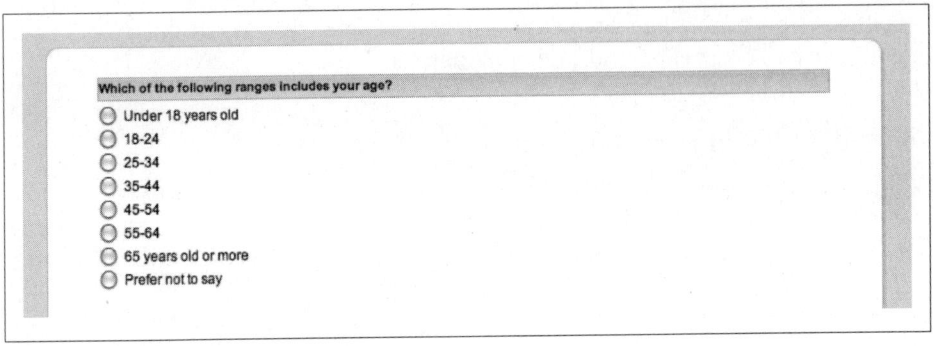

Figure 7-6. A demographic survey question

Surfographics

We call environmental data that relates or affects someone's online activity *surfographics*. In other words, anything that describes a user's environment while they surf—like accessibility tools, or multiple computers per household—are surfographics characteristics. It's extremely important to track, and is often overlooked. It includes use and behavioral data that can affect the accuracy and usefulness of your other web monitoring data.

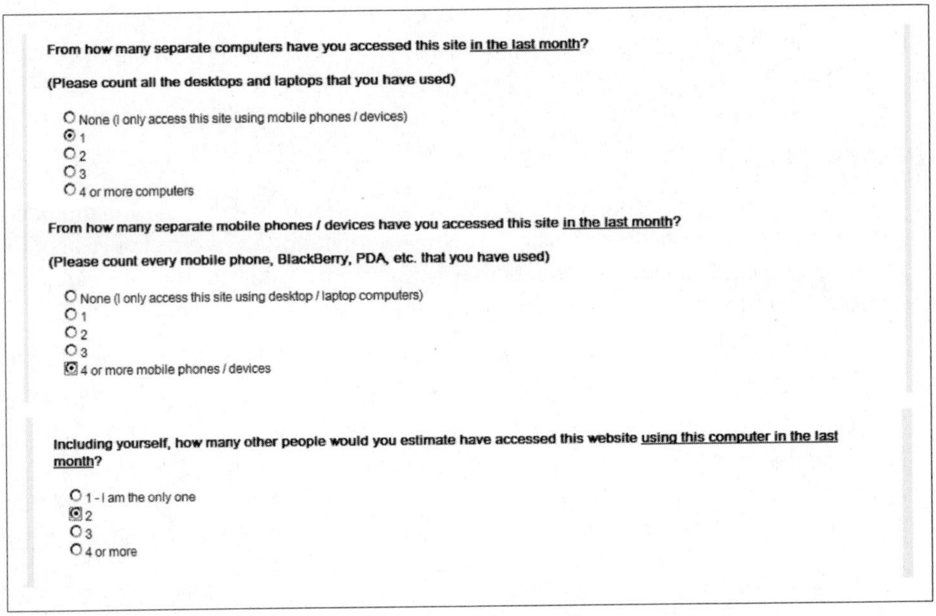

Figure 7-7. Surfographic questions on computer use help improve unique visitor counts

Cookie disambiguation

If you're running a media site, one of your most important metrics is audience size—how many people visit your website. Most organizations count this by measuring the number of unique cookies that request data from the site on a given day, but those numbers are wrong. And it's not just deleted cookies that skew unique visitor count—one person with several computers, or one computer with several users, can interfere with a proper assessment of readership.

If you use a survey like the one in Figure 7-7 to collect data on how many computers visitors have or how many people share a computer, you can adjust your readership numbers accordingly.

Familiarity with web technologies

Some visitors are extremely comfortable with the Web, while others only discovered email last week. You'll get dramatically different results to questions like those in Figure 7-8 depending on the respondent's experience with the Internet, and it's important to tie this data to the rest of their responses.

Figure 7-8. Surfographic questions designed to measure the respondent's use of web technologies

This is also important information for web designers and usability testers, as something that seems obvious to one segment of your audience may be opaque to another.

Collection of Visit Mechanics Unavailable Elsewhere

Your web analytics tools should show you where your visitors came from. Some traffic sources can't be tracked, however, and in these cases it's acceptable to ask visitors how they found you, as shown in Figure 7-9.

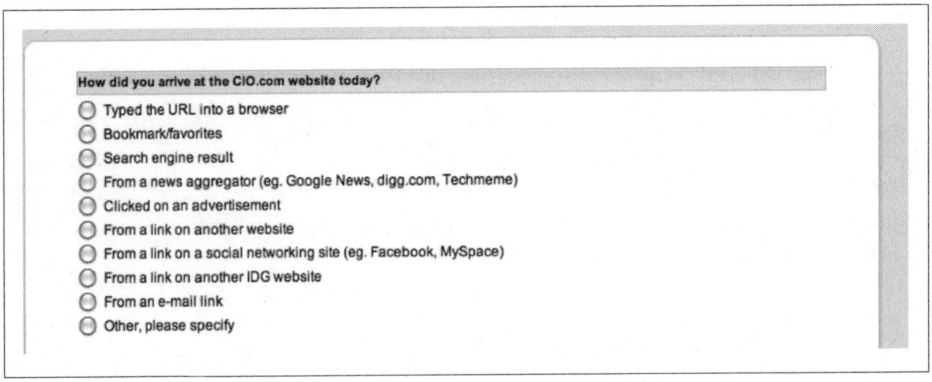

How did you arrive at the CIO.com website today?
- Typed the URL into a browser
- Bookmark/favorites
- Search engine result
- From a news aggregator (eg. Google News, digg.com, Techmeme)
- Clicked on an advertisement
- From a link on another website
- From a link on a social networking site (eg. Facebook, MySpace)
- From a link on another IDG website
- From an e-mail link
- Other, please specify

Figure 7-9. One important use of VOC is to determine the source of a visit, particularly ones in which referring URLs aren't available

For example, a visit that began with a desktop email client or a desktop microblogging client will lack a referring URL—both appear to have been typed into a browser. And yet one started with an email message from a friend, while the other stemmed from a community discussion on Twitter.

You should still compare this data to analytics information—if a visit came from a search engine, but the visitor claims it was from a social network, you've learned that this particular visitor doesn't pay a lot of attention to how he finds out about destinations on the Web (or tends to lie!), and you can view the rest of his results with that in mind.

Jonathan Levitt of VOC provider iPerceptions notes that his firm sees "two distinct measurements of visit mechanics: the path to the site (i.e., 'direct referrer') and the medium that most influenced the desire to visit (i.e., 'print media,' 'broadcast", and so on)." In other words, where someone came from and what motivated him to visit may not be the same thing.

It's also a good idea to find out where your visitors hang out, so you can be sure to include those destinations in your community monitoring strategy with a question like the one in Figure 7-10. With the rapid growth of social networks, it's important to keep your list of sites current so users can quickly indicate the ones they visit.

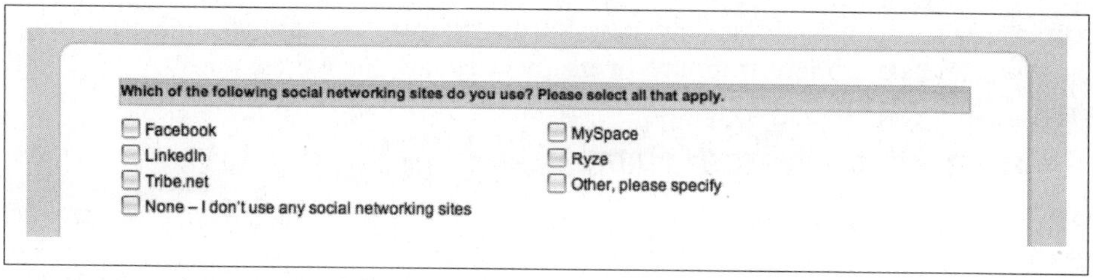

Figure 7-10. Surveying respondents for their social network use

What VOC Isn't

Now that we've addressed some of the things VOC tries to do, let's look at some things it *isn't*.

Many of the criticisms leveled at VOC—that it's unscientific or doesn't yield insights that can be applied to all visitors—aren't really fair. VOC doesn't make these claims, but many web operators who've misused VOC give it a bad name.

Here's what VOC never set out to be.

It's Not a Substitute for Other Forms of Collection

VOC is not an excuse to abandon all other forms of collection. In many cases, VOC surveys ask visitors questions that are unnecessary, because the answers can be found elsewhere. This is a sign that an organization's web monitoring tools are siloed: the people who know about performance aren't talking to the folks who run analytics, who in turn aren't sharing data with the people in usability.

Unnecessary questions reduce survey completion rates because the longer a VOC survey is, the more likely people are to abandon it rather than giving you the insights you need. So if you can get an answer somewhere else, don't waste your respondents' time—they'll only give you three to five minutes of it.

It's Not Representative of Your User Base

Perhaps the biggest criticism leveled at visitor surveys is their sampling bias. It's true that only a certain kind of visitor will respond to a survey, however good the invitation. While larger samples can mitigate sampling error, the answers you get still won't be representative of your user base. You're less likely to get responses from power users who are in a hurry, and even then, they're probably only going to offer feedback in certain situations. And you're more likely to hear from zealots and outliers.

This criticism misses the point: VOC should capture insights that you may be able to investigate. In the travel site case mentioned above, a few responses saying that visitors were just checking availability prompted the site operators to research further and confirm that this was, in fact, the case for many of their customers. Then, through analytics and experimentation, they were able to adapt their sites.

It's Not an Alternative to a Community

The best way to understand the needs, emotions, and aspirations of your target market is to visit it where it lives—in Facebook groups, chat rooms, Twitter feeds, news aggregators, and blog comment threads. You can use VOC to find out where your market hangs out, or to dig deeper into something you hear online, but you need to marinate in your community to really understand it.

It's Not a Substitute for Enrollment

If you need to constantly poll your market to understand its needs, you should convince visitors to let you contact them through email or RSS feeds. This lets you go back to them several times with additional questions and build your own panel of respondents. Remember, however, that enrolled respondents—and friends you interact with on social networks—are more loyal and "tainted" with opinions. After all, they liked you enough to enroll. So while it's good to survey them, you still need to examine newcomers.

In other words, VOC works best on visitors who you don't yet know, and who are new to the site, as soon as they've formed an opinion of you. Ask them too soon, and they won't have visited your site; ask them too late, and they may have left forever.

Four Ways to Understand Users

VOC technology is deceptively straightforward: ask your visitors questions and analyze the results. Below the surface, however, those questions are the result of a process of research and planning, sample selection, invitation, clustering, segmentation, and reporting.

There are four ways to understand the voice of your customers, as shown in Figure 7-11.

Focus groups
> A moderator guides discussion among several topics with a sample of the target market. These provide a great way to understand a customer base that's readily available. Focus groups let you see nonverbal cues, and a skilled moderator can steer the conversation in new directions you may not have considered, based on the participants' feedback.

Online surveys
> This is what most people mean when they talk about VOC. Visitors are intercepted during their visits and are asked if they will participate in a survey. While not as

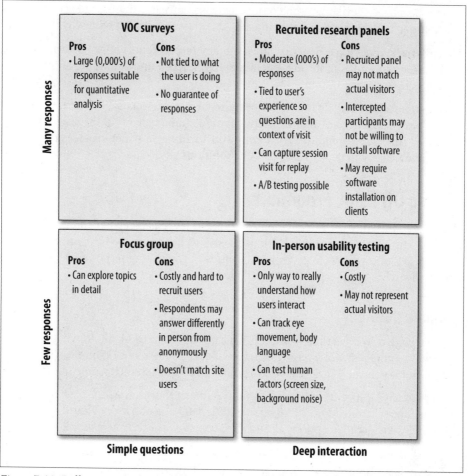

Figure 7-11. *Different methods of surveying a site's target market*

open-ended as a focus group, online surveys allow you to solicit feedback from hundreds of participants who are familiar with the actual website experience.

Usability analysis

This involves watching a participant as she uses the site. This borders on web interaction analytics (WIA) and may even include eye tracking and other tools to understand how people use a site. Usability analysts often record participants' faces and voices during the session and encourage them to provide feedback.

Recruited research panel services

These combine VOC surveys with interaction capture and run studies across a recruited panel that may include hundreds of users. Participants download a plugin that tracks their online experiences and occasionally asks them questions related to the site as they surf. While more costly and invasive than simple surveys, this approach allows a much better integration of what users did with why they did it.

The approach you choose depends on how many respondents you want to survey and on whether you want simple answers or a more in-depth understanding of their experiences as they browse the site.

All site operators should do focus group studies and in-person usability testing. All you need is patience, an observant eye, an open mind, and a belief that the only thing that matters is the user's opinion. When you want to survey a wider cross-section of the market, however, you need to look to VOC solutions, and in particular, online surveys.

Kicking Off a VOC Program

Listening to the voice of your customers requires a few clear steps: planning the study, avoiding known pitfalls, asking the right questions, designing the navigation, integrating the VOC into your website, trying the study out, selecting respondents, collecting data, and analyzing the results.

Planning the Study

Before you give your customers a voice, it's a good idea to think about what you want to learn. If you're just looking for general feedback, a simple button on the website will suffice. But if you want insights, you're going to have to ask them some questions.

It's best to have a specific question in mind. Avinash Kausik suggests four basic questions (which he's rolled into a free service, called 4Q and shown in Figure 7-12, along with VOC provider iPerceptions at *http://4q.iperceptions.com/FAQs.aspx*).

- Satisfaction: "How would you rate your site experience overall?"
- Intention: "What was the primary purpose of your visit?"
- Accomplishment: "Were you able to complete the purpose of your visit today?"
- Details: "What do you value most about this site?" or "Why weren't you able to achieve the purpose of your visit?" Of course, visitors can't answer this until the end of the visit.

These four questions get to the heart of VOC: Why did visitors do what they did?

The goals of the study

If you just want to understand what motivates visitors, some open-ended motivational questions like 4Q's are good. Most of the time, however, you'll have a specific question you want answered, for example, why are people abandoning the purchase? Are people buying for themselves or someone else?

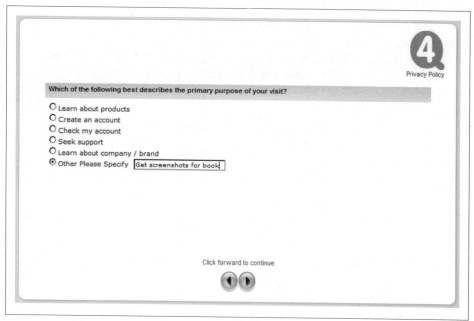

Figure 7-12. A question from iPerceptions' 4Q service

Imagine that you want to know whether people heard about your website from radio or television advertising. You can either give respondents an open-ended question and analyze the response text for patterns and insights (such as the number of times "Radio" and "TV" appear), or you can give visitors multiple-choice answers ("Radio," "TV") and get data that's easier to calculate, but that may miss some responses ("Blimp").

Consider the hotel booking abandonment issue we described at the start of the chapter. One way the travel site operator might have found the answer is by asking an open-ended question, such as, "Why did you visit the site today?" There's a chance that some people will respond, "I was just checking hotel availability and had no plans to purchase."

On the other hand, the operator might ask several questions that, taken together, suggest a motivation, such as, "Did you plan to purchase something today?" and "Are you checking hotel availability?" The problem with this second approach is that you need a hypothesis. Until you have a clear idea of what to ask, open-ended questions are better. Armed with a hypothesis, you can switch to a multiple-choice format to quantify and validate or repudiate it.

Once you know what you're trying to learn, you need to decide which survey approach will answer your questions. A simple questionnaire may suffice, particularly if you're trying to find the answers to known questions.

Often, however, you'll want to segment respondents for further insight:

Before/after comparison
If you want to know whether you made things better or worse by making changes to your site, you need to collect satisfaction data prior to a change, then compare it with results collected after the change.

Demographic comparison
If you want to find out which segments of your user community prefer a particular element of the site, you'll need to collect demographic data (for segmentation) alongside satisfaction data. This may have already been collected for you by services like Quantcast, but you will still need the data so you can segment responses in multiple dimensions.

A/B comparison of two choices
If you want to know which version of your website your visitors prefer, you need the mechanics to show them different versions, then solicit their responses. For more in-depth comparisons, you may want to use a usability testing service and ask respondents which version they prefer, being sure to randomize the version they see first to avoid learner bias.

Comparing the impact of website latency on satisfaction
A good way to understand the economics of performance and capacity planning is to ask users about their perceived experience with page latency and correlate those experiences with actual web latency measurements. For this to work, you need to associate RUM data with the VOC record.

Comparing experience based on actions
You may want to compare user experiences based on how the user interacted with the site. For example, did visitors who used the search tool have a better experience than those who browsed by directory? To do this type of comparison, you need to invite visitors to participate in the study up front, then adjust the questions you ask them based on what they did on the site itself. This means associating web analytics data with VOC survey questions. Some analytics vendors have integrated VOC data, and some VOC providers overlay their results with free analytics tools using plug-ins like Greasemonkey to make it easier to tie survey results to outcomes.

The Kinds of Questions to Ask

Your VOC study will include several kinds of questions: those that segment respondents, those that solicit feedback on something visitors saw, those that check whether visitors perceived elements of the site, those that try to capture the visitor's state of mind, and those that look for general feedback.

Why We Ask Questions We Know the Answers to

Many of the questions in VOC surveys can be answered through other sources. For example, you can usually tell where a visitor is located by his IP address. If this is the case, you shouldn't bother asking visitors directly for their IP addresses.

Sadly, however, many VOC tools ask questions whose answers lie in data collected by other monitoring tools. This is a consequence of the state of web monitoring: pieces of the visitor portrait live in analytics, RUM, WIA, and VOC tools, and are seldom linked properly, forcing analysts to ask redundant questions and use up precious seconds in a survey.

A much more elegant way to answer some of these questions—and not bother the respondent—is to monitor visitors using browser-side JavaScript. For example, you might track whether a visitor subscribes to your RSS feed in a cookie, and include this information in the survey response. This can be technically difficult to implement with hosted services, however, which is why many sites waste visitors' time with questions about activity.

While visitors typically don't shy away from answering such questions, we still think it's a lousy idea to ask questions you already know the answers to, and you should avoid doing it whenever possible by implementing better integration with other monitoring technologies.

Segmentation

Segmentation questions are used to group responses demographically or psychographically, as well as to disqualify respondents. They often solicit personal information, but shouldn't be discriminatory. Any personal data you collect may be subject to a legal review or legislation depending on how it is associated with the visitor's identity.

Segmentation questions can include basic demographics ("How old are you?") or more detailed behavioral questions ("How often do you enter your credit card online?") Recall that questions around online behavior are known as *surfographics*.

You can also segment visitors according to whether they have other interactions with your company, for example, whether they visit your brick-and-mortar retail outlet or subscribe to your print magazine (Figure 7-13).

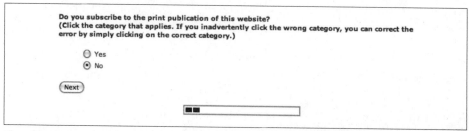

Figure 7-13. Correlating brick-and-mortar behaviors with online behaviors

Evaluation

Once you know which segment the respondent belongs to, you can ask her to evaluate the site. Pairing questions like, "Was it easy to find what you wanted?" with follow-up open-ended responses ("Why or why not?") will give you a good indication of the user's goals.

Another useful question to ask is, "How likely is it that you would recommend us to a friend or colleague?" This question is the basis of the Net Promoter Score, and is thought to strongly reflect your visitors' true intentions and feelings about your business or site. According to research by General Electric, Net Promoter scores correlate closely with a company's revenue growth (*www.businessweek.com/magazine/content/06_05/b3969090.htm*).

If the user has completed a transaction as part of a visit, you can ask her about it, but you need to wait until after she has completed the transaction before gathering feedback. There are three ways to accomplish this:

1. Intercept visitors after they have completed the transaction, so that only those who have performed the action you want them to evaluate will get the survey.
2. Use a script on the transaction completion page to let the VOC service know that this visitor needs to be given a different survey.
3. Ask visitors to self-identify ("Did you buy something today?") and have the VOC service adjust subsequent questions in the survey accordingly.

Recall

What if visitors left your site because they didn't see what you wanted them to? You may have had the perfect offer for them, but they simply didn't notice.

If you have actions or content within the site, you can test a visitor's recall once his session has expired. "Did you notice the ad for a car?," for example, will tell you a lot about why the visitor didn't click on a particular ad. You can also test a visitor's recall of marketing campaigns to see whether he heard and remembered advertisements he saw elsewhere.

While you may think content is highly visible, don't be fooled: in thousands of web usability tests that tracked eye movement (conducted by the Nielsen/Norman Group), so few participants glanced at pictures on the site that the numbers were too small to report. If you're not seeing conversions, determine whether visitors even noticed the offer.

General feedback and exploration

If you're trying to glean insights from your visitors, asking them what products or services they'd like to see from you is a good start. You can also verify the clarity of your marketing messages by asking how *they* would explain what you do to others.

Mindset

You can ask motivational questions ("Why did you visit the site?"), as well as those that tell you where a user has been prior to visiting ("Are you surfing with a specific goal in mind, or just browsing the Web casually?")

Designing the Study's Navigation

Once you've got your questions, you need to think about how to structure the survey. Your survey will have four main segments—introduction, demographics and segmentation, questions and opinions, and conclusion.

During the *introduction* you set the expectations for the respondent. If you want to hide the purpose of the study because you don't want to bias respondents, explain that you'll reveal it at the end of the survey. The introduction also states terms of service, contact information, and expected duration.

The *demographics* questions are those you'll use to segment responses. You may want to ask first (particularly if you want to do a lot of analysis), but if you're just doing research to try to capture visitor sentiment, you may want to wait until later, since you'll get more answers to questions placed earlier in the survey.

The *questions and opinions* section is the core of your survey. This is where you provide ratings and ask open-ended questions of participants. It includes recall, mindset, and evaluation questions, as well as general feedback.

In the *conclusion*, thank the visitor, obtain contact information if needed, and close the survey. If you're offering to share the results with the respondents, here's where you tell them when to expect those results. You may also wish to ask if you can contact them again in future.

Branching logic

You may want to vary the questions you ask during the questions-and-opinions section based on the demographics you collect about each visitor. For example, if you asked, "Is this your first time visiting the site?" you may have a different set of questions than those you'd ask repeat visitors.

Ideally, you'll know these conditions beforehand—your analytics tool knows whether they are returning visitors, or whether they bought something, for example. However, that analytics data may not be available in your VOC product. Some VOC tools support branching logic to selectively ask questions as the survey progresses.

Randomizing questions

You may want to randomize question order. This will ensure that all questions get responses even with respondent dropout (though, as noted, if you're seeing dropout, your survey is likely too long).

Another reason for randomizing the order of questions is to eliminate learned bias: respondents may answer one question differently depending on which questions came before it, and randomizing question order is a good way to control this effect. Learned bias is a common problem with usability tests. When test subjects see two versions of a site in A/B comparisons, they learn from their mistakes on the first version and perform better on the second version, so testers know to try A first sometimes and B first at other times. You should do the same with your questions if your VOC tool will let you.

Control questions

One way to check whether answers are accurate is to ask a related question elsewhere in the survey. You might, for example, ask whether a respondent bought something online in the last year, and then later ask how comfortable he is using a credit card online. You should see a strong correlation between the two, which you will verify during initial testing.

Using control questions like these, you can identify responses that don't seem rational and either eliminate them or downplay them in the final study. This kind of control model is especially important if you're offering a reward for survey completion, since you'll get some respondents who are simply completing the survey to get the reward, and you need to eliminate them.

Why Surveys Fail

There are plenty of pitfalls for anyone assembling a survey. The Society for Technical Communication suggests several steps in gathering questions from stakeholders at *http://www.stcsig.org/usability/newsletter/0301-surveybloopers.html*. Even if you follow these excellent suggestions, however, there's plenty to trip up the unsuspecting web analyst. Figure 7-14 shows the eight main reasons why users didn't complete surveys, according to Katja Lozar Manfreda and Vasja Vehovar in an article titled "Survey Design Features Influencing Response Rates in Web Surveys" (*http://www.websm.org/uploadi/editor/Lozar_Vehovar_2001_Survey_design.pdf*).

Since it's hard enough to get the right people to answer surveys in the first place, here are some things you should try *not* to do. The top two results are simply boredom: don't ask people too many questions.

Don't ask for frequency of visits

Asking visitors how often they visit a site (Figure 7-15) is bound to lead to inaccuracy—people don't have good recall of this kind of data. Your analytics tools have a much better record of this information.

If you can't get data from your analytics tool into your VOC tool, however, you'll need to collect this so you can segment responses by visit frequency.

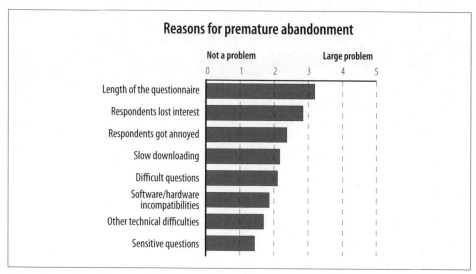

Figure 7-14. The eight main reasons people don't complete surveys

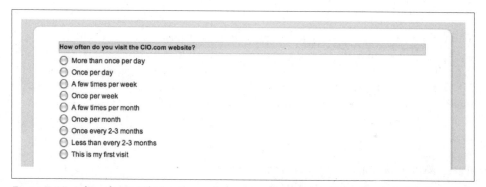

Figure 7-15. Asking for visit frequency can be a waste of a question if you can get it elsewhere

Don't ask about subscription rates

If the respondent is a newsletter recipient, you should know that from her cookie or the personalization of her login. Linking email recipients to their visits is vital if you want to understand how your users get to your site. In other words, you should already have this data from your analytics of referring URLs and returning visitors, or from your newsletter emailing service.

Don't ask what they just did

Another thing we shouldn't have to ask is what a visitor just did on his current visit (Figure 7-16). In the precious few questions you have, you need to focus on *why* someone did something, not *what* he did. VOC shouldn't be a band-aid for bad web analytics.

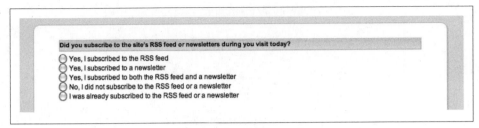

Figure 7-16. Things you should be able to work out without asking your visitors

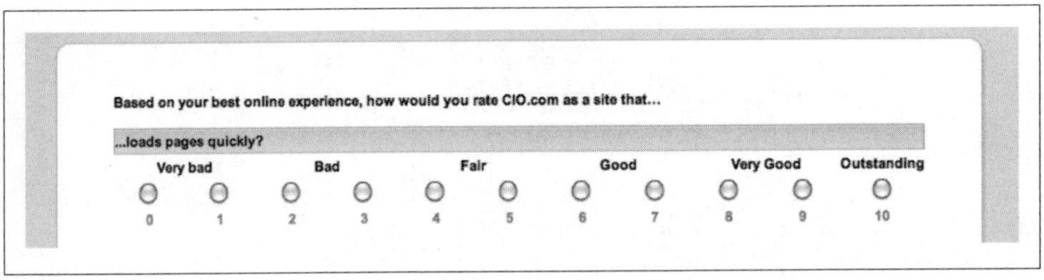

Based on your best online experience, how would you rate CIO.com as a site that...

...loads pages quickly?

Very bad		Bad		Fair		Good		Very Good	Outstanding
0	1	2	3	4	5	6	7	8	9	10

Figure 7-17. A Likert scale asking visitors how quickly they think a site loads

This is different from the recall questions we looked at earlier. Asking whether someone saw an ad helps you to understand abandonment. There's no other analytical tool that can tell whether a visitor saw an ad. However, asking a user if she clicked a particular button is wasteful, because we have other ways of tracking that. Don't mistake recall questions that look at visitor perception with those that ask about visitor actions. You shouldn't ask people to recall things that can be more accurately recorded by analytical tools.

Don't use VOC to determine web performance

When it comes to data like web performance, you should be measuring it yourself. Asking users whether performance is acceptable is somewhat useful, because it gives you an idea of their expectations, but if you're not correlating this to the performance they actually received, it's meaningless.

In the example shown in Figure 7-17, we can find out whether users felt performance was acceptable. But were they on a LAN connection or a dial-up modem? Correlating this data to performance would be incredibly useful, since it would give us an idea of what acceptable delay was and help us with capacity planning. On its own, however, it's a waste of your visitors' time—and yours.

Don't ask about loyalty

Analyzing loyalty through VOC is only slightly useful. It's one thing to know the site encourages visitors to return. It's far more meaningful to see if that encouragement

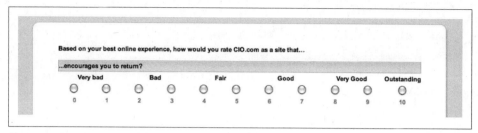

Figure 7-18. A question to assess whether respondents are likely to return

Figure 7-19. Determining a visitor's country of origin can be better done through IP address lookup, making the survey shorter

worked (as indicated by web analytics' loyalty data) than to ask questions like the one in Figure 7-18.

Don't ask for demographic data that you can get from other sources

It's easy to resolve a visitor's IP address to her city of origin (Figure 7-19); it's positively trivial to ask her what country she's from. Asking this kind of question in a VOC survey is a waste of everybody's time unless you simply can't get it elsewhere. With modern geographic lookups of IP addresses, however, there's no excuse for this.

Don't ask questions they can't answer

Your visitors may not remember everything about their visit, so asking them specific questions about it (like the one in Figure 7-20) won't yield accurate information.

You might be tempted to prompt visitors beforehand, for example, by saying, "During your visit today, pay attention to the ads you see, as we'll be asking you questions about them." If you do this, however, the answers you get won't apply to other visitors. You'll discourage respondents, increase survey abandonment, and get data that isn't useful.

While we're at it, this is the perfect time to air one of our pet peeves: don't ask visitors about their impressions of a visit before they've visited the site!

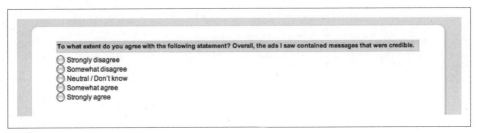

Figure 7-20. How do visitors answer if they didn't notice the ads on the website?

Don't ask too many questions

It's better to have a few questions answered by many respondents than a large number of incomplete questionnaires. This might seem counterintuitive—after all, if you front-load the survey with useful questions, you'll get lots of answers for those and few for the rest of the survey.

One of the most useful things about surveys is correlation. If you have incomplete surveys, you can't correlate responses to one another and derive the insights you need ("People from China find performance is too slow!" or "Women dislike the advertising on the site").

When it comes to surveys, fewer questions is always better. Pick a question you want the answer to, and get the answer. You can run the other surveys later.

Don't ask a question that won't reveal anything

If you have a question that everyone will answer the same way, don't waste your visitors' time asking it. When you trial your survey, see if there are questions that everyone gives the same answer to. If there are, consider whether asking it will give you any useful data. Perhaps you need to reword it for respondents to give you a variety of responses.

Always give visitors an out

As designers of the website, we seldom imagine all the answers someone might give. We're simply too familiar with the website to ask good questions. When designing a survey, give respondents an answer for every situation. This can sometimes be subtle.

In Figure 7-21, a visitor may have never had a relationship with the site before this visit. To respond that they have no relationship, they need to choose "Other" and then enter "none." It's as though the survey's designers couldn't conceive of somebody having no relationship with them.

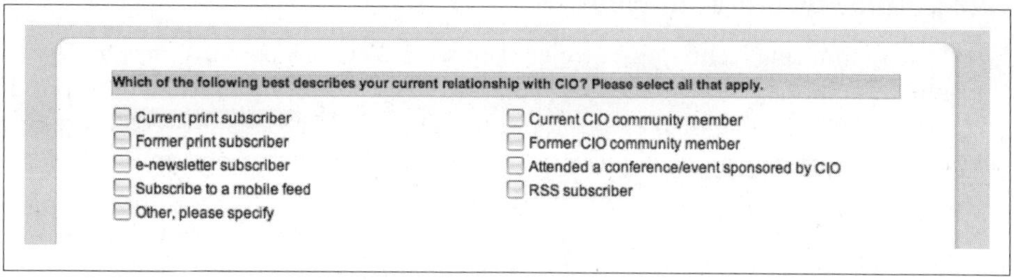

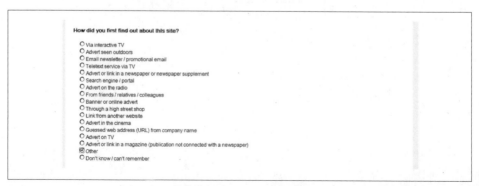

Which of the following best describes your current relationship with CIO? Please select all that apply.

☐ Current print subscriber   ☐ Current CIO community member
☐ Former print subscriber   ☐ Former CIO community member
☐ e-newsletter subscriber   ☐ Attended a conference/event sponsored by CIO
☐ Subscribe to a mobile feed   ☐ RSS subscriber
☐ Other, please specify

Figure 7-21. There's no way for a respondent to say, "I don't have a relationship"

Discourage lingering

Users tend to dwell over their answers, wanting to go back and edit them. You want to discourage this behavior, partly because it means they'll take too long to complete the survey and partly because their initial answer is likely to be correct.

There's another reason to discourage reflection. Later in the survey, you may ask other questions that are there specifically to test for consistency with previous answers. If respondents can go back and edit their earlier responses, it defeats your ability to cross-reference these control questions and determine whether a respondent was actually answering consistently. Consumer behavior researchers use these control questions to discount some responses when analyzing studies.

Some VOC services automatically advance the survey to the next question when the respondent clicks an answer. While this doesn't give the respondent the ability to correct a mistake before submitting, it also keeps the survey moving along briskly. We're fans of brisk.

Provide room for subjective feedback when no answer applies

In cases where one of the options is "other," you should provide a text box for respondents to type their answers. Otherwise, you may be completely ignorant of an important category of response, as is the case in Figure 7-22.

How did you first find out about this site?

○ Via interactive TV
○ Advert seen outdoors
○ Email newsletter / promotional email
○ Teletext service via TV
○ Advert or link in a newspaper or newspaper supplement
○ Search engine / portal
○ Advert on the radio
○ From friends / relatives / colleagues
○ Banner or online advert
○ Through a high street shop
○ Link from another website
○ Advert in the cinema
○ Guessed web address (URL) from company name
○ Advert on TV
○ Advert or link in a magazine (publication not connected with a newspaper)
◉ Other
○ Don't know / can't remember

Figure 7-22. In this case, there is no room for subjective feedback

Integrating VOC into Your Website

Of all the monitoring technologies we've seen so far, VOC is perhaps the simplest to implement. In the case of "intercepted" surveys, where visitors are invited to respond, you simply embed a line of JavaScript within the page that loads the VOC service's scripting logic. Here's an example for iPerceptions' launcher script on *www.bitcur rent.com*:

```
<!-- Begin: 4q.iperceptions.com --><script
 src="http://4qinvite.4q.iperceptions.com/I.aspx?sdfc=b2523cb8-1064-
5ca1c10e-a480-421d-a186-3
fac395d08db&lID=1&loc=4q-web1" type="text/javascript"
 defer="defer" ></script><!-- End: 4q.iperceptions.com -->
```

The steps involved in a VOC submission are outlined in Figure 7-23.

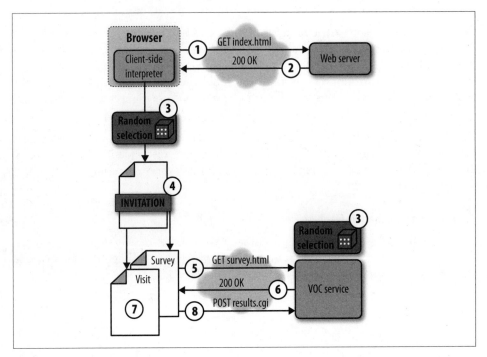

Figure 7-23. Steps involved in selecting, inviting, and submitting a VOC survey

They are as follows:

1. The browser requests a page from the web server.
2. The web server returns the page containing the JavaScript that will launch the survey.

```
<!--Start Kampyle Exit-Popup Code-->
<script type="text/javascript">
var k_push_vars = {
        "view_percentage": 30,
        "popup_font_color": "#000000",
        "popup_background": "#D4E2F0",
        "header": "Your feedback is important to us!",
        "question": "Would you be willing to give us a short (30 seconds) feedback?",
        "footer": "Thank you for helping us improve our website",
        "yes": "Yes",
        "no": "No",
        "text_direction": "ltr",
        "images_dir": "http://cf.kampyle.com/",
        "yes_background": "#76AC78",
        "no_background": "#8D9B86",
        "site_code": 9144165
}
</script>
<script type="text/javascript" src="http://cf.kampyle.com/k_push.js"></script>
<!--End Kampyle Exit-Popup Code-->
```

Figure 7-24. JavaScript for a Kampyle VOC pop up; the view_percentage value determines what percentage of visitors are invited

3. A randomizing function, either within the JavaScript (Figure 7-24), which is easier to implement, or through the VOC service, which adds more steps to the process but affords greater control, determines whether the visitor should receive the invitation.

4. Selected visitors are shown the invitation. Note that visitors can be intercepted at the start, middle, or end of a visit.

5. When a respondent agrees to participate in a survey, the JavaScript requests the VOC survey from the service.

6. The survey loads in a new window, often behind the current visit. In some cases, the visitor completes the survey immediately.

7. The visitor finishes the visit.

8. The visitor completes the survey, which is sent back to the VOC service for analysis.

If you don't want a separate interception, you can embed a survey form from services such as Wufoo, SurveyMonkey, or even Google Forms within your site.

Trying the Study

Before you run the study, you need to try it with a test audience that resembles your intended respondents in order to identify any problems with the survey before you roll it out. This can save you time and money. There's a second reason, too: if your actual responses differ significantly from those of your trial group, it's a sign that your respondents may not be a part of the market you thought they were.

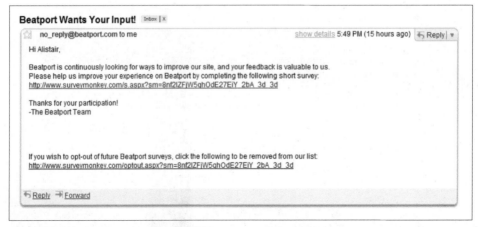

Figure 7-25. A recruitment mail for a VOC survey to existing users

The dry run can be conducted on paper or over the phone, but it's always best to try it as an actual web survey. When trying the survey, check that the test group understands what you are asking—if test respondents express confusion over how to answer the question, adjust it. Finally, time how long it takes for them to complete the survey; if it's more than a couple of minutes, it's too long. Split it into several studies.

Choosing Respondents

Once you're sure the study asks the right questions in the right way, it's time to find people to complete it. You can either recruit respondents or intercept visitors.

Recruitment

When you recruit respondents, you send a message out to people asking them for feedback, either through an email campaign or a paid research firm.

Let's look at how Beatport.com, an online music store, recruits survey respondents. The first step is to send a message like the one in Figure 7-25 to known users. In this particular case, there's no mention of a reward. The message is simple, text-based, and provides opt-out information.

On clicking the link, the respondent goes to a hosted survey like the one shown in Figure 7-26.

Once the survey is completed, the respondent is sent to the page containing free downloads (shown in Figure 7-27). The surveyor didn't mention this in the invitation, so it's unlikely that it helped improve return rate. Nevertheless, by rewarding known respondents with free content, the company is building brand loyalty with its customers and increasing the chances that respondents will complete future surveys.

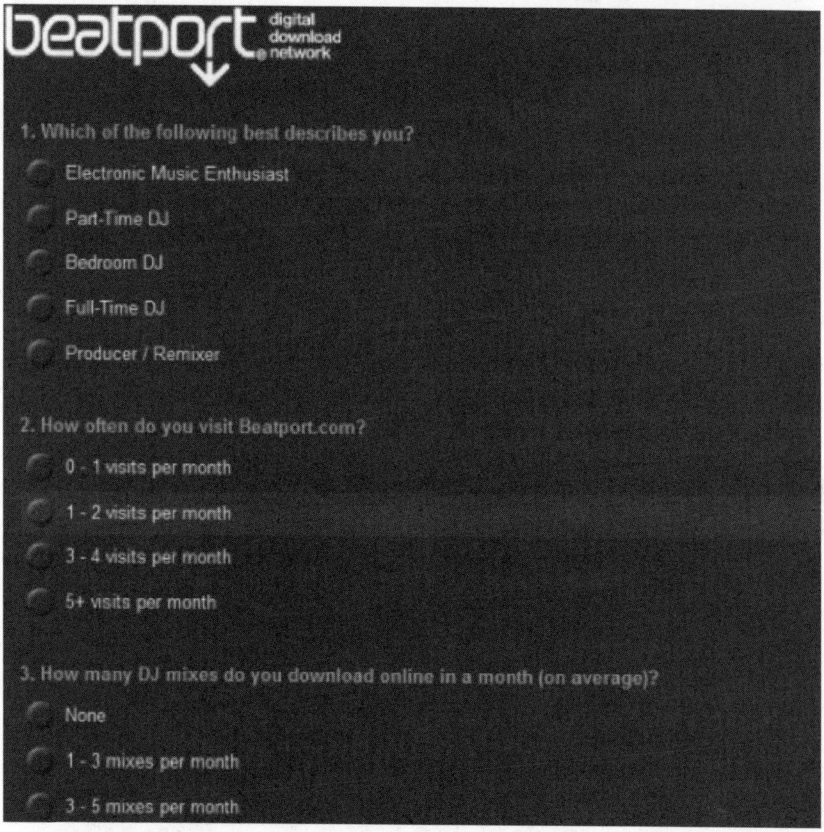

Figure 7-26. A hosted web survey on Beatport.com

What's notable about this reward is that it's embedded in the shopping cart, as shown in Figure 7-28, so the survey may actually prompt visitors to browse for additional content and make more purchases. In effect, this survey can be used as a strategy to invite back enrolled users who haven't visited the site in a while. The company has also built viral features into the landing page, encouraging the spread of content as a result of survey response.

In this example, the application lets visitors send recommendations to their friends. This is recruitment that ultimately turns into an e-commerce opportunity and viral distribution. It's a near perfect example of Dave McClure's Pirate Metrics: acquisition, activation, retention, referral, and revenue, wrapped in a VOC study.

The Beatport example benefits from a known mailing list of existing users. You probably won't have it this good. In many cases, you'll be using a third-party list and you may have to pay for a panel of respondents.

Figure 7-27. A free download page shown to survey respondents on Beatport.com

Figure 7-28. The Beatport reward is integrated into the shopping cart, making it not only a survey, but also a sales opportunity

Both mail campaigns and paid panels have limitations, however.

If you try to *recruit respondents through email*, you'll need to buy a mailing list that's segmented to your target audience (teachers in Florida, for example), and you can expect most of your mails to bounce. A 2003 study (*www.supersurvey.com/papers/su persurvey_white_paper_response_rates.pdf*) showed a 13.5 percent response rate for online survey invitations, with higher response for smaller, more targeted mailings. The good news is that half of the responses in the study came in within 17 hours of sending out the invite, so you'll get your answers quickly. If you're emailing your enrolled customers, you'll have a much better response rate, but you won't be able to study new visitors who haven't yet signed up

On the other hand, if you're using a *paid research panel service*, you'll get all the responses you can pay for, but you may find that the respondents—who are drawn from the service's paid panelists—aren't as good a fit for your target audience. For example, you may have to settle for teachers nationwide rather than just those in Florida.

Either way, recruitment gets you new answers relatively quickly. If you use a third-party mailing list, the people who answer your questions won't necessarily be your customers. To capture the real voice of your customers, you need to intercept them as they use your site.

Interception

When you intercept respondents, you ask them to participate in the survey either by interrupting them or by building questions into the transaction itself. While interception often happens when visitors first arrive, you can intercept them anywhere—after a checkout, for example, or when they've taken a particular action.

You intercept in several ways:

- Through an *invitation to participate*, such as the one shown in Figure 7-29, which, if accepted, results in a new browser window containing a survey. When the visitor has ended his visit, he can complete the survey.
- By collecting data on *a form built into the site itself*. This model is common on blogs and other content sites with a sidebar format suitable for asking a small number of questions.
- Through *additional questions during a checkout process*. While this might seem natural and noninvasive, it may impact the conversion rate on the site, so you must use it judiciously.

Self-selection and feedback buttons

Feedback buttons (Figure 7-30) are another way to intercept users by self-selection, but they won't generate a statistically representative cross-section of visitors. This means you're more likely to hear from users who are delighted or disgruntled.

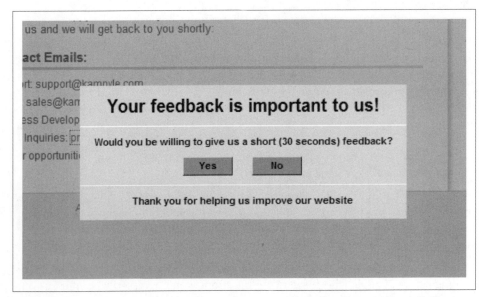

Figure 7-29. A simple interception from Kampyle generated by the JavaScript in Figure 7-24

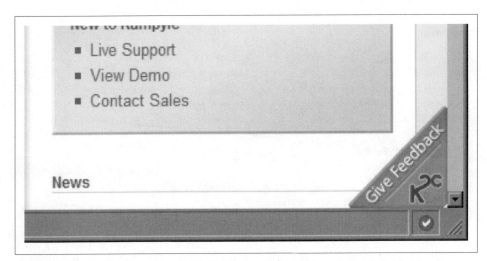

Figure 7-30. A feedback button on a site running Kampyle's VOC service

You should provide visitors with the ability to leave feedback, but not at the expense of inviting them to share their impressions with you, so don't just rely on feedback buttons.

When capturing feedback, anecdotal evidence suggests that simple visual ratings systems, such as the one in Figure 7-31, produce better responses with a greater completion rate than numerical data, at least online.

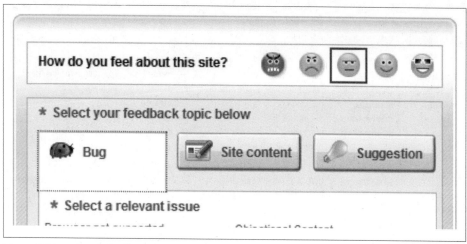

Figure 7-31. A simple feedback form

An overview of VOC methods

Before we continue, let's look at the various methods of capturing customer responses that we've seen so far summarized in Table 7-1.

Table 7-1. Various methods of capturing customer responses, expected response rates, and usefulness

	How it works	Suitable for	Example	Response rate	Correlation to no. of visitors
Interception					
Invitation to participate	Ask visitors if, once they're done with their visits, they're willing to answer a survey	Understanding visitors in depth	Demographic survey	5–15 percent	High (as long as respondent bias is controlled)
Inline form	Ask questions on the site itself	Surveying all visitors with a vote or similar question	Ask whether visitors want to see more content of this sort; Great for page-level feedback	1–2 percent (few visitors will see the survey or respond, unless it has community and ranking features associated with it)	High (respondents are real visitors), but not random, which affects generalization
Questions in checkout	Ask questions as part of a transaction	Asking specific questions to a subset of visitors who com-	Satisfaction with a checkout process	25–50 percent (higher if included in transaction process, but may affect conversion rates)	High (respondents completed a task tied to an outcome before answering)

	How it works	Suitable for	Example	Response rate	Correlation to no. of visitors
		plete a partic- ular action			
Recruitment					
Mass mail- ing to target market	Send invita- tions to a mail- ing list that matches target demographics	Getting a broader un- derstanding of a market that doesn't currently visit your site	Market re- search prior to launch	2–10 percent (impersonal messages and spam blocking lead to poor re- sponse rate)	Low (unless mailing list is highly targeted)
Mailout to enrolled visitors	Send survey in- vitation to en- rolled customers	Getting a large number of responses from loyal visitors in a short time frame	Community feedback and testing on a new release	5–20 percent (personalized messages and an engaged audi- ence yield better response rates)	High (80 percent of respond- ents complete the survey)
Paid panel	Pay panelists who match demographics to respond, sometimes with a capture of their visits	Detailed VOC and WIA from predictable, controlled audience	Understanding customer mindset during abandonment	100 percent (paid to com- plete the task and provide feedback)	Medium (depends on whether the research panel represents your target demographic)

Deciding Who to Ask

How many people do you want to survey? The simple answer is, of course, as many as possible. One commonly cited reason for surveying many people is to ensure that data is statistically significant—the more respondents, the more confidence you have in your results.

Remember, however, that your survey is already biased based on the people who responded to it. Much of what you collect won't be applicable to your site's visitors as a whole, anyway. You still need a large sample, but not because it will accurately model your entire market. The real reasons for wanting many responses are twofold: to validate patterns, and to segment.

You need to know which responses aren't complete outliers, and for this, you need dozens, even hundreds, of answers. Many of the clustering and visualization tools on the market require a large number of inputs to function properly; if you only have a few responses, you may as well read responses by hand.

The more important reason for a large sample size is segmentation. Once you get the results of the original survey, you'll probably have other questions. Imagine that you're looking at visitor satisfaction with the site after a change. You notice that it has improved significantly overall, but there's a large range in responses.

Should you be content with that data? Of course not. You should wonder whether there's a hidden pattern to your responses—did men prefer the change, but not women? Did it work better for younger visitors, but not older ones? If you have a large enough sample of responses, you can segment the results in ways you didn't foresee when you created the study.

We'll look at analysis later in this chapter. For now, know that more results are better.

On the other hand, you have a limited number of visitors, and you don't want to distract all of them with a questionnaire. This is the dilemma of VOC: how do you strike a balance between being receptive to feedback and not annoying your audience?

You must spread out your surveys with the use of a daily quota. The quota dictates how many surveys you'll invite visitors to each day, up to your target number of surveys. It's based on the estimated volume of visitors and the response rate to invitations.

Imagine, for example, that you want 20 responses a day and your site has 1,000 unique visitors a day. You also know that only 10 percent of visitors that you invite actually agree to take the survey. You therefore need to invite every fifth visitor to participate in the survey, as shown in Table 7-2.

Table 7-2. Estimating invite interval for an intercepted VOC survey

Estimated visitors	Response rate	Daily quota	Invite interval
1,000	10%	20	5

Some VOC services have sophisticated algorithms for managing quotas based on fluctuating traffic levels and varying response rates, deciding on the fly whether to intercept a visitor. Others set a daily invite interval and stick to it, leaving you to guess at traffic and conversion as best you can. And the simplest of systems store a percentage in the JavaScript for invitations, hoping that you'll get enough responses.

Private Panels

If you're using a recruited paid panel, your respondents may install software that will augment their responses, for example, by asking them questions as they surf the site. Some hosted services have access to panels of users willing to participate in research. This allows you to preselect respondents and target a particular demographic (such as males between 25 and 30 in the U.S.), but doesn't tell you what your actual visitors' motivations are. As a result, private panels are more useful for usability and navigation testing than for understanding actual visitor mindsets.

Disqualifying Certain Visitor Types

While each of your visitors has voice, you may not want to hear all of them. Should you be interested in the mindset of a particular segment, you'll need to qualify who gets a survey, based on a particular demographic (age, gender, and so on), a respondent's personal background (past use of this or other sites), or surfographic data (ownership of a particular product, comfort using the Web).

When possible, disqualify respondents based on technical information within the page rather than through questions. For example, if you only want to survey your "Gold" customers, put a cookie into the web session flagging those visitors as Gold status and modify the JavaScript that launches the invite so that it invites only Gold customers to take the survey.

Encouraging Participation

When you're running a VOC study, you will be interested in three things:

The response rate
How many visitors receive an invitation (either recruited by email or intercepted by pop up) and accept it, arriving at the survey.

The start rate
How many visitors, once presented with the survey, start it. Some visitors will forget they agreed to participate in the survey and close it, while others may change their minds upon learning more about it.

The completion rate
How many visitors who, having started the survey, finish it.

Your goal, of course, is to maximize all three of these.

Getting Great Response Rates

One of the biggest challenges in web surveys is soliciting participation from the right respondents. To begin with, invitations to participate in a web survey get fewer responses than other forms of survey. In 2007, Lozar Manfreda et al. conducted a met-astudy that looked across 45 separate studies. Their paper, titled "Web Surveys versus Other Survey Modes – A Meta-Analysis Comparing Response Rates" (*International Journal of Market Research* 50, no. 1; 79–104), showed that, on average, web-based surveys got an 11 percent lower response rate than other survey modes such as phone or email.

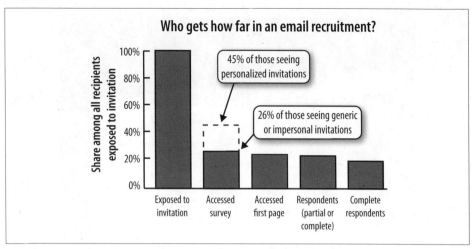

Figure 7-32. Percent of visitors who responded to and completed surveys

A higher response rate means that your survey results are more representative of your visitors in general. It also means fewer interrupted visitors (since you don't have to ask as many in order to get the number of responses you want). So, you should strive for the best possible response rate. Unfortunately, that's not always possible. An earlier study by the same group indicated that 74 percent of people exposed to general invitations and 55 percent of people who received individual invitations never accessed the survey.

Once visitors responded to the invitation, however, 83 percent started to answer the survey and 80 percent completed it, as shown in Figure 7-32 (*http://www.icis.dk/ICIS_papers/C2_4_3.pdf*). Our discussions with several VOC tool providers confirmed that most studies see similar response rates.

If you're inviting visitors to participate, you need to make the invitation appealing and give them a reason to devote some time to the effort. Simply being polite works wonders. Smart VOC surveyors also personalize their email messages, state the purpose of the survey up front, send reminder email messages, and ensure that formats and design are simple.

A lightbox approach that overlays the offer on the web page, like the one shown in Figure 7-33, can be particularly good at focusing the visitor's attention. It's also an opportunity to give a brief, clear message about your organization that may help reinforce your brand.

Depending on your target audience, a more impactful and eye-catching approach, such as the one shown in Figure 7-34, may work better. It is your responsibility to test and see which method works best for your audience.

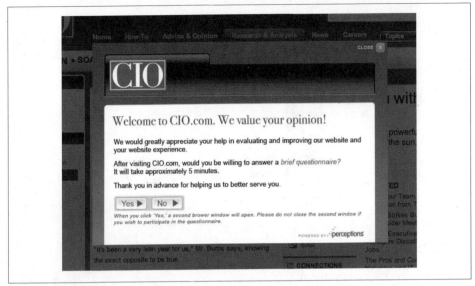

Figure 7-33. A lightbox approach is noticeable and forces visitors to respond, rather than simply letting them close a pop up or new window

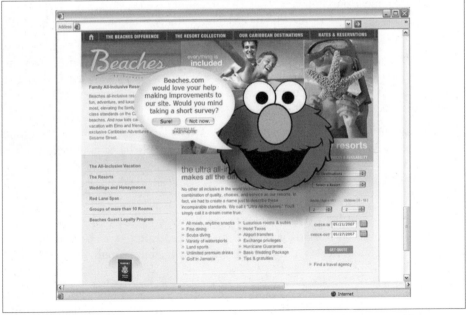

Figure 7-34. Invitations should be targeted to audience and brand tone whenever possible

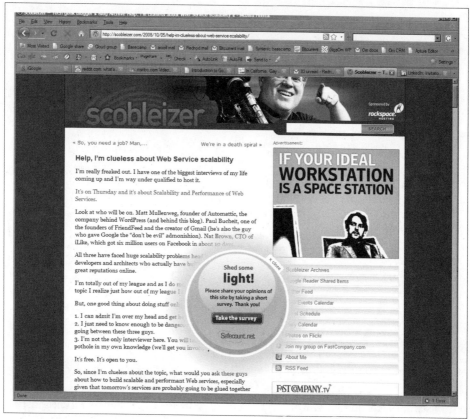

Figure 7-35. The use of an overlay is an elegant method to encourage users to begin surveys, as it still allows users to read a portion of the site

As Figure 7-35 shows, another advantage of overlaid invitations is that visitors can see some of the site behind the invitation at the same time, so they associate the survey in front with the brand and content behind it.

The risks of rewards

Should you reward respondents for participation? The web operators we've spoken with suggest that rewards can increase response rates by 15 to 20 percent, particularly for consumer surveys.

In business-to-business surveys, rewards that are perceived as bribes may have a negative effect. One popular way to avoid this perception is to offer to share the survey's results with respondents, who may have an interest in some of the outcomes.

Rewards may increase the total number of respondents, but may also reduce the number of responses that fall within your desired respondent demographic. So, if you're going to reward participation, be sure you ask control questions that disqualify

Associated Newspapers - Survey Prize Draw Rules

1. Only one entry will be accepted per person.

2. This prize draw is only open to people aged 18 or over. The prize winners will be selected at random by Survey Interactive from all entries received.

3. Only the winners will be contacted, via email.

4. Prizes must be taken as stated and cannot be deferred.

5. There are no cash alternatives and prizes are non transferable.

6. Associated Northcliffe Digital and Survey Interactive do not accept any responsibility for late or lost entries due to the Internet. Proof of sending is not proof of receipt. The decision of Associated Northcliffe Digital is final in every situation, including any not covered above and no correspondence will be entered into.

7. Entrants to prize draws will be deemed to have accepted these rules and to agree to be bound by them. These rules are governed by the laws of England and Wales.

8. We will not pass on your personal details to any other organisation without your permission, except for the purpose of awarding your prize if necessary.

9. This prize draw is not open to employees or contractors of the participating websites or any person directly or indirectly involved with those organisations or running of this prize draw including direct family members.

10. Entries will not be accepted after 23:59pm on 07/12/2008.

11. Should you have any questions about the prize draw please contact the Insight team at Associated Northcliffe Digital on +44 (0)20 7752 8400 or write to Northcliffe House, 2 Derry Street, Kensington London W8 5TT.

Figure 7-36. Clearly stating the terms of any reward is essential

responses from respondents who aren't part of your target audience and are just on your site to win a prize.

The type of reward you offer may vary depending on survey length, too. A 2004 study on response rate and response quality found that "vouchers seem to be the most effective incentive in long questionnaires, while lotteries are more efficient in short surveys. A follow-up study revealed that lotteries with small prizes but a higher chance of winning, are most effective in increasing the response rate.[*]

If you're going to offer rewards, it's important to state the rules and regulations as part of the survey conclusion, as shown in Figure 7-36.

In particular, you may have to comply with state regulations on prizes and contests, and should tell respondents that you will only contact them in the event that they win.

---

[*] Deutskens, E.; de Ruyter, K.; Wetzels, M.; Oosterveld, P.; "Response Rate and Response Quality of Internet-Based Surveys: An Experimental Study", *Marketing Letters* 15, no. 1 (February 2004): 21–36(16).

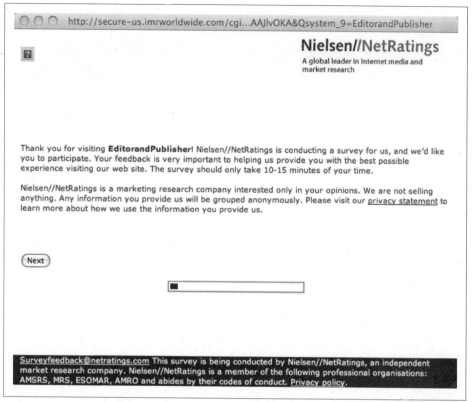

Figure 7-37. A progress bar is an important part of properly setting respondent expectations

Setting Expectations

It's vital that you tell respondents what to expect early in the process. This includes the purpose of the survey, the way in which data will be used or shared, and what steps you're taking to protect respondents' privacy. It should also include the estimated time to complete the survey and clear details on how to contact the operators of the survey, as shown in Figure 7-37. VOC vendor iPerceptions noted that when it first added a progress bar to surveys, there was a significant increase in completion rate.

Permission to Follow Up

You should always ask whether you can follow up with a respondent, using a question like the one in Figure 7-38, either because you may need further clarification on responses or because you may want to include the respondent in subsequent surveys that rank improvement on the site.

A secondary KPI for your site might be the percentage of VOC respondents who agree to a follow-up and become part of your recruited panel.

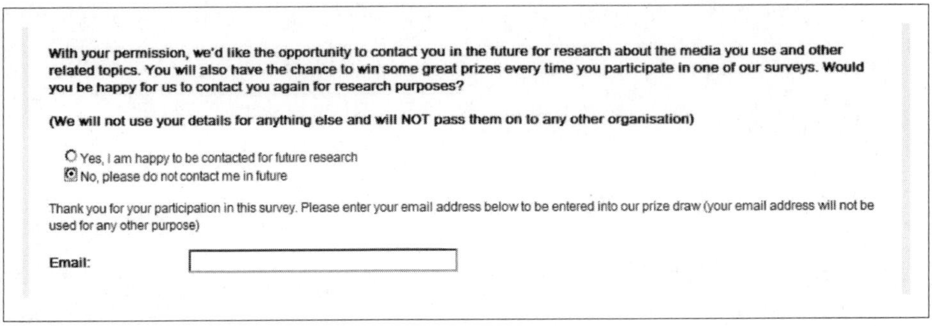

With your permission, we'd like the opportunity to contact you in the future for research about the media you use and other related topics. You will also have the chance to win some great prizes every time you participate in one of our surveys. Would you be happy for us to contact you again for research purposes?

(We will not use your details for anything else and will NOT pass them on to any other organisation)

○ Yes, I am happy to be contacted for future research
◉ No, please do not contact me in future

Thank you for your participation in this survey. Please enter your email address below to be entered into our prize draw (your email address will not be used for any other purpose)

Email: [   ]

Figure 7-38. An example of a follow-up request at the end of a survey

Improving Your Results

If you aren't getting the results you want from your VOC research, you need to tune your approach. Doing so requires many of the techniques we discussed when looking at web analytics. In many ways, the VOC conversion funnel should be treated as a transaction like any other on your site, and you should monitor and optimize it to ensure you get good information. Here are some problems you may encounter, and suggestions for addressing them.

High recruitment bounce rates

If your email invitations to respondents are bouncing, you need to make sure your messages don't look like spam. Using plain-text email messages (instead of image-heavy, multiple-object messages) works well. Similarly, make sure the message is very simple and short, with only a single URL. Try to make the title of the message direct, and address it to the individual, but don't use words like "survey" or "invitation" in the title—and use them only sparingly in the body of the message.

Low recruitment response rates

Invitation recipients are 19 percent more likely to respond to a personal, targeted message aimed at them individually than to a generic one. If you're targeting your message and still not getting answers, you can consider a reward or try to explain more about why taking the survey will help the recipient somehow. Appeal to recipients' altruism—making the Web a better place—and explain why they, in particular, have been selected.

Poor interception response rates

You're asking people to respond when they visit your site, but they're not taking you up on the offer. You need to optimize your invitation—experiment with several messages and compare the results. Change the point at which the invitation is shown, the segments you target, or the rewards that you're offering. If all else fails, either recruit

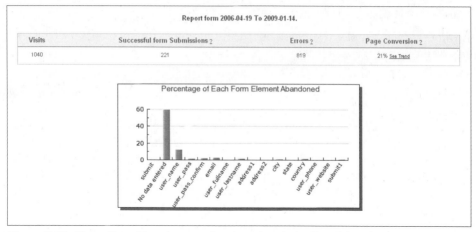

Figure 7-39. Using FormAlive form analysis to identify bottlenecks and troubleshoot survey completion issues

respondents or resort to placing questions within the checkout process—just be sure not to hurt conversions.

Poor start rates

If people are accepting your offer to participate in the survey, but are not starting the survey, you may want to begin it immediately rather than suggesting that they first finish their visit to the site. You may also be able to use JavaScript to detect when the page is closed and bring the survey window to the front, asking visitors if they wish to close the browser or navigate away from the page without completing the survey.

Poor completion rates

If users are starting the survey but not finishing it, make it shorter. Long surveys are the main cause of abandonment, and incomplete responses can't be properly correlated with one another.

Some of the WIA tools we saw in the previous chapter can analyze form completion rates, since they see keystrokes and mouse movements within a page of a form to determine where users are spending the most time, which forms they're refilling, and which fields they are leaving blank.

This kind of analysis, shown in Figure 7-39, suggests where problems occurred or where the visitor gave up. You're less likely to be able to integrate WIA monitoring from one vendor with a VOC survey from another, however, so this may only apply to forms you're operating yourself.

Large number of disqualified responses

If you're getting completed surveys, but the results aren't useful, there may be one of several problems:

You're not asking the right people
> Demographic data from the survey suggests that you're hearing from visitors that aren't your intended audience. Because respondents don't match the group you're hoping to analyze, you need to disqualify earlier in the process. If you're recruiting participants through email, you need to change your mailing list or ask qualifying questions in the email itself. If you're using a paid panel, you need to urge the vendor to adjust the makeup of its panel. If you're providing a reward, try removing it and see if this leads to better, albeit fewer, results.

You're not asking the right questions
> If you're not getting a picture of what customers are saying, you may be constraining their responses too much. It's okay to have a hypothesis and test it, but make sure there is room for some open-ended answers (for example, a respondent may write, "Your colors are too bright"). Then analyze the open-ended responses for patterns ("10% of respondents say colors are too bright"), and consider turning them into a more structured question ("Our colors are too bright. Agree/disagree") in future surveys.

The users don't understand the questions
> If you're getting inconsistent responses, some of the respondents may simply not understand what's being asked of them. See if there's a pattern of misunderstanding ("More men under 30 answer question five incorrectly") and retest the questionnaire with that segment to find out where the misunderstanding lies. Often, identifying misunderstandings or confusing terminology can improve your organization's marketing communications as a whole. If, in the process of optimizing your VOC studies, you identify jargon or words that your market doesn't understand, be sure to tell the marketing department!

Once you've got data you can use, it's time to analyze it.

Analyzing the Data

Once you've collected enough responses, you can start analyzing the data. When respondents can only pick one answer from a selection—usually communicated through a radio button or a drop-down list— you simply count the number of responses. This can be tricky, however, as simple answers can often hide more complicated patterns. In other words, distributions can often be more valuable than aggregations. Thankfully, much of the heavy lifting can be done for you by using the algorithms provided by existing VOC vendors.

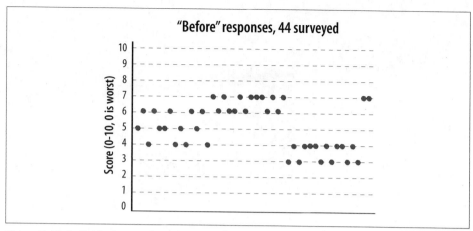

Figure 7-40. Individual scores from 44 respondents prior to a site change

The importance of segmentation and representation

It's critical to consider your results in the context of segments of visitors that responded to your survey. Let's look at a hypothetical example of how segmentation can show very different results.

A media site wanted to improve the excitement generated by its visual content. Knowing that results are only useful if compared to something, it first surveyed visitors before making the change to its site. It intercepted 44 respondents, asking them for their impressions of the website on a scale of 1 to 10, where 1 was "bad" and 10 was "good." The results are shown in Figure 7-40.

The organization then changed the site, adding glamorous visuals and more movement and multimedia. Again, it collected 44 responses from visitors rating the site. These results are shown in Figure 7-42 (shown later).

At first glance, the results were encouraging. The rating climbed from an average of 5.15 to an average of 6.31—an 11.6 percent improvement. The web operator had also decided to ask two segmentation questions: respondent age and gender, When segmented along these dimensions, the results were very different.

As Figure 7-41 shows, while the site changes had improved the rating given by younger visitors, it had been less effective for older visitors.

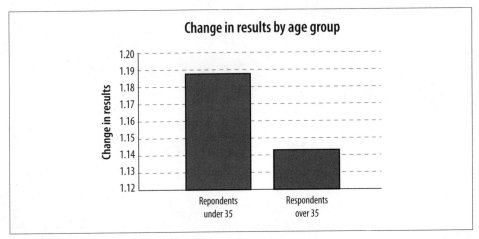

Figure 7-41. Change in site rating segmented by age group

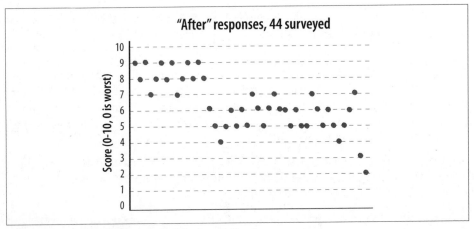

Figure 7-42. Individual scores from 44 respondents after a site change

Perhaps most importantly, the results showed that while the changes improved men's ratings of the site, they had actually lowered ratings from female visitors, as shown in Figure 7-43.

On closer analysis of the data, the company determined that the respondents had not been evenly represented across gender and age. 57 percent of respondents were female and 64 percent were under 35.

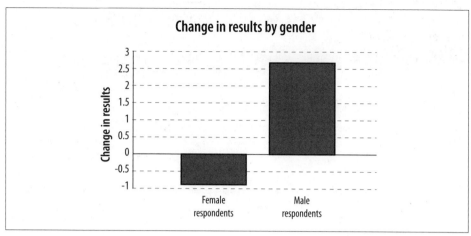

Figure 7-43. Change in site rating segmented by gender

This is a fairly straightforward example, but it should show you the importance of data exploration and segmentation. If the site sought the approval of young males, the change might have been a success, but if it was targeting older women, the change was a disaster, as summarized in Figure 7-44.

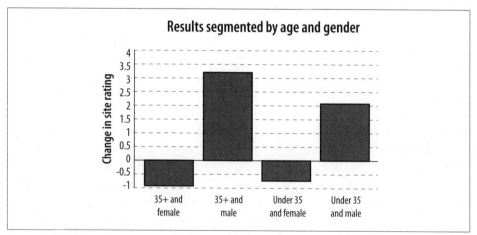

Figure 7-44. Relative change in site rating for four segments of respondents

While a full discussion of statistical analysis is beyond the scope of this book—and most commercial packages offer tools to help with this—here are some things to bear in mind.

Analyzing integer data

The easiest data to analyze is whatever you've collected numerically. This may be a value you've asked visitors to give you ("How many other social networks do you belong to?") or rating data ("On a scale of 1 to 5, where 1 is dismal and 5 is awesome, how good is this site?").

Every time you analyze numbers you need to calculate some basic statistics (Table 7-3).

Table 7-3. Some basic statistical terms

Measurement	What it means	How it's calculated	Concerns and uses
Mean (or average)	The average of all responses	Add the answers and divide by the responses	Averages can be misleading—a few outliers can strongly influence them, so it's common to trim off the highest and lowest values to get a better sense of an average
Median (50th percentile)	The number that "splits" all the responses in half	Sort the responses and find the halfway point	Better for finding middle ground than an average when there are large outliers
Mode (most common)	The response that happens the most	Count how many times each value or category is seen	Helps you understand what's most common in responses; unlike means and medians, modes apply to categorical data, for example, "the most common name in the survey is Smith"
Standard deviation (dispersal)	How dispersed the answers are	Add up how far each response is from the mean, square the results, find the mean of that square, and take the root of it	If all respondents provided the same answer, there would be no deviation; if answers varied considerably, standard deviation would be high; standard deviation is a measure of uncertainty

Wikipedia maintains a comprehensive discussion of statistics and provides more detailed information on some of these terms at *http://en.wikipedia.org/wiki/Portal:Statistics*. Figure 7-45 shows how the four statistical metrics describe the distribution of the 44 responses we looked at earlier.

Most commercial software will calculate means and standard deviations, as well as segmenting across various categories of data you've collected. Figure 7-46 shows an example of this.

In addition to aggregate analysis of visitors' responses, most tools will let you view an individual response as well (as shown in Figure 7-47). As we've seen, looking at the individual responses and understanding their distribution can be far more revealing than just looking at averages.

It's important to be able to move from aggregate trends—overall customer satisfaction, main reasons for complaints, top motivations, and so on—to individual responses. This way, you can understand patterns while still gleaning the open-ended insights for which VOC is so important.

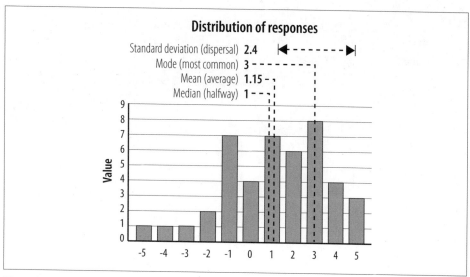

Figure 7-45. How various statistical terms describe a data distribution

Displaying numerical breakdowns

To visualize multiple-choice data, you can simply show scores for each answer or a "top five" list. Figure 7-48 shows more complex data—in this case, a comparison of two brands' rankings along with a statistical description of the results.

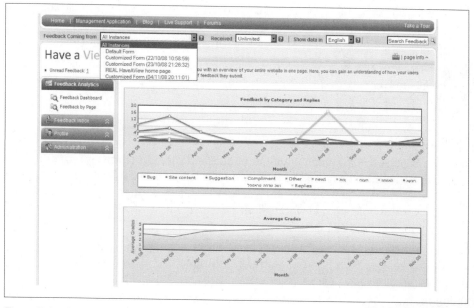

Figure 7-46. Statistics on visitor grading of websites within Kampyle shows averages over time

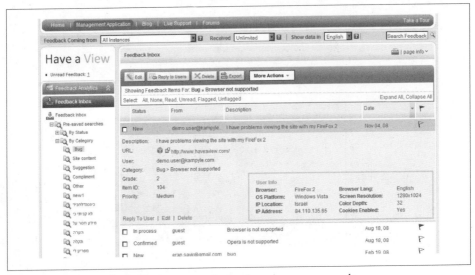

Figure 7-47. A view of individual responses that make up an average grade

Ordinal data

If you ask respondents for a list of responses in order, you're collecting ordinal data. For example, you might say, "List your five favorite ice cream flavors, with the most favorite first."

While this data is useful, you may want to weight it to ensure the best visualization. You may give a weight of 5 to the first answer, 4 to the second, and so on. The result is a weighted scoring of preference that you can add up across respondents.

Displaying many possible answers can be challenging. New visualizations such as the tag cloud shown in Figure 7-49 can help you analyze data for patterns and trends that might not be as obvious.

Open-ended data

If you're asking users to submit open-ended data and you have relatively few responses, you owe it to yourself to read them individually. If you have large numbers of open-ended responses, you may want to analyze only those whose quantitative responses are unusual, for example, people who had an extremely unsatisfying experience or those who could not complete a task.

Some VOC services offer parsing tools that will look at important words (nouns and verbs) and analyze them, presenting them as ordinal data based on how often they occur, or in tag clouds. Advances in natural language parsing and semantic interpretation promise to make it even easier to extract sentiment and meaning from a large number of responses.

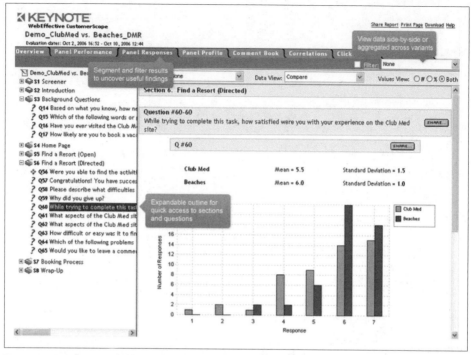

Figure 7-48. Keynote's WebEffective provides statistical data and distributions by question

Integrating VOC Data with Other Analytics

Once you've extracted VOC data, you may want to share it with other data sources. Some web analytics vendors offer data warehouses in which VOC responses and performance metrics can be analyzed alongside analytics.

Depending on the service you're using, user activity may have already been captured along with responses, as shown in Figure 7-50. This is the case for some recruited panels in which respondents install software to track their visits. If you're using such a service, you can see who a visitor was, what she did, and her responses within the same page.

If you're manipulating large amounts of VOC data, you may want to export it to a spreadsheet or to statistical software, as Figure 7-51 shows. Many tools and survey sites can export data as comma-separated value (CSV) files suitable for importing into such software packages.

If there's one theme throughout this book, it's the convergence of web visibility tools. VOC is no exception.

There are examples of VOC companies that integrate with WIA and web analytics platforms, but this is something that we'll look at in more detail in Chapter 17.

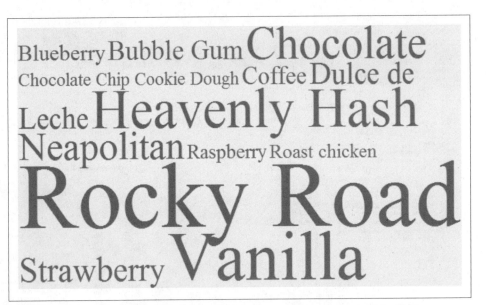

Figure 7-49. A tag cloud visualizing frequency of responses within unstructured feedback

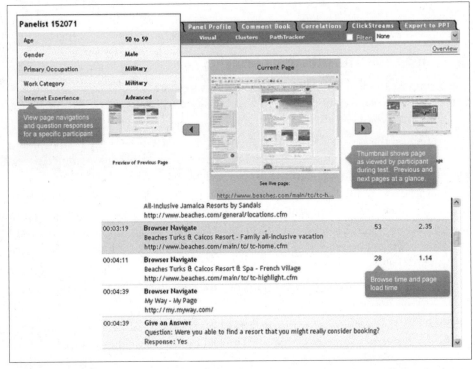

Figure 7-50. Integration of page replay (WIA) and survey responses (VOC) in a single tool

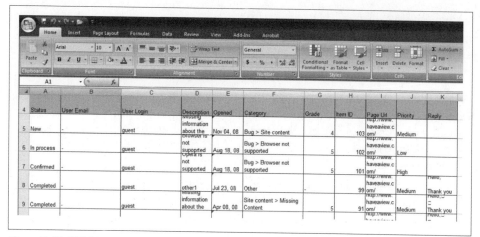

Figure 7-51. VOC responses exported to a spreadsheet for further analysis

Advantages, Concerns, and Caveats

VOC is an essential tool for understanding your visitors' mindsets and their reactions to the changes you make. However, it's not a substitute for other forms of tracking and analysis, and it requires rigorous statistical analysis of data distributions to avoid misinterpretation. Results may be biased toward a subset of your visitors rather than reflecting the opinions of all your users.

Learning What to Try Next

If you can harness VOC, you have a real advantage: *you learn what to try next*. Discover a possible improvement, and you know what to measure with analytics, EUEM, and WIA. If you can adopt a mindset of quick iteration that collects and analyzes the voice of your customers, you'll be much more likely to succeed online.

Becoming Less About Understanding, More About Evaluating Effectiveness

VOC is a poor substitute for actually interacting with your customers. As the Web shifts from a one-to-many monologue to a many-to-many conversation, there are less intrusive, less misleading ways to determine what your target audience is after. We'll look at this in Chapters 11 through 14 of this book, when we turn our attention toward online communities.

VOC won't be marginalized, however. It will be disseminated differently—through Twitter threads and Facebook groups. There will be a human element to it, as we invite social contacts to give us more detailed information resulting from a survey.

Community managers will have several predefined surveys running at all times and will guide vocal community members toward them.

You May Have to Ask Redundant Questions

We've admonished operators for asking questions that can be answered elsewhere. Doing so is lazy and often less accurate than cross-referencing other data you already have. But we would be remiss if we didn't recognize that there is a good reason this still happens despite a desire to keep surveys short and completion rates up: *data compartmentalization.*

In some cases, you may be legally prevented from associating VOC data (such as race, gender, or age) with a visitor's identity (analytics, WIA). If this is the case, you have no choice but to collect data within the VOC system, even if it already exists elsewhere, because you can't associate the respondent with analytics and EUEM tools without breaking the law. If so, we forgive you for asking visitors things you already know—just try not to do it too much.

Voice of the Customer Maturity Model

Maturity level	Level 1	Level 2	Level 3	Level 4	Level 5
Focus	Technology: Make sure things are alive	Local site: Make sure people on my site do what I want them to	Visitor acquisition: Make sure the Internet sends people to my site	Systematic engagement: Make sure my relationship with my visitors and the Internet continues to grow	Web strategy: Make sure my business is aligned with the Internet age
Who?	Operations	Merchandising manager	Campaign manager/SEO	Product manager	CEO/GM
VOC	"Contact us" buttons and on-site feedback; emphasis on satisfaction	Surveys within the site via opt-in invitations; emphasis on loyalty	Engaging the public Internet (chatrooms, social sites, etc.) and analyzing key topics and discussions; emphasis on word-of-mouth and virality	Customer collaboration in product and service design; user engagement; emphasis on lifetime value creation, giving the user a sense of ownership	Consumer feedback tied in to corporate planning through quantitative analysis of VOC and community data; customer as a collaborator in the growth of the company

Web Performance and End User Experience

Analytics, WIA, and VOC has shown you what visitors are doing and why they're doing it. But there's another equally important question to answer: could they do it? Visitors expect a responsive online experience that's reliable. Sites that are slow, or that are often down, can't serve their visitors. Measuring this is the domain of End User Experience Management. Part III contains the following chapters:

- Chapter 8, *Could They Do It?: End User Experience Management*
- Chapter 9, *Could They Do It?: Synthetic Monitoring*
- Chapter 10, *Could They Do It?: Real User Monitoring*

Could They Do It?: End User Experience Management

Web analytics is your best insight into your website, your brand, and your customers online. But if the site's not working properly, there simply won't be any activity to watch. Sites that aren't reachable can't sell things, and slow sites lose users. Google estimates that for every additional 500 milliseconds of delay, their site loses 20% of their traffic (*http://web2.sys-con.com/node/804850*).

Web applications break in strange and wonderful ways, many of which are beyond your control. Monitoring performance and uptime is usually the job of a web operations team, and it's usually an IT function. And that's the start of the problem.

In most companies, the skills needed to run a website are often broken up into distinct silos. One group is responsible for designing the application, another for testing it, another for managing the platforms and infrastructure, another for analyzing what happens on it, and another for understanding how web activity relates to community opinion.

This separation of roles might make sense on an organizational chart, but for web applications, it's dysfunctional. Web analytics and web performance are two sides of the same coin. They should be in the same business unit and should be judged by metrics like user satisfaction, conversion rates, and uptime. The teams responsible for the website's design, analytics, community, and support need to work alongside the web operations teams that are responsible for making sure the infrastructure, servers, and networks are healthy.

Web operators rely on two main types of tools to make sure their sites are functioning properly. The first, called synthetic testing, simulates visitor requests at regular intervals to make sure the website is working. The second, real user monitoring (RUM), analyzes actual user traffic to measure responsiveness and detect errors.

What's User Experience? What's Not?

A visitor's experience is the performance, availability, and correctness of the site they visit. For the purpose of this book, we'll define it as *a measure of how accurately a user's visit reflects the visit its designers intended*. We refer to the task of managing this experience as End User Experience Management, or EUEM.

In this book, we make the distinction between EUEM and usability. Usability, which we measure with WIA and usability testing, looks at how users tried to use the site. If visitors did something wrong—if they couldn't find a particular button, or if they entered their hat size in the price field—they weren't doing something the designers intended. While that issue is, of course, vitally important to web operators, it's a usability problem.

Here, we're concerned with how well the website delivers the experience the designers intended. Problems may include pages that took a long time to reach a visitor or rendered too slowly in the visitor's browser, websites that broke because third-party components or plug-ins didn't work, and content that wasn't correct or wasn't shown on the screen properly.

At its simplest, EUEM is about answering the question, "Is this site down, or is it just me?" Every site operator has experienced a moment of sudden dread when they try to visit their own site and find that it's not working. If you implement EUEM correctly, you'll know as soon as there's a problem, and you'll know who's affected.

ITIL and Apdex: IT Best Practices

Before we get further in this discussion, we need to talk about standards. The IT Information Library (ITIL) is a set of recommendations and best practices for operating IT services (*www.tso.co.uk*). It includes a dauntingly comprehensive set of processes and terms, but its fundamental message is simple and straightforward: *any IT service exists to perform a business function*.

ITIL is all about running IT as a service. If you're in web operations, your "customers" are really the rest of the business, and you're doing your job when the services the business needs are available as expected.

Terms like "performance" and "availability" have many meanings to many people. ITIL defines availability as "the ability of a service to perform its agreed function when required, as a function of reliability, maintainability, serviceability, performance, and security." In other words, availability is the percent of the time that you did what you said you would.

Availability decreases when the service you run—in this case, a website—isn't working properly. That may be because of an incident, which is "an unplanned interruption to an IT service" or "a reduction in the quality of an IT service." It may also be because

---

the site is slow. Slowness simply means that the response time—the time taken to complete an operation or transaction—exceeds an agreed-upon threshold.

Other factors, such as a security breach or an inability to update the site, all affect the availability of an IT service. In this book, we're going to focus primarily on what ITIL calls Response Time and Incident Management, both of which affect the availability of the web service. For web operators, the task of "day-to-day capacity management activities, including threshold detection, performance analysis and tuning, and implementing changes related to performance and capacity" is called *performance management*.

Though ITIL makes good sense, it's a bit formal for many startups that are focused on delivering products to markets before their funding runs out. If you're looking for something a bit less intimidating, consider Apdex (Application Performance Index), which is a formula for scoring application performance and availability that's surprisingly simple, and can be used for comparison purposes.

Originally conceived by Peter Sevcik, Apdex is now an industry initiative supported by several vendors. An Apdex score for an online application is a measurement of how often the application's performance and availability is acceptable.

Here's how it works: every transaction in an application (such as the delivery of a page to a browser) has a performance goal, such as, "This page should load in two seconds." Every page load is then scored against this goal. If the page is delivered within the goal, the visitor was "satisfied." If the page was delivered up to four times slower than the goal, the visitor was "tolerating." If it took more than four times longer—or simply wasn't delivered at all because of an error—the visitor was "frustrated."

To calculate an Apdex score, you use a simple formula, shown in Figure 8-1.

$$Score = \left\{ \frac{(Satisfied) + (Tolerating/2)}{All} \right\}$$

Figure 8-1. Calculating an Apdex score

In other words, you add up all the satisfied measurements, 50% of all the tolerating measurements, and 0% of all the frustrated ones, then divide by the total number of measurements.

 http://www.apdex.org/overview.html has information on how to calculate an Apdex score, as well as vendors who support the metric. One of its key strengths is the ability to "roll up" scores across many different sites, pages, and applications to communicate performance and availability at different levels of an organization.

Apdex is useful for several reasons. You can take several Apdex scores for different pages or functions on a website, and roll them up into a total score, while retaining individual "satisfied" thresholds. It's also a consistent way to score a site's health, whether you have only a few measurements—a test every five minutes, for example— or many thousands of measurements every second.

A Note on Terminology

While ITIL's terms are useful, we're going to use terms that are more common in the web operations world. When we refer to *availability* here, we're talking about uptime —*the amount of time that the site can be reached, even if it's relatively unresponsive*. This is similar to ITIL's concept of incidents and problems—if you see an error, availability goes down. If your site's always up, its availability is 100%.

When we refer to *performance*, on the other hand, we mean web infrastructure performance, specifically *the responsiveness of an application as experienced by the end user*. While we often talk about average performance, averages themselves aren't very useful. We care more about how many of the measurements we've taken exceed a threshold or about what the worst-served visitors are experiencing.

Why Care About Performance and Availability?

You may have formal contracts with your site's users that dictate how much of the time your site will be working and how quickly it will handle certain requests. Even if you don't have a formal SLA in place, you still have an implied one. Your visitors have expectations. They'll get frustrated if your site is much slower than that of your competitors, or if your performance is inconsistent and varies wildly. On the other hand, they'll be more tolerant of delay if they're confident that you'll give them the information they're after or if you've been recommended by others.

Failing to meet those expectations hurts your business:

- A site that's unresponsive or plagued by incidents and unpredictable availability has lower conversion rates.

- Sites that deliver a consistently poor end user experience are less likely to attract a loyal following. Poor site performance may also affect perception of your company's brand or reputation.

- You may be liable for damages if you can't handle transactions promptly, particularly if you're in a heavily regulated industry such as finance or healthcare.

- Poor performance may cost you money. If you have a formal contract with users, you may be liable for refunds or service credits. Slow or unavailable sites also encourage customers to find other channels, such as phone support or retail outlets, that cost your organization far more than handling requests via the Web. Once

---

visitors try those channels, they may stick with them, costing you even more money.

There are six fundamental reasons companies measure the performance of their sites:

- To establish baselines
- To detect and repair errors
- To measure the effectiveness of a change
- To determine the impact of an outage
- To resolve disputes with users
- To estimate how much capacity will be needed in the future

Establish agreed-upon baselines

You need to determine what "normal" performance and uptime are, partly so others can tell whether you're doing your job and partly so you can set thresholds to warn you when something is wrong.

The developer who built your website had an idea of what performance should be like. She was using an idealized environment—her desktop—connecting to a server that was otherwise idle. To understand the composition and efficiency of a web page's design, developers rely on tools like Firebug and YSlow (Figure 8-2).

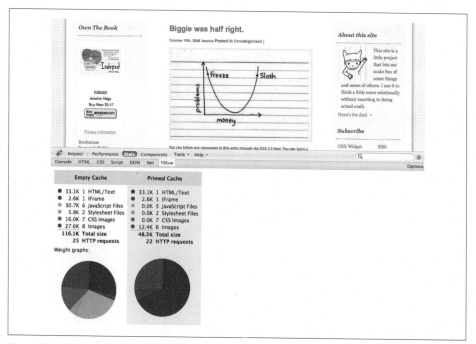

Figure 8-2. YSlow shows the elements of latency for a page

Once launched, websites seldom perform as expected. Changes in traffic, Internet conditions, and visitors' specific hardware and networks all affect how quickly a page loads. The performance you see during release testing won't match what users experience on a busy day, and to understand the impact of load you need to monitor the application at all times of the day, every day of the week. You also need to consider a variety of user environments. Knowing that your site is slow today is interesting, but knowing that it's slow every week at 3:00 A.M. is something you can act on.

Your analytics data can show you when peak traffic occurs, and can even identify times when particularly large or processor-intensive pages are being requested. Some sites, like online flower stores, university enrollment platforms, and tax filing portals see a hundredfold increase in traffic on just a few days of the year, and are comparatively idle the remainder of the year.

If you don't know what those expectations should be, a great place to start is to try your competitors' sites to see what performance they offer. Without agreement on what's "fast enough" and when maintenance is allowed, you're aiming at a target that the rest of the company can move whenever it wants to.

Detect and repair errors to reduce downtime

The most obvious day-to-day use of website monitoring technologies is to detect problems with the site before you hear about them from users. This improves total uptime and reduces user frustration.

Many companies that deploy monitoring tools are shocked by what they first see. They discover that users have been working around problems and suffering through slowdowns the site's operators knew nothing about. Because many of the commercial EUEM applications and services help you to visualize performance and availability, it's easy to communicate these problems with other departments and get the support and capital to fix them.

Measure the effectiveness of a change

Web applications are always changing (or at least, they should be if you're heeding our advice on web analytics and experimentation). Often, companies will find a version of a site that tests well with visitors but has an adverse long-term impact on EUEM or capacity. There's a trade-off to make, which is one of the reasons that user experience is now a concern for marketers as well as technologists.

Even if you're not altering content and software yourself, your server and hardware vendors may be sending you upgrades, or your service provider may be changing routing. Like it or not, you'll be overseeing hundreds of changes. Without monitoring performance before and after the changes, you'll have no idea whether things got better or worse.

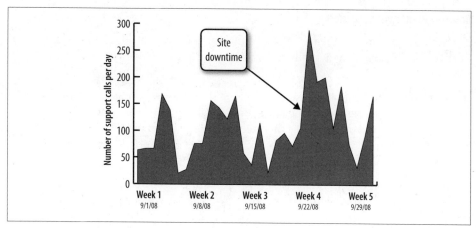

Figure 8-3. Site downtime caused support cases to dramatically increase Monday, September 22, 2008

EUEM tools answer a crucial question behind any change to content, code, equipment, or service providers: *did the change make performance better?* "Better" may mean faster, or less costly, or more reliable. Ultimately, however, you need to tie the change back to business outcomes like conversion, stickiness, and adoption. Dynamic web content updates need to be a part of your organization's change, configuration, and release management processes.

Know the impact of an outage

No matter how well you're doing your job, something will break. In the early days of web operations, sites went completely off the air. Today, we've developed technologies like load balancing and redundancy for websites to prevent most site-wide errors. Instead, errors are transient, affecting some users but not others.

When things break, you need to know the impact of the outage. How many users were affected? When the site returned, was there a flood of returning traffic that slowed things down? You need to track health so you can tie outages to changes in user behavior, such as a spike in phone banking when the web portal gets too slow (Figure 8-3).

In the IT world, the concept of Business Service Management (BSM) encourages us to treat every IT system as a business service. Hardware and network metrics need to be tied to business terms such as lost sales, abandoned shopping carts, and reduced contributions.

Resolve disputes with end users

There's no substitute for the truth. By monitoring your site through tests or user monitoring, you'll know what actually transpired. In disputes with customers, you're able to replace anecdotes with accountability.

If you have contractual obligations with your subscribers, you can prove something wasn't your fault and avoid having to issue a refund. If you're a SaaS provider, you have to deliver uptime and performance that's acceptable or your customers will cancel their contracts and you'll see increased churn, which will reduce revenues and mean you have to spend more money on sales.

Dispute resolution can go even further. By capturing a transcript of what happened, you have a record that may hold up in court. Some heavily regulated industries rely on these records. Consider, for example, an insurance portal. If a buyer purchases fire insurance, then suffers flooding damages and claims, "I actually bought flood insurance but the server got it wrong," having a copy of the user's session is extremely handy.

Estimate future capacity requirements

So you're monitoring the site, validating changes smoothly, optimizing campaigns, fixing problems as they occur, and resolving arguments with newfound finesse and aplomb. Great—your business is probably growing. And that means it's time to add more capacity.

How do you know when you'll need more servers? Because performance and availability can be related to traffic loads, as was the case for Twitter's Fail Whale (Figure 8-4), you can use performance data and web analytics to understand how soon you'll need new physical or virtual machines. EUEM tools let you profile your application to understand where bottlenecks occur and to see which tiers of your infrastructure need additional capacity.

Things That Affect End User Experience

These days, there are so many moving parts in a web application that analyzing end user experience can be tough. Despite all the mashups, plug-ins, mobile browsers, and rich content, however, if you understand the fundamental things that make up a web transaction, you're well on your way to measuring the experience of your website's end users.

At its core, EUEM can be broken down into two components: availability and performance.

Figure 8-4. Twitter.com's fail whale, a familiar sight during the service's growing pains

Availability

As visitors interact with your site, some of them will have problems. These fall into two categories: *hard* errors and *soft* errors. Hard errors are the obvious ones—the application has broken, and the visitor sees a message to that effect. This includes the ubiquitous 404 message, errors saying that the database isn't working, an abruptly terminated connection, or a sudden period of network congestion. They're easy to find if you know what you're looking for because there's a clear record of the error happening.

Soft errors are those in which the application doesn't do what you intended, for example, a site that forces users to log in twice, a navigational path that's broken, a form that won't let users submit the right data, or a page whose buttons don't work. Soft errors are more difficult to detect and more costly to diagnose because they confuse users without making it clear that the application is to blame.

In the end, errors can happen with the application, the network, or the infrastructure. With the many technical disciplines involved, resolving an outage can often mean getting everyone on a conference call to argue about why their part of the site isn't to blame. Better monitoring avoids these calls, or at least ends them quickly. This is why organizations need to start thinking of their websites as end-to-end systems, focusing on end user experience, and making web operations an integral part of a BSM strategy.

Performance problems

Your visitors will experience widely varying levels of performance. There are dozens of factors that affect the latency of a website:

- *Something specific to the user*, such as a slow connection or an old computer. This is a client-side issue best handled by customer service, site requirements, and user expectations.

- *Slowdowns on the Internet*, such as periods of congestion or connections from far away. This is a service management issue that can be addressed by choosing the right service providers, using content delivery networks (CDNs), and having a good geographic distribution strategy across your data centers.

- *Server software that's inherently slow* because it's doing something time-consuming, such as generating a report. This is a performance tuning issue that can be addressed with software engineering.

- *Infrastructure that's insufficient to handle the current traffic load* because many users are competing for resources. This is a capacity issue that can be addressed with additional servers or equipment.

- *Application issues* that are inherent to the kind of website you're running, the application server you're using, or the dependencies your application has on other systems.

The ultimate measurement of performance is how long it takes for the user to interact with the application. Slow performance isn't the only problem, however. Inconsistent performance can be particularly bad, because users learn to expect fast performance and assume your site is broken when delays occur, which amplifies abandonment.

Now that we know why performance and availability matter, let's look at how a web session works and where things go wrong. We're going to give you a working knowledge of web protocols, but we're going to keep it specific to HTTP and focus on things you need to know. If you're interested in learning more about networking, we strongly recommend Richard Stevens' book *TCP Illustrated* (Addison-Wesley Professional). For a better understanding of HTTP, check out *HTTP: The Definitive Guide* by David Gourley et al. (O'Reilly).

The Anatomy of a Web Session

Getting a single object, such as an image, from a web server to a visitor's browser is the work of many components that find, assemble, deliver, and present the content. They include:

- A *DNS server* that transforms the URL of a website (*www.example.com*) into an Internet Protocol (IP) address (10.2.3.4).

- *Internet service providers* (ISPs) that manage routes across the Internet.

- *Routers and switches* between the client and the data center that forward packets.
- A *load balancer* that insulates servers from the Web on many larger websites.
- *Servers and other web infrastructure* that respond to requests for content. These include web servers that assemble content, application servers that handle dynamic content, and database servers that store, forward, and delete content.

That's just for a single object. It gets more complicated for whole pages, and more so for visits.

Sessions, Pages, Objects, and Visits

You'll hear us talk a lot about sessions, pages, and objects as we describe web activity. Here's what we mean:

Session
> A connection between two computers established for the purpose of sending data, usually over Transmission Control Protocol (TCP).

Object
> A file retrieved from a web server. This may be a web page (*index.html*), a component of a page (*image.gif*), or a standalone object (*document.pdf* or *download.zip*). Log-based analytics tools often refer to this as a "hit."

Container object
> An object that includes references to other objects within it. A typical example is a file like *index.html* that has images, stylesheets, and JavaScript files within it.

Component object
> An object that's assigned to a container. An image embedded in a page is a component of that page. Note that some components (particularly stylesheets) can also be containers of other objects (such as background images).

Page
> A container that contains many objects (such as *index.html*, *search.php*, and so on) displayed to a user. This is what analytics tools call a "page view."

Transaction
> A series of pages across which a visitor performs a particular action, such as purchasing a book or inviting a friend.

Visit
> The extent of a visitor's interaction with a website, consisting of one or more pages (and possibly one or more transactions). Visits are normally considered terminated after 30 minutes, although this can vary tremendously depending on the type of site you operate. Analytics tools call this a "unique visit." Some tools call a visit a session, but in order to avoid confusion between visitor sessions and TCP sessions, we'll use "visit" to refer to a visitor's interaction with a sequence of pages.

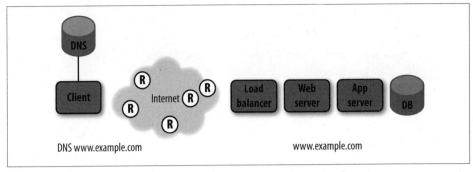

Figure 8-5. Requesting an IP address for www.example.com from a local DNS server

Finding the Destination

When you type a URL into a browser, the browser first needs to find out the IP address of the website. It does this with the Domain Name Service (DNS) protocol, which ultimately resolves the name into one or more addresses and sends the address back to your browser, as shown in Figure 8-5.

You can run your own DNS lookups by typing **nslookup** and the domain name you want to resolve at most command prompts. To launch a command prompt on a PC, click Start→Run, then type **command.com**. On a Mac running OS X, click the Spotlight icon in the top righthand corner of the screen (or press Command-Space) and type **Terminal**, then choose the Terminal application from the Applications list that appears.

In the following example, the IP address for yahoo.com is provided by a DNS server (cns01.eastlink.ca) as 68.180.206.184.

```
macbook:~ alistair$ nslookup yahoo.com
Server:     24.222.0.94 (cns01.eastlink.ca)
Address:    24.222.0.94#53

Non-authoritative answer:
Name:    yahoo.com
Address: 68.180.206.184
```

Here's what you need to know about DNS:

- The DNS lookup happens when the visitor first visits your site. The client's computer or his local ISP's DNS server may keep a copy of the address it receives to avoid having to look it up repeatedly.

- DNS can add to delay, particularly if your site is a mashup that has data from many places, since each new data source triggers another DNS query.

- If DNS doesn't work, users can't get to your site unless they type in the IP address—in fact, asking them to type in the IP address directly is one way to see whether the problem is with DNS.

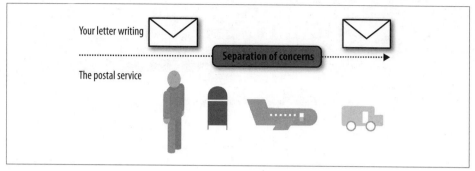

Figure 8-6. Separation of concerns between letter writing and postal services

 Typing an IP address instead of a URL may not work either. If a web server is hosting multiple websites on one machine, the server needs to know which website users are looking for. Therefore, if a user requests a site using the server's IP without including the hostname in the request, he won't get the site he's looking for.

- There are many DNS servers involved in a lookup, so you may find that your site is available from some places and not others when a certain region or provider is having DNS issues.

Establishing a Connection

Armed with the IP address of the destination site, your browser establishes a connection across the Internet using TCP, which is a protocol—a set of rules—designed to send data between two machines. TCP runs atop IP (Internet Protocol), which is the main protocol for the Internet. TCP and IP are two layers of protocols.

The concept of protocol layers is central to how the Internet functions. When the Internet was designed, its architects didn't build a big set of rules for how everything should work. Instead, they built several simpler, smaller sets of rules. One set focuses on how to get traffic from one end of a wire to another. A second set focuses on how to address and deliver chunks of data, and another looks at how to stitch together those chunks into a "pipe" from one end of a network to another.

Here's an analogy to help you understand the Web's underlying protocols.

Think for a moment about the postal system—address an envelope properly, drop it in a mailbox, and it will come out the other end. You don't need to worry about the trucks, planes, and mail carriers in between (Figure 8-6). Just follow the rules, and it'll work as expected.

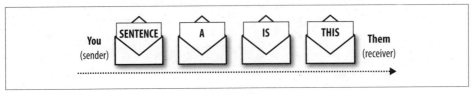

Figure 8-7. Sending a message with one word per envelope

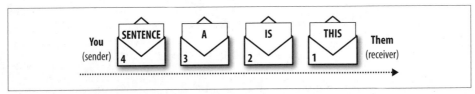

Figure 8-8. Adding a sequence number to the envelopes

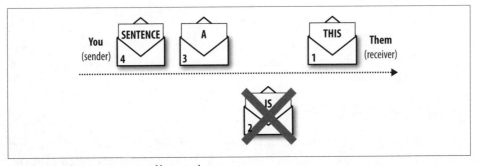

Figure 8-9. The consequences of losing a letter

Now imagine trying to write a message to someone, but having to use a new envelope for each word, as shown in Figure 8-7.

It wouldn't work. You'd have no control over the order in which the envelopes were delivered. If several envelopes arrived on the same day, or if they arrived out of order, the receiver wouldn't know in what order to open them.

To resolve this issue, you and your reader would need to agree on some rules. Perhaps you'd say, "I'll put a number on each envelope, and you can read them in the order they're numbered," (Figure 8-8).

We still have a problem: what if a letter is lost? As Figure 8-9 shows, your reader, stuck at letter 1, would be waiting for lost letter 2 to show up while the remaining letters continue to arrive.

So you'd also need a rule that said, "If you get a lot of letters after a missing one, tell me, and I'll resend the missing letter." The conversation between sender and receiver looks more complex, as Figure 8-10 illustrates, but it's also much more reliable.

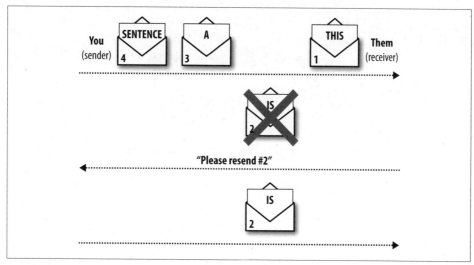

Figure 8-10. Receiver recovering from a lost letter

You might even say, "Let me know when you've received my letters," which would help you to understand how long they were taking to be delivered. If you didn't get an acknowledgment of delivery for a particular letter, you could assume it was lost, and resend it.

A set of rules like the ones we've just defined makes up a protocol. Notice that you don't need to concern yourself with how the postal service works in order to have your one-word-per-letter conversation. You don't care whether the letters are delivered by car, plane, carrier pigeon, or unladen swallow. Similarly, the postal service isn't concerned with the rules of your conversation. There is a *separation of concerns* between you and the postal service.

Put another way, your letter conversation is a "layer" of protocols, and the postal service is another "layer." You have a small amount of agreed-upon interaction with the postal layer: if you address the letter properly, attach postage, and get it in a mailbox, they'll deliver it.

In the same way, the Internet's layers have a separation of concerns. IP is analogous to the postal service, delivering envelopes of data. For your computer to set up a one-to-one conversation with a server across IP, it needs a set of rules—and that's TCP. It controls the transmission of data (that's why it's called the Transmission Control Protocol). The pattern of setting up a connection between two machines is so commonplace that we usually refer to it as the TCP/IP protocol stack. The connection between the two machines is the *TCP session*, shown in Figure 8-11.

The modern Internet is extremely reliable, but it still loses data from time to time. When it does, TCP detects the loss and resends what was lost. Just as your conversation with

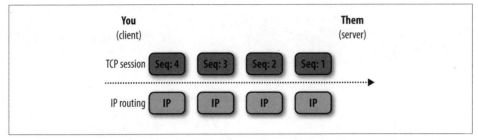

Figure 8-11. The combination of TCP and IP creates end-to-end sessions between a client and a server

a friend would slow down briefly while you resent a lost envelope, so packet loss on the Internet increases delay.

How TCP works

Because IP doesn't guarantee the order in which packets will be delivered—or even whether they'll arrive at all—TCP manages things like the sequence of delivery, retransmission, and the rate of transmission.

TCP also allows us to time the network connection, since we can measure how much time elapses between when we send a particular packet and when the receiver acknowledges receipt. This comes in handy when we're analyzing real user traffic and diagnosing problems.

You don't need to understand TCP in detail, but you should know the following:

- TCP creates an end-to-end connection between your browser and a server.
- If the network loses data, TCP will fix it, but the network will be slower.
- TCP uses a fraction of the available bandwidth to handle things like sequence numbers, sacrificing some efficiency for the sake of reliability.
- TCP makes it possible to measure network latency by analyzing the time between when you send a packet and when the recipient acknowledges it. This is what allows network monitoring equipment to report on end user experience.
- By hiding the network's complexity, TCP makes it easy for the builders of the Internet to create browsers, web servers, and other online applications we use today.

Deciding which port to use

A modern computer may have many TCP connections active at once. You may have an email client, a web browser, and a shared drive. All of them use TCP. In fact, when you surf several websites, you have TCP connections to each of them. And those sites have TCP connections to thousands of visitors' computers. To keep track of all these connections, TCP gives each of these sockets, or TCP ports, numbers.

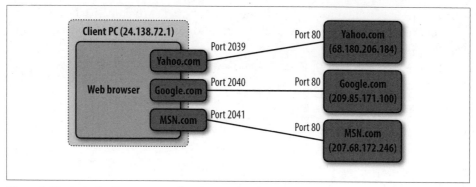

Figure 8-12. A single client connected to three web servers

For the really common applications, such as the Web, email, and file sharing, the Internet community has agreed on standard numbers. The Web is usually port 80, and encrypted web connections are usually port 443. However, you can connect to any port number on which a web server is running and ask it for content, so sometimes the web port is 8080, or 8000; it varies.

Four things uniquely identify a TCP session: the client and server IP addresses, and the TCP port numbers being used. In Figure 8-12, the TCP session to Yahoo.com is identified as 24.138.72.1:2039 to 68.180.206.184:80. No other pair of computers on the Internet has that same combination of addresses and ports at that time.

Setting up the connection

Your browser knows the IP address to which it wants to connect. And it knows it wants to connect to port 80 on the server because it wants to access that server's web content. The TCP layer establishes a session between the client running the browser and port 80 on the server whose IP it received from the initial DNS lookup. After a brief exchange of information known as a three-way handshake, there's a TCP session in place. Anything your browser puts into this session will come out at the other end, and vice versa.

You can try this out yourself. Open a command prompt and type the "telnet" command, followed by the domain name of a website and the port number, as follows:

```
macbook:~ alistair$ telnet www.bitcurrent.com 80
```

You'll see a message saying you're connected to the server, and the server will wait for your request.

```
Trying 67.205.65.12...
Connected to bitcurrent.com.
Escape character is '^]'.
```

You can request the home page of the site by typing GET, followed by a forward slash signifying the root of the site:

```
GET /
```

If everything's working properly, you'll see a flurry of HTML come back. This is the container object for the home page of the site, and it's what a browser starts with to display a web page.

```
<!DOCTYPE html PUBLIC "-//W3C//DTD XHTML 1.0
 Transitional//EN" "http://www.w3.org/TR/xhtml1/DTD/xhtml1-transitional.dtd">
<html xmlns="http://www.w3.org/1999/xhtml">
<head profile="http://gmpg.org/xfn/11">
<script type="text/javascript" src="http://www.bitcurrent.com/
wp-content/themes/grid_focus_public/js/perftracker.js"></script>
<script>
```

After a while, you'll see the end of the HTML for the page, followed by a message showing that the connection has been closed.

```
</body>
</html>
Connection closed by foreign host.
```

In theory, you could surf the Web this way, but it wouldn't be much fun. Fortunately, browsers hide all of this complexity from end users.

Securing the Connection

There's one more thing to consider before your browser starts requesting web pages. If what you're about to send is confidential, you may want to encrypt it. If you request a page from a secure website (prefixed by https://), your browser and the server will use a protocol called the Secure Sockets Layer (SSL) to encrypt the link.

Again, you don't need to understand SSL. Here's what you do need to know:

- It makes the rest of your message impossible to read, which can make it harder for you to troubleshoot problems with a sniffer or to deploy inline collection for analytics.
- The server has to do some work setting up the connection and encrypting traffic, both of which consume server resources.
- SSL doesn't just secure the link, it also proves that the server is what it claims to be, because it holds a certificate that has been verified by a trusted third party.
- The browser may not cache some encrypted content, making pages load more slowly as the same object is retrieved with every page.
- It makes the little yellow padlock come on.

Retrieving an Object

Now you're finally ready to retrieve and interpret some web content using HTTP. Just as IP handles the routing of chunks of data and TCP simulates a connection between two computers, so HTTP focuses on requesting and retrieving objects from a server. The individual layers of communication are shown in Figure 8-13.

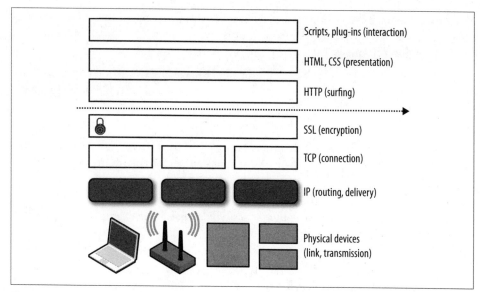

Figure 8-13. Layers of communication working together to deliver a website to a visitor

Your browser retrieves pages through a series of simple requests and responses. The browser asks for an object (*index.html*) by name. The server replies that yes, in fact, it has this object (200 OK) and sends it.

There's an important caveat here. In many cases, your browser may have already visited the website in question. Much of the content on a web page doesn't change that often. Menu bars, images, videos, and layout information are only modified occasionally. If your browser had to download every object every time it visited a site, not only would your bandwidth bill be higher, but the pages would also load more slowly. So before asking the server for an object, your browser checks to see if it has a copy in memory on your PC. If it does—and if the copy isn't stale—it won't bother asking for it. We'll look at caching in more detail shortly.

There are seven standard HTTP commands, called *methods*, that a browser can send to a server. Of these, only three are commonly used on most sites:

- GET asks for an object by name.
- POST sends data, such as the contents of a form, to the server from the client.
- HEAD asks the server to send only the descriptive information about the object, but not the object itself.

Other commands, such as PUT, TRACE, OPTIONS, and DELETE, were seldom used until recently, but PUT and DELETE have newfound popularity in Ajax-based websites.

Your browser also tells the server things about itself, such as what kind of browser it is, whether it can accept compressed documents, which kinds of objects it can display, and which language the visitor speaks. The server can use some of this information to respond in the best way possible, for example, with a page specifically designed for Internet Explorer or one written in the visitor's native language.

The request to the server can be as simple as GET page.html, but it's common for a browser to send additional information to the server along with the request. This information can come in several formats. Common ones are:

Cookies
>This is a string of characters that the server gave to the browser on a previous visit. This allows the server to recognize you, so it can personalize content or welcome you back.

Information in the URL structure itself
>A request for *www.example.com/user=bob/index.html*, for example, will pass the username "bob" to the server.

A URI query parameter
>URI query parameters follow a "?" in the URL. For example, an HTTP request for *http://www.youtube.com/watch?v=Yu_moia-oVI&fmt=22* contains two parameters. The first is the video number (v=Yu_moia-oVI) and the second is the format (fmt=22, for high definition).

Remember that we said your browser only uses a cached copy of content if it's not stale? Your browser will check to see if it has a fresh copy of content, and then retrieve objects only if they're more recent than the ones stored locally. This is called a *conditional GET request*: the server only sends an object if the one it has is more recent than the one the browser has, which saves on bandwidth without forcing you to see old content.

This is one of the reasons why, when testing site performance, you need to specify whether the test should simulate a first-time or a returning user. If a browser already has much of the content, performance will be much better because the browser already has a copy of many parts of the page.

The initial response: HTTP status codes

If everything is running smoothly, the server will respond with an HTTP status code.

```
HTTP/1.0 200 OK
```

This is a quick message to acknowledge the request and tell the client what to expect. Status codes fall into four groups:

- The *200* group indicates everything is fine and the response will follow.
- The *300* group means "go look elsewhere." This is known as a *redirect*, and it can happen temporarily or permanently. It's used to distribute load across servers and data centers, or to send visitors to a location where the data they need is available. Instead of responding with a "200 OK" message, the server sends back a redirection to another web server.

the page you requested does not exist

help | blog | stats | feedback | bookmarklets | socialite | buttons | widget | code | mobile | store | advertise

WIRED.com - WIRED How-To

Use of this site constitutes acceptance of our User Agreement and Privacy Policy. (c) 2009 CondeNet, Inc. All rights reserved.

Figure 8-14. A 404 apology page offering suggested links to the visitor

- The *400* group indicates client errors. The client may have asked for something that doesn't exist, or the client may not have the correct permissions to see the object it has requested. While you might think that 400 errors are a visitor's problem, they can be the result of broken links you need to fix. Figure 8-14 shows an example of a 404 apology page.

- The *500* group indicates server errors. These are serious issues that can occur because of application logic, broken backend systems, and so on. Sometimes these errors can produce custom pages to inform visitors of the problem or to reassure them that something is being done to fix the issue, as shown in Figure 8-15.

418 I'm a Teapot

On April 1, 1998, Larry Masinter, currently principal scientist at Adobe, wrote an RFC for an HTTP extension. The Rationale and Scope reads as follows:

"There is coffee all over the world. Increasingly, in a world in which computing is ubiquitous, the computists want to make coffee. Coffee brewing is an art, but the distributed intelligence of the web-connected world transcends art. Thus, there is a strong, dark, rich requirement for a protocol designed espressoly [sic] for the brewing of coffee. Coffee is brewed using coffee pots. Networked coffee pots require a control protocol if they are to be controlled."

The RFC goes on to describe the 418 status code in detail at *http://tools.ietf.org/html/rfc2324*.

Figure 8-15. An application error apology page on a Wordpress server

Object metadata: Describing what's being sent

Just before it sends the object, the server gives us some details about the object the browser is about to receive. This can include:

- The size of the object
- When it was last modified (which is essential to know in order to cache objects efficiently)
- The type of content—a picture, a stylesheet, or a movie, and so on—so your browser knows how to display it
- The server that's sending the answer
- Whether it's OK to cache the object, and for how long to use it without checking for a fresh object from the server
- A cookie the client can use in subsequent requests to remind the server who it is
- Any compression that's used to reduce the size of the object
- Other data, such as the privacy policy (P3P) of the site

Here's an example of metadata following an HTTP 200 OK response:

```
HTTP/1.x 200 OK
Date: Mon, 13 Oct 2008 04:31:29 GMT
Server: Apache/2.2.4 (Unix) mod_ssl/2.2.4 OpenSSL/0.9.7e
Vary: Host
Last-Modified: Wed, 13 Jun 2007 19:15:36 GMT
Etag: "e36-70534600"
Accept-Ranges: bytes
Content-Length: 3638
Keep-Alive: timeout=5, max=100
Connection: Keep-Alive
```

```
Content-Type: image/x-icon
P3P: policyref="http://p3p.yahoo.com/w3c/p3p.xml", CP="CAO DSP COR
  CUR ADM DEV TAI PSA PSD IVAi IVDi CONi TELo OTPi OUR DELi SAMi
  OTRi UNRi PUBi IND PHY ONL UNI PUR FIN COM NAV INT DEM CNT STA POL HEA PRE GOV"
```

Each line is a *response header*. Following the headers, the server provides the object the client requested. One of the reasons for HTTP's widespread success is the flexibility this affords: the server may provide many headers that your browser simply ignores. This has allowed browser developers and server vendors to innovate independently of one another, rather than having to release browsers and servers at the same time.

Preparing the response

Now the server has to send the object your browser requested. This may be something stored in the server's memory, in which case it's a relatively simple matter of retrieving it and stuffing it into the existing TCP session. Your browser will do the rest.

On most modern sites, the web page contains some amount of dynamic content. This means the server has to think about the answer a bit. This server-side processing—what we'll refer to here as *host time*—is a major source of poor performance. Assembling a single object may require the web server to talk to other servers and databases, and then combine everything into a single object for the user.

Once the object is prepared, the server sends it to your browser.

Sending the response

The other major source of poor performance comes from network latency (what we'll refer to as *network time*). From the moment the server puts the object into the TCP connection until the moment the last byte of that object comes out at the other end, the clock is ticking. Large objects take longer to send; if anything's lost along the way, or if the sender and receiver are far apart, they'll take longer still.

Your browser receives the object. If the request was for a standalone object, that would be the end of it: your browser would shut down the TCP connection through a sequence of special TCP messages, and you'd be able to access the received object (i.e., the web page).

Getting a Page

Pages contain many things, such as pictures, video, and Flash, as well as formatting information (stylesheets) and programming logic (JavaScript). So your browser now has to finish the job by retrieving the components of the container object it received.

Modern browsers are eager to display pages quickly. Your browser doesn't even wait for the container object to load completely—as it's flowing in, the browser is greedily analyzing it to find references to other objects it needs to retrieve. Once it sees additional objects listed in that page, it can go ahead and request them from the server, too. It can

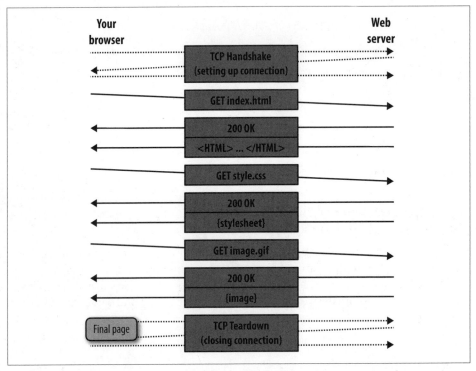

Figure 8-16. A bounce diagram showing the retrieval of a page and its component objects

also start executing any JavaScript that's on the page and launching plug-ins, such as Java, ActiveX, or Flash, that it will need to display content.

With the client and server communicating so much, it would be a waste to tear down that TCP session just to reestablish it. Version 0.9 of HTTP initially behaved in this way. If your browser needed more objects (and it usually does), it had to reconnect to the server. Not only was this slow, it also kept the servers busy setting up and tearing down connections. The Internet's standards body moved quickly to correct this.

In version 1.1 of the HTTP protocol, the TCP connection is kept alive in anticipation of additional requests. Figure 8-16 shows this process.

Doing lots of things at once: Parallelism

The examples we've seen so far are serial—that is, the browser and the web server are requesting and sending one object at a time. In reality, web transactions run in parallel. Here's why: if you're on a one-lane road, it's hard to overtake someone. Add a second lane, and suddenly traffic flows freely. The same thing is true with network connections. A single TCP session means that a really big object can stop smaller objects from getting through quickly. This is known as *head-of-line blocking*, and is similar to a single slow truck delaying dozens of faster cars.

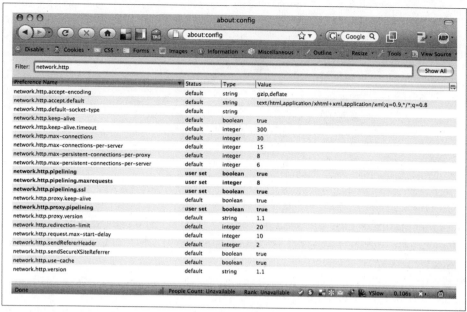

Figure 8-17. The About:config panel in Firefox, showing connection properties that users can adjust to change connection parallelism

Web browsers resolve this issue by establishing multiple TCP sessions to each server. By having two or more sessions open, the client can retrieve several objects concurrently. Some browsers, such as Firefox, let users configure many of these properties themselves, as shown in Figure 8-17. For more information about parallelization and pipelining, see Mozilla's FAQ entry at *www.mozilla.org/projects/netlib/http/pipelining-faq.html*.

TCP parallelism makes browsers work better, but it also makes your job of measuring things more difficult. Because many HTTP transactions are occurring simultaneously, it's harder to estimate performance. Many web operations tools use *cascade* diagrams like the one in Figure 8-18 to illustrate the performance of a test or a page request. Notice that several objects (*script.js* and *style.css*, for example) are being retrieved simultaneously.

Interpreting the page

Early browsers simply displayed what they received. With the introduction of Java-Script in 1993, developers started to embed simple programs within the container page. Initially, they focused on making pages more dynamic with functions such as rollover images, but these quickly grew into highly interactive web pages.

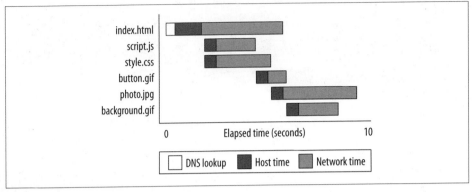

Figure 8-18. A simple cascade diagram for a page

Since then, JavaScript has become a powerful programming language in its own right. Interactive sites like Google Maps rely heavily on browser scripting to make their applications run quickly on the client.

At the time a page is loaded, the browser interprets the portions that are written in JavaScript. Because this may involve rewriting portions of the page that were just loaded, JavasScript needs to run first. That's right: a page can contain a program that actually modifies the page itself.

Why Should JavaScript Rewrite a Page?

There are times when it makes sense to have JavaScript actually rewrite much of the page that loaded it. If a page contains large quantities of repetitive data that can be generated programmatically, it may be faster and less taxing on network connections to have the browser manufacture the page.

Consider, for example, a page with a 100 × 100 cell table in it. Rather than sending the HTML for the table (which could contain 10,000 lines of `<td></td>` markup in it and would take a long time to deliver), JavaScript can create the HTML for that table once the page has arrived at the client.

Compared to the 10,000 cells of the original table, the code is relatively small. And by getting the browser to do some of the heavy lifting, the server frees itself up to do other things. If you use this approach, you must test it carefully and use it sparingly; in fact, with compression, the advantages of such an approach can be negligible at best.

Assembling the objects

Once your browser has the page and enough information about the objects it contains, it can display them to the user. Note that the browser doesn't need all the objects on the page to show you the page—if your browser knows how big an image will be, it can begin displaying the page, leaving a space where the image will go. Including image sizes within the container page means the browser can show you the page sooner.

A timeline of page milestones

There are several important milestones in the loading of a page, and these are the things you'll be monitoring with synthetic and RUM tools. They don't always happen in this order, and their order depends on how a particular site was built.

Initial action
> This is the action that started it all—either typing in a URL or clicking on a link. It's the basis for all page measurement.

Initial DNS response time
> If the browser is visiting a site for the first time, there must be a DNS lookup.

SSL negotiation
> If the page is encrypted, the browser and server need to set up the encrypted SSL channel before transmitting data.

Container first byte
> This is when the start of the container object reaches the browser. The time from the initial action to the first byte of the container can serve as a rough estimate of how much time the server took to process the request.

Component objects requested
> At this point, the browser knows enough to start asking for embedded elements, stylesheets, and JavaScript from the server.

Component first byte
> This is when the start of the component reaches the browser.

Third-party content DNS lookups
> If any of those component objects come from other sites or subdomains, the browser may have to look them up, too.

Start processing instructions
> This is when the browser starts interpreting JavaScript, possibly triggering other retrievals or page modification. This may block other page activity until the scripts are finished.

Page starts rendering
> This is the point at which the browser has rendered the page sufficiently for a user to start interacting with it. This is the critical milestone that governs end user experience.

All objects loaded
> Some objects, such as those below the viewable screen, may continue to load even after the user starts interacting with the page.

Entire page loaded, including backup resources
> To improve usability, scripts on the page may intentionally fetch some data after the page is completed.

In-page retrieval (type-ahead search, etc.)
Many applications may retrieve additional components in response to user interactions. Microsoft's Live Maps, for example, retrieves new map tiles as users scroll.

Getting a Series of Pages

Few visits last for a single page. Each user action may trigger another page load, with a few differences from what happened when the initial page was loaded:

DNS lookups are unnecessary
You shouldn't need to look up the IP address again through DNS.

Some content will be in the local cache
There will hopefully be plenty of up-to-date content in the local cache, so there will be less to retrieve.

New cookies to send
The browser may have new cookies to send this time around, particularly if this was the first time the user visited the site.

Wrinkles: Why It's Not Always That Easy

What we've just outlined is a relatively straightforward process. If everything goes as planned, the page will load properly. Unfortunately, things often go awry. Much of the work of EUEM is in smoothing out the many wrinkles that can happen along the way.

DNS Latency

DNS can take time to resolve, or fail to resolve entirely. If users can't get your IP address, they can't visit you. One of the primary tasks of a web operations team is to manage the propagation of IP addresses through DNS—not just making sure they're fast, but also that they're correct and that they can be changed relatively smoothly.

On some systems, you can use the `dig` command line utility to measure DNS latency (the `dig` utility is included in the Windows distribution of BIND, available at *www.isc.org/software/bind*).

```
macbook:~ alistair$ dig www.brainpark.com

; <<>> DiG 9.4.2-P2 <<>> www.brainpark.com
;; global options:  printcmd
;; Got answer:
;; ->>HEADER<<- opcode: QUERY, status: NOERROR, id: 23382
;; flags: qr rd ra; QUERY: 1, ANSWER: 2, AUTHORITY: 0, ADDITIONAL: 0

;; QUESTION SECTION:
;www.brainpark.com.        IN    A

;; ANSWER SECTION:
```

```
www.brainpark.com. 1800 IN   CNAME  brainpark.com.
brainpark.com.     1800 IN   A      174.132.128.251

;; Query time: 98 msec
;; SERVER: 24.222.0.94#53(24.222.0.94)
;; WHEN: Sat Feb  7 16:25:00 2009
;; MSG SIZE  rcvd: 65
```

In this example, a DNS lookup for Brainpark.com (*http://brainpark.com*) returned a single IP address, and took 98 milliseconds.

DNS Latency for the Practical Man

If you do anything over the Internet, whether it's email or web browsing or pirating movies, DNS lookups are part of it. If you can't look up domain names, you can't surf the Web, because your browser doesn't know where to go. If your email server can't lookup domain names, it just holds on to your mail until it can. While it's supercritical to be able to do DNS lookups at all, there's a lot of redundancy built in, so it works almost all the time. So the next question is, "How fast does the DNS need to be?"

(The answer is going to be 20 milliseconds—that's 1/50 of a second, or less. But let's take the scenic route to explain how we get there.)

To keep a user happy, the answer depends on the activity.

Sometimes, it just doesn't matter. If you are going to spend hours downloading a multigigabit HD movie, any reasonable DNS lookup or lookups are just lost in that large transfer time. In general, whenever DNS is a small part of the transaction, it isn't an issue.

For web browsing, a trivial web page uses one DNS lookup, so if my DNS service has a 1/10 of a second latency (the time between sending a DNS query and getting the corresponding response) there's a 1/10 of a second, or 100 millisecond, DNS time tax. But that's a 1900s-style web page. Today, a really simple web page probably uses about 10 DNS lookups—a few for the content, and many more for the advertisements and other inserts. If each lookup takes 1/10 of a second, 10 might seem to add 1 second to the time it takes to view a page.

However, today's web browsers are clever and do the DNS lookups in parallel, usually up to some limit. So 10 lookups might be done as 2 convoys of 5 each, adding 2/10 of a second. That's barely noticeable compared to the time it takes to send the data, and probably not annoying. For a complicated Facebook page, there might be 200 DNS lookups, adding 4 seconds, and that's into the annoying range. Of course, once the DNS lookups get cached (i.e., remembered in a local copy) you stop paying for the lookups. So the good news is once your browser locates the ad servers, you only wait for the ads, rather than the DNS.

If I'm sending email, I may not care if the email server has slow DNS, since it rarely matters if there's a delay of minutes—of course, the systems administrator may care a lot if his email server gets clogged or he has to buy more servers. The typical email takes around a dozen DNS lookups for every hop it takes between sender and receiver, and three to fours hops is not unusual, so it adds up. Spammers generate a lot of DNS

---

lookups, even for mail that only goes a hop or two before being discarded. Not to be outdone, antiqspammers can use 5–10 DNS lookups to vet email by checking to see if the sender is on any of multiple "blacklists" of spammers.

So 20 ms is really a rule of thumb that seems to keep users happy and not cost service providers too much in terms of DNS server hardware and software.

But when ISPs want to sell their service as blindingly fast, even faster is better. An idle, reasonably hot PC these days answers queries in a very small number of milliseconds if it has good software (some is slow), so the time it takes to do a query is often dominated by waiting in line behind other users' queries at the server, or just how long it takes to move through the routers and cables between the user's machine and the DNS server. But the good news is that as the network's speed increases and processors get more powerful, it's possible to generate faster and cheaper DNS queries to keep up with the applications' growing appetites.

—Paul Mockapetris,
author of DNS and the chairman of Nominum

There are a number of DNS testing tools, and managed DNS services like UltraDNS, that can help you monitor DNS performance. Most synthetic testing tools will test the DNS component of page retrieval as part of their normal testing.

Multiple Possible Sources

Most big websites are hosted in several places. This is done for two main reasons. First, it spreads out the load and reduces the impact of a major outage. Second, it puts computers closer to users, which reduces the network delay when accessing websites.

Sometimes, the DNS response of your site will include several IP addresses. If the first address doesn't work, your browser can, in theory, try the others. Google, for instance, has four IP addresses that a browser can use:

```
macbook:~ alistair$ nslookup www.google.com
Server:   24.222.0.94
Address: 24.222.0.94#53

Non-authoritative answer:
www.google.com  canonical name = www.l.google.com.
Name:    www.l.google.com
Address: 64.233.169.147
Name:    www.l.google.com
Address: 64.233.169.103
Name:    www.l.google.com
Address: 64.233.169.104
Name:    www.l.google.com
Address: 64.233.169.99
```

In practice, however, users won't wait long enough for the first site to time out—they'll have already gone elsewhere or tried to reload the page. DNS is the first line of defense against regional failure and global delay. This takes two forms: Global Server Load Balancing (GSLB) and CDNs.

Global Server Load Balancing

Instead of sending everyone on the planet to a single destination, large sites can send visitors to the servers closest to them. In GSLB, the DNS server decides which of the data centers is "best" for the visitor. This depends on several factors:

Whether the data center has the information the visitor needs
> Not every piece of content is available in every data center. There may also be legal constraints or pricing policies that require you to send certain users to certain locations.

The delay between the visitor and the data center
> This may be based on geography or on network round-trip time.

The responsiveness of the servers in each data center
> There's no point in sending a user to a data center that's nearby if the servers there are unacceptably slow.

GSLB doesn't always rely on DNS, however. Sometimes the web server needs to first determine what content a visitor wants, then redirect the visitor's browser to the optimal data center. The server can't rely on DNS-based GSLB, because it has to first receive the request from the browser, interpret it, then respond with an HTTP message sending the visitor elsewhere. This is a popular method because it overcomes the staleness and address caching issues that plague DNS-only GSLB. However, it introduces a redirect delay into the start of each session since the client must now request a page from another server (often with an additional DNS query).

Content delivery networks

You might want multiple locations for your data, both for resiliency and for better network performance, but it's expensive to deploy infrastructure in several locations around the world.

CDN companies are willing to help—for a price. Think of them as alternate networks that are tuned for fast delivery and circumvent many of the bottlenecks normal traffic faces. CDNs don't actually have parallel networks to the Internet. CDNs have two main strategies to speed up the Internet for their paying customers:

- *Get data close to users* to reduce the amount of information that must cross the Internet.
- *Speed Internet traffic up* as much as possible for the data that must cross the network.

Some CDN providers focus on making sure that all the content users need is at the edge; others focus on making sure that data travels quickly over the wide area network (WAN); and some blend the two functions.

To get data close to users, a CDN will:

- Handle the DNS lookup process on behalf of its subscribers, directing visitors to the closest place from which to serve data.
- Cache data at the edge, moving content that doesn't change often—such as images or MP3s—closer to users, where it can be served more quickly.
- Execute some processing on machines that the CDN owns, close to visitors. This is known as an *edge-side include*, and it's useful when a website needs to assemble data from several sources within a page.

To speed up Internet traffic, a CDN will:

- Choose better paths across the network, changing routes more rapidly based on better measurements than the Internet as a whole.
- Optimize for specific applications (such as video or web traffic) by tweaking certain parameters of protocols rather than offering one-size-fits-all handling (as the Internet does).
- Use the latest HTTP protocols to send multiple requests at once and compress the results.
- Send redundant packets along different paths to overcome sporadic packet loss.
- Bypass congestion points by connecting to multiple carriers and service providers rather than relying on the carriers to send users' traffic from one to another.
- Split single data streams at the edge to reduce the amount of traffic at the core, usually for streaming content.

A CDN takes over your DNS, and (hopefully) speeds it up along with the rest of your site. Many of the techniques just outlined mitigate or overcome the performance and availability bottlenecks that we're about to see—so if performance is important to your business, you need a CDN. While CDNs were once a daunting prospect requiring long-term contracts, recent developments by cloud computing firms like Amazon are turning them into a pay-as-you-go acceleration utility for web applications.

Slow Networks

Network latency is the result of two things: the time it takes to complete a trip across the network (round-trip time), and the number of times, or turns, the application needs to traverse it. Simply put, turns multiplied by round trips, equals network latency. All networked applications can be analyzed in this way. To improve your application's performance you need to either reduce the round-trip time or reduce the number of turns.

How Fast Can Networks Get?

There's an upper limit to web performance: the speed of light. That's as fast as data can travel, and it means a packet sent from New York will arrive in Las Vegas no sooner than 13 milliseconds later.

Or is it?

Quantum effects suggest that we may be able to send data instantly. We want to avoid being laughed at in the future, so we'll just point to *www.physorg.com/news137937526.html* and say we told you so.

Network connections seldom travel at the speed of light (see the sidebar "How Fast Can Networks Get?"), and there's lots of room for improvement in the effect networks have on end user experience.

An extremely simple object request, such as a GET for a tiny image, would take exactly one round trip to retrieve, assuming the server responded instantaneously. On a LAN, this is a matter of milliseconds. On an Internet connection, it's far greater. Let's compare a device on a LAN to one in China by running a ping test from both locations.

```
$ ping 192.168.1.1
PING 192.168.1.1 (192.168.1.1): 56 data bytes
64 bytes from 192.168.1.1: icmp_seq=0 ttl=64 time=2.250 ms
64 bytes from 192.168.1.1: icmp_seq=1 ttl=64 time=1.995 ms
64 bytes from 192.168.1.1: icmp_seq=2 ttl=64 time=1.974 ms
--- 192.168.1.1 ping statistics ---
3 packets transmitted, 3 packets received, 0% packet loss
round-trip min/avg/max/stddev = 1.974/2.073/2.250/0.125 ms

$ ping yahoo.cn
PING yahoo.cn (202.165.102.247): 56 data bytes
64 bytes from 202.165.102.247: icmp_seq=0 ttl=44 time=367.263 ms
64 bytes from 202.165.102.247: icmp_seq=1 ttl=44 time=428.287 ms
64 bytes from 202.165.102.247: icmp_seq=2 ttl=44 time=399.339 ms
64 bytes from 202.165.102.247: icmp_seq=3 ttl=44 time=382.795 ms
--- yahoo.cn ping statistics ---
4 packets transmitted, 4 packets received, 0% packet loss
round-trip min/avg/max/stddev = 367.263/394.421/428.287/22.604 ms
```

The round-trip time to China is 190 times greater than the round-trip time to a local device.

In addition to geographic distance, many other factors can affect round-trip time, such as low bandwidth connections, intermediate devices, and congestion.

Low bandwidth

A 14.4 Kbps modem would send a 5 MB file between two computers in about six minutes; the same object would take only five *milliseconds* over a dedicated Gigabit Ethernet network.

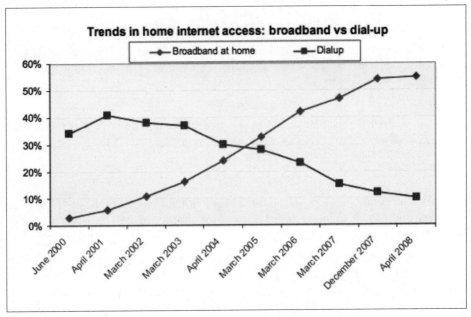

Figure 8-19. Consumer Internet connections are getting faster

Increasingly, we're surfing on faster connections using faster machines, as shown in Figure 8-19 (the Pew Internet and American Live Project, Home Broadband Adoption 2008 survey, *http://www.pewinternet.org/Reports/2008/Home-Broadband-2008.aspx*).

Despite our faster connections, performance issues plague the Internet at both the edge and the core. You can estimate how much of your population has which kind of bandwidth—and tweak your site design accordingly—using analytics or RUM tools, because the speed at which the web page loads will usually be subject to the bottleneck of the visitor's connection. This is known as the "last mile" of the connection, and it's generally the slowest.

Hops—devices between your browser and your destination

A broadband connection is no guarantee of a fast Internet. As a packet travels to the server, it traverses several routers, each of which knows where to send it next. Pairs of routers are connected by wires, sometimes in the same room, sometimes on opposite shores of an ocean. When a site slows down, it's often because one of these connections is slow. A command-line utility called *traceroute* can show you what's going on.

Traceroute Is a Hack

To prevent packets from circulating forever and filling up the tubes of the Internet, every packet on the Internet keeps track of how many routers have forwarded it. Each time a router forwards a packet, it decreases the hop count of that packet by one. When the hop count reaches zero, the router deems the destination unreachable and sends a message (called using an ICMP TIME_EXCEED) back to the sender.

Traceroute exploits this feature by fiddling with the hop count of packets it sends. The first packet has a hop count of one, which the first router decreases. Realizing the hop count is now zero, the router sends back a message to the sender pronouncing the destination unreachable.

Traceroute now knows the IP address of the first router—after all, it just received a message from it. In addition to looking up the IP address of the router to find out its name (if it has one), traceroute can now increment the hop count by one and discover the second router. It continues this process until it gets a response from the destination, as shown in Figure 8-20.

Traceroute can use several different methods, depending on which protocol you want to test. Probes can be sent with UDP, TCP, or ping (ICMP) protocols. Depending on how intermediate devices treat different protocols, you may get different results. For example, a firewall may block UDP packets but allow pings to pass, meaning a UDP-based traceroute won't discover devices behind the firewall.

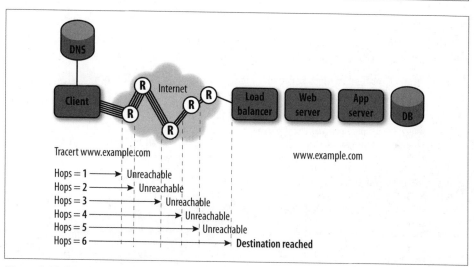

Figure 8-20. How traceroute discovers the routers between itself and a destination on the Internet

Let's look at an example of traceroute in action. First, we'll run a test from a home network located in Montreal to *www.brainpark.com*.

```
macbook:~ sean$ traceroute www.brainpark.com
traceroute to brainpark.com (208.71.139.130), 64 hops max, 40 byte packets
 1  192.168.0.1 (192.168.0.1)  1.102 ms  0.693 ms  0.566 ms
 2  10.183.128.1 (10.183.128.1)  5.655 ms  6.787 ms  8.134 ms
 3  24.200.227.77 (24.200.227.77)  5.510 ms  5.719 ms  9.674 ms
 4  24.200.250.82 (24.200.250.82)  8.187 ms  6.630 ms  5.733 ms
 5  ia-cnnu-bb04-ge13-1-0.vtl.net (216.113.122.14)  7.195 ms  7.696 ms  9.712 ms
```

```
 6  POS1-0.PEERB-MTRLPQ.IP.GROUPTELECOM.NET (66.59.191.157)  5.706
    ms  5.773 ms  6.117 ms
 7  GE4-0.WANA-MTRLPQ.IP.GROUPTELECOM.NET (66.59.191.137)  7.513 ms
    5.870 ms  5.863 ms
 8  POS4-0.WANA-TOROONXN.IP.GROUPTELECOM.NET (66.59.191.226)  14.216
    ms  16.317 ms  14.184 ms
 9  POS5-0.PEERA-CHCGIL.IP.GROUPTELECOM.NET (66.59.191.106)  26.245 ms
    28.390 ms  26.594 ms
10  GT-360NETWORKS.PEERA-CHCGIL.IP.GROUPTELECOM.NET (66.59.191.90)
    53.759 ms  52.112 ms  52.016 ms
11  lau1-core-01.360.net (66.62.7.1)  93.512 ms  93.913 ms  93.927 ms
12  pdx1-core-01.360.net (66.62.3.13)  214.342 ms  101.376 ms  199.833 ms
13  slc1-core-01.360.net (66.62.3.9)  88.250 ms  88.970 ms  87.604 ms
14  slc1-edge-01.360.net (66.62.5.3)  87.971 ms  87.835 ms  88.154 ms
15  66.62.56.26 (66.62.56.26)  101.184 ms  100.683 ms  99.282 ms
16  qwk.net (208.71.136.2)  95.589 ms  96.070 ms  95.583 ms
17  mail4.qwknetllc.com (208.71.139.130)  95.969 ms  94.018 ms  96.740 ms
```

In the resulting output, the first column shows the hop count—the number of routers between the client and the server. The second column shows the router at that hop, identified by an IP address and possibly a DNS address. This can sometimes provide useful clues (for example, POS4-0.WANA-TOROONXN.IP.GROUPTELECOM.NET suggests that this is a router in Toronto, Ontario, belonging to Group Telecom).

The third column shows the timing of three probes sent to that router. The last row in the output shows the final gateway to the destination address, which is 17 hops away.

If we try the same test from a server in Hong Kong, however, latency and hop count change significantly:

```
traceroute to brainpark.com (208.71.139.130), 30 hops max, 40 byte packets
 1  tmhs3506-v1.pacific.net.hk (202.14.67.252)  0.708 ms  0.611 ms
 2  v206.tmhc2.pacific.net.hk (202.64.154.65)  0.336 ms  0.310 ms
 3  v102.wtcc2.pacific.net.hk (202.64.5.6)  100.642 ms  0.972 ms
 4  202.64.3.142 (202.64.3.142)  0.579 ms  0.531 ms
 5  PNT-0039.GW1.HKG4.asianetcom.net (203.192.178.137)  0.710 ms  0.694 ms
 6  gi13-1.gw2.hkg3.asianetcom.net (122.152.184.42)  0.559 ms  0.608 ms
 7  po2-1-1.cr3.hkg3.asianetcom.net (202.147.16.73)  1.898 ms  1.917 ms
 8  po14-2-0.cr1.nrt1.asianetcom.net (202.147.0.169)  54.458 ms  54.375 ms
 9  po2-1-0.gw3.lax1.asianetcom.net (202.147.0.162)  180.820 ms  180.301 ms
10  gi1-31.mpd01.lax05.atlas.cogentco.com (154.54.11.133)  179.917 ms  257.576 ms
11  te4-2.ccr02.lax01.atlas.cogentco.com (154.54.6.189)  188.953 ms  189.052 ms
12  te4-1.ccr02.sjc01.atlas.cogentco.com (154.54.5.69)  191.762 ms  193.605 ms
13  te3-3.ccr02.sfo01.atlas.cogentco.com (154.54.2.125)  185.090 ms  185.235 ms
14  te4-3.mpd02.den01.atlas.cogentco.com (154.54.5.197)  217.569 ms  223.162 ms
15  te4-1.mpd01.den01.atlas.cogentco.com (154.54.0.109)  210.145 ms  209.567 ms
16  360-networks.demarc.cogentco.com (38.104.26.74)  213.263 ms  211.191 ms
17  66.62.4.65 (66.62.4.65)  223.749 ms  223.591 ms
18  66.62.3.21 (66.62.3.21)  223.494 ms  227.203 ms
19  66.62.5.3 (66.62.5.3)  226.369 ms  222.972 ms
20  66.62.56.26 (66.62.56.26)  239.029 ms  237.787 ms
21  208.71.136.2 (208.71.136.2)  233.600 ms  235.684 ms
22  mail4.qwknetllc.com (208.71.139.130)  223.938 ms  223.219 ms
```

As the output shows, not only does the latency of a single packet climb from 94 milliseconds to 220 milliseconds, but there are also five additional hops needed to reach the destination. Each of those links and each of those routers introduces additional delay into the network round-trip time.

Congestion and packet loss

Much of the Internet can't move as many packets as its users would like. Peer-to-peer traffic and spam further clog links. The routers have to do something with all of this excess traffic, and more often than not, they simply discard it, relying on TCP to recover from the loss.

Loss of data generally means that the packet must be resent, a process known as *re-transmission*. We say "generally" because many UDP-based applications such as voice and video can handle a certain amount of loss. In certain situations (like real-time updates), delivery is more important than getting 100 percent of the data. For a voice call, it is better to have a brief moment of static than to introduce more and more delay as the call progresses.

As we've seen, the TCP stack handles retransmission of data. There's a bit more to it, though. Consider a highway on which it takes one minute to drive from point A to point B. If there's only one car driving back and forth on the highway, you can send at most one carload every two minutes across the highway. Double the number of cars, and you double the number of carloads per minute. But there's a limit to this: at some point, the highway becomes congested, and the introduction of additional cars slows down all cars on the highway. Maybe there's an accident or a stalled car; whatever the case, you need to reduce the number of cars in transit. In other words, there's an optimal number of cars in transit that maximizes the carloads per minute.

TCP tries to send as much data as possible without being greedy. It, too, is seeking the optimal number of things in transit. To do this, it first sends one packet and waits for a response. Then it sends a couple, and waits for the response. It continues to do this —putting more and more packets "in flight" between the sender and receiver—until something goes wrong. In practice, different computers and devices have different initial numbers of packets in flight and different ramp-up rates to squeeze more performance out of their Internet connections, but the basic functionality is the same.

Packet loss has three effects:

- TCP resends the packet, which consumes bandwidth.
- The sending TCP stack reduces the number of packets in flight, reducing the overall bandwidth of the connection.
- The receiving TCP stack, which is trying to deliver packets in the right order, has to hold on to all other packets until the right one arrives, which means it can't deliver the rest of the content until the arrival of the required packet.

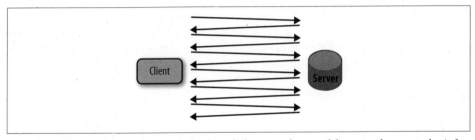

Figure 8-21. A high number of turns with relatively low LAN latency delivers performance that's fast enough

In other words, packet loss is bad. The bigger your pages, the more congested your networks, the greater the distance between sender and receiver, and the greater the number of objects, the higher the chances that loss will happen.

While you need to do everything you can to minimize the impact of round-trip time—most notably, getting your content close to your end users and making sure it's as small as possible—most of the delay in web applications comes from turns. It's the back and forth repetition that comes from retrieving many objects that really slows down websites.

Turns: Back and forth is the real problem

In early client/server computing, the client and the server lived on the same LAN, as shown in Figure 8-21. Performance was decent; the client queried the database, and the database responded.

Things were also predictable. If you knew the round-trip time, the number of turns, and the lossiness of your network, you had a fairly good idea of how significant the network delay would be.

When we moved these applications to WANs, however, they died. Most of them had been tested on fast connections to nearby servers, and the speed of those connections hid the constant chatter. On a slower WAN link, those hundreds of back and forth conversations slowed things down interminably, making otherwise decent applications ponderous and unwieldy, as Figure 8-22 shows.

The early web-centric design model replaced the client-server model. The browser was a dumb edge to a smart core, where the Web, application, and database talked amongst themselves and presented a ready-made page to the browser. This reduced the number of turns across the slow WAN, as shown in Figure 8-23, and gave visitors acceptable performance.

With the creation of larger pages, richer content, stylesheets, scripts, and third-party mashups, Internet applications are once again sliding toward a world filled with back and forth chatter (Figure 8-24).

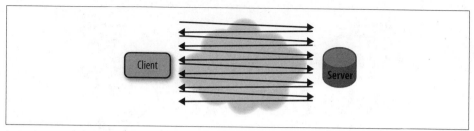

Figure 8-22. A high number of turns, coupled with high WAN latency, make the application unacceptably slow

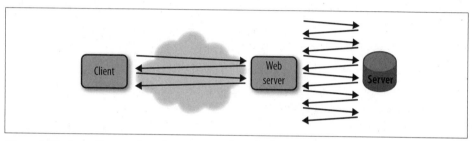

Figure 8-23. A smaller number of turns on a relatively slow WAN offered acceptable performance for early web applications

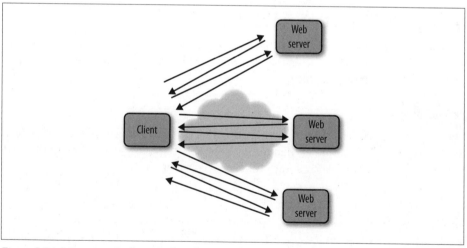

Figure 8-24. Many sources for a single web page mean modern web applications and mashups are at risk of poor performance

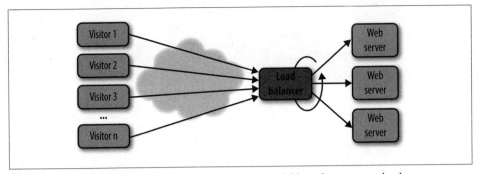

Figure 8-25. Load balancers distribute load across all available web servers in the data center on a per-visitor basis

So there are several factors behind network latency:

- The size of the content, which dictates how "big" the page is
- The number of objects and, by association, the number of turns you'll need to make across the Internet connection
- The available bandwidth
- The amount of packet loss and retransmisson that interferes with delivery
- The round-trip time between the browser and the web server

Fiddling with Things: The Load Balancer

Most websites of any significant size have a load balancer between them and the outside world. The load balancer shields the website from the Internet and attempts to make the site faster and more reliable.

Distributing load

One of the original reasons for using load balancers was to balance the load across many servers equitably. Rather than running one very big web server, companies could buy several smaller ones and "spray" traffic across them.

The simplest way to share load is to iterate incoming visits across web servers in a round-robin model, as shown in Figure 8-25. However the round-robin model doesn't take into account other factors, such as the work that each server is doing or the fact that some servers may be more powerful than others. Today, load balancers use more sophisticated algorithms, choosing the server with the fewest users on it or the one that responds most quickly.

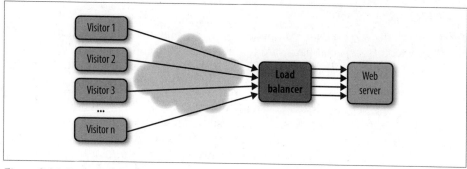

Figure 8-26. By consolidating incoming requests and sending them across a few, more efficient, connections to servers, load balancers offload network processing

Avoiding broken machines

The other basic reason for balancing load is reliability. By having several machines, your website has redundancy. If one server fails, the load balancer can remove it from rotation and insulate your visitors from the outage.

Address translation

The load balancer has a public IP address, while the servers behind it can have private ones. This makes a site harder to hack, since the attacker can't connect directly to a private IP address across the public Internet. It also means you don't need as many IP addresses overall.

Consolidating requests

A large part of the processing that web servers do involves network signaling—managing the many TCP sessions they have at any given moment. Both the client and the server have to keep track of packet sequencing, retransmission, and so on. For a client, this is easy—it's only handling a few connections. For a server, however, it can be deadly. Servers may be tracking thousands of TCP sessions, and this can consume more resources than the actual website itself.

Load balancers can help. They can offload these thousands of small network connections and instead maintain only a few connections to each server.

As Figure 8-26 shows, the web server has only a few connections to deal with. The load balancer can use a technique called *pipelining* to send several requests to the server at once, further increasing the efficiency of this connection. Note, however, that pipelining makes it much harder to monitor network connections, since the usual model of request and response isn't as easy to follow when several requests are sent simultaneously.

Compressing and encrypting responses

Encrypting traffic takes a lot of computation—something that's better left to the load balancer, which has special hardware to encrypt as efficiently as possible. This can include encrypting and decrypting traffic, as well as compressing data and even adjusting caching parameters to reduce the number of requests.

In the case of encryption, the load balancer needs to decrypt the traffic so it can look at the HTTP headers (which would otherwise be hidden) and decide which server should handle them, based on cookie and content. You don't need to understand the inner workings of a load balancer to monitor this, but it is important to know where the data is and isn't encrypted. If you sniff the wire between the load balancer and the web servers, you may see traffic in plain text that will be encrypted and compressed before it's sent across the Internet, so internal measurements of page size may not match those of the pages that reach the users.

Server Issues

There's only so much a load balancer can do to help with performance. While networking, protocol inefficiencies, and general-purpose tasks like redirection and encryption can often be tackled by hardware upgrades, bandwidth purchases, and the use of a CDN, application and server problems are more complex and may require rewriting or rearchitecting an application.

Server OS: Handling networking functions

The first part of a server to handle a browser's request is the operating system. It's a traffic cop for all TCP/IP connections, and may have other functions (such as logging or security) that consume processing power. Different operating systems will have different performance characteristics. There are many server monitoring tools that can provide insight into OS performance (Figure 8-27).

Web service: Presentation layer, file retrieval

Just as a desktop operating system runs applications (for example, a word processor), so a server operating system runs services (such as web, authentication, or database). The first part of any web application cluster is the actual HTTP service. Two of the dominant web servers on the Internet today are Apache and Microsoft's IIS web server, but hosted services like Google's web stack are behind a growing number of websites (shown in Figure 8-28).

If the TCP client on your desktop has its counterpart in the TCP stack of the server, the HTTP in your web browser (e.g., Firefox) has its counterpart in a web service (e.g., Apache). The web service handles the HTTP status codes and persistence needed to deliver objects properly. It gets those objects from memory, from disk, or from application processing.

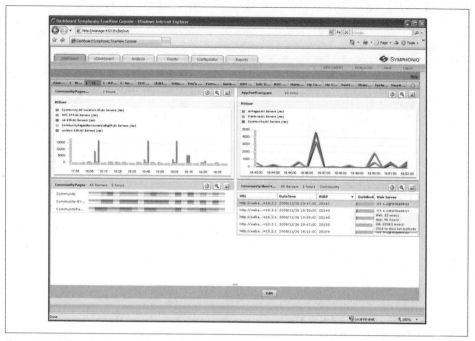

Figure 8-27. Symphoniq correlates OS, App Server, and database performance information with end user experience

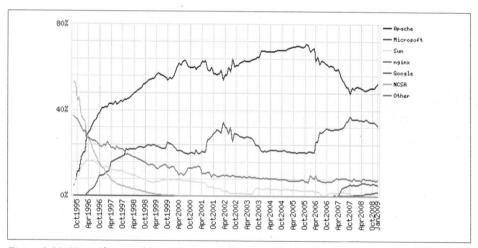

Figure 8-28. Netcraft's monthly "web server survey" results (http://news.netcraft.com/archives/web _server_survey.html)

When it comes to measuring just the web server's performance, the key metric is how long it takes to get back an HTTP response for a trivial object. If we request a tiny image from the web server, the server shouldn't have to think about the request much—it

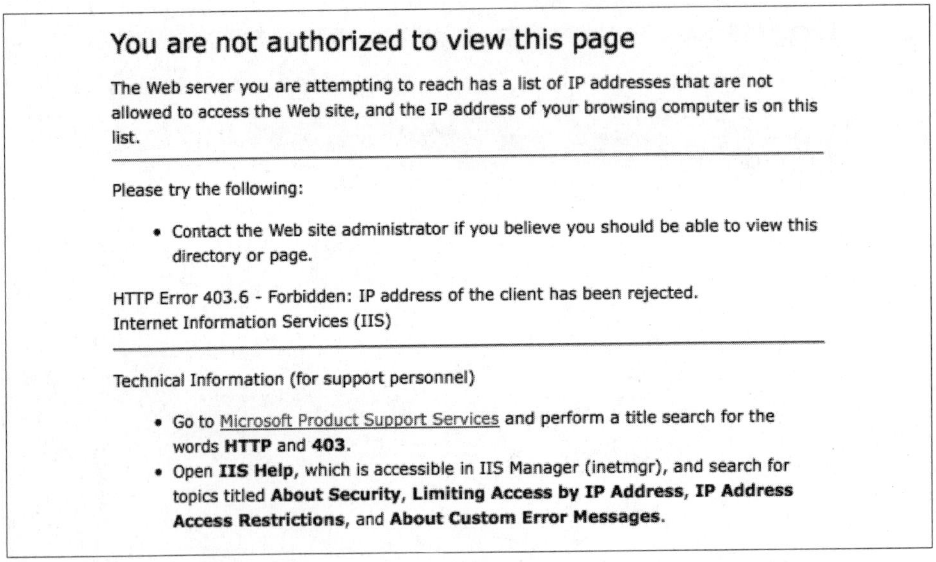

You are not authorized to view this page

The Web server you are attempting to reach has a list of IP addresses that are not
allowed to access the Web site, and the IP address of your browsing computer is on this
list.

Please try the following:

- Contact the Web site administrator if you believe you should be able to view this
  directory or page.

HTTP Error 403.6 - Forbidden: IP address of the client has been rejected.
Internet Information Services (IIS)

Technical Information (for support personnel)

- Go to Microsoft Product Support Services and perform a title search for the
  words **HTTP** and **403**.
- Open **IIS Help**, which is accessible in IIS Manager (inetmgr), and search for
  topics titled **About Security, Limiting Access by IP Address, IP Address
  Access Restrictions**, and **About Custom Error Messages**.

Figure 8-29. A 403 error

should just deliver the image. And because the image is tiny, we won't have several
turns of the network to complicate things. The time it takes to retrieve a small object,
less the network round-trip time, is roughly the delay that the web service has
introduced.

We may also encounter problems with the client's request. While these are often mal-
formed requests from browsers, they can also include blocked IP address ranges, as
shown in Figure 8-29, or invalid visitor credentials. Problems like these are particularly
hard to detect without looking at logfiles, leading to the all-too-common "it works for
me" defense from IT teams.

Dynamic tier: Application logic

A web service that handles only static content isn't very common. Typically, the web
service has an application service behind it, either on the same server or on a separate
tier of servers. This service runs a program each time a request arrives and sends the
output of that program back to the web tier for delivery.

There are a number of popular application servers online today—many more than there
are web servers. In the Microsoft world, IIS is both a web server and an application
server, and it focuses on generating Active Server Pages (the *.aspx* extension of many
sites). Similarly, Java is a common server language that generates pages with a *.jsp*
extension. A website can be written in any language, but some, like PHP, Ruby, and
Python, are increasingly common.

You may be more familiar with turnkey applications such as WordPress, Drupal, Mediawiki, or PHPBB. All of these are application servers in their own right. Many of these

tools come with built-in logging and problem recording that's delivered within the web server logs, and they provide more detailed information about errors than web servers do on their own. Figure 8-30 shows a series of errors reported by the WordPress platform.

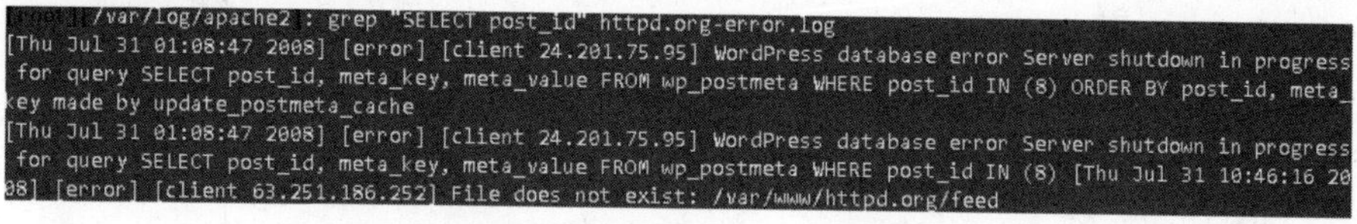

```
[root]:/var/log/apache2 : grep "SELECT post_id" httpd.org-error.log
[Thu Jul 31 01:08:47 2008] [error] [client 24.201.75.95] WordPress database error Server shutdown in progress
for query SELECT post_id, meta_key, meta_value FROM wp_postmeta WHERE post_id IN (8) ORDER BY post_id, meta_
key made by update_postmeta_cache
[Thu Jul 31 01:08:47 2008] [error] [client 24.201.75.95] WordPress database error Server shutdown in progress
for query SELECT post_id, meta_key, meta_value FROM wp_postmeta WHERE post_id IN (8) [Thu Jul 31 10:46:16 20
08] [error] [client 63.251.186.252] File does not exist: /var/www/httpd.org/feed
```

Figure 8-30. WordPress errors in Apache server logs

Depending on the application server you're using and the verbosity of logging that you configure, you'll get varying degrees of visibility into its health. Remember, however, that logging consumes processing and storage resources—if you enable logging to better understand a problem, you may introduce even more delays than you had in the first place. Some large sites simply can't turn on logging because of the impact on web performance and the volume of data it creates.

From a performance standpoint, application servers cause a lot of delay. You can measure the application's delay by comparing the "trivial" web request you use to test the web service tier with the responsiveness of an application request.

Storage tier: Data, state persistence

Behind the application servers lie databases that the application tier can query to retrieve data. Databases store information such as product catalogs and social graphs; they may also store application state.

Databases are often the cause of host latency. The database may be busy doing something else (such as rebuilding an index or clearing out data). It may have many requests for the same data, it may be dealing with inefficient queries, or it may just be digging through many records. There are so many ways databases can go wrong that database tuning is a dark art that fills entire books (see IBM's list of known causes of slow database performance at *http://www-01.ibm.com/support/docview.wss?uid=swg21174563*).

Some sites have huge amounts of data, as in the case of search engines that use other storage technologies in which shards of data are spread out across many machines, each independent of the others. This shared-nothing approach, which is used in Google's filesystem and other cloud computing models, scales more smoothly because it avoids contention that can happen with traditional databases.

Sometimes a database will return an error to the application server. The application server should recognize the error and deal with it gracefully; if it does so, you need to collect the error as part of your monitoring even if the error wasn't shown to the user.

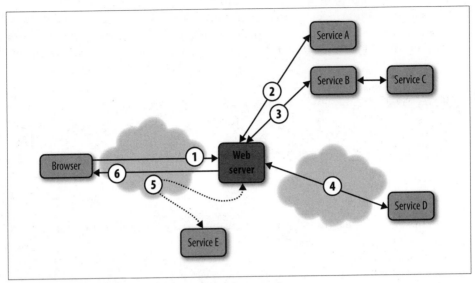

Figure 8-31. Communication model in an SOA-driven architecture

To determine the impact database latency has on overall site performance, compare the performance of pages that hit the database to those that are returned by the application server tier without consulting the database. Some web operators create custom pages that specifically test database functions such as writing data or retrieving information, and use this as part of their testing strategy.

Third-party components: Web services, transaction processing

With an increasingly distributed web, your application server may also talk to backend systems you don't control. For example, during credit card validation, your server may have to check with a merchant bank as well as fraud detection systems to validate the transaction before confirming a purchase.

Designing applications using a set of predefined services—known as a service-oriented architecture, or SOA—focuses on assembling building blocks that are interoperable through well-defined interfaces. This makes maintaining and upgrading the application easier. For example, a developer could replace one currency conversion service with another and not have to change the rest of the application.

However, SOA creates a great deal of machine-to-machine communication that's hard to troubleshoot. Figure 8-31 shows an SOA-driven backend for a website.

The browser sends a request (1) to the web server, which places backend calls (2) to other web services. Some of those web services (3) may themselves rely on other services; some may be located across dedicated wide-area connections (4) that increase overall latency; still others may be connected across the Internet (5) with all of the unpredictability that entails. Eventually, the server can send the response to the browser (6), but what appears to be host latency is in fact the result of third-party service delays.

Remote services can respond in unexpected ways, and it's up to your application servers to deal with these conditions gracefully. If the application relies on third-party services, you also need to consider the impact of the network between your data center and the third party.

Permissions and authentication

Even if everything is working smoothly, you may often encounter problems with authentication. Most content in an application has permissions associated with it. A surprising number of web problems stem from having the wrong permissions—a server that's not allowed to retrieve a certain object, for example, or a blog application that can't write to a particular directory.

The relationship between workload and performance

Sites that get suddenly busy often get suddenly slow. There is a correlation between traffic volume and web performance, but it's not a linear one, so it can be difficult to detect until it's too late. A server that can handle 1,000 hits per second may be fine at 900 hits per second, but at 1,001 hits a second, traffic is arriving faster than the server can deal with it, and a backlog quickly forms. If all else is equal, the greater the traffic to a site, the slower the responsiveness and the higher the number of incidents.

One of the most challenging aspects of web performance management is scaling. Because websites are connected to the entire world, their traffic levels can vary significantly from day to day. This is especially true for seasonal businesses, for example, a tax filing website at tax time or a costume store at Halloween.

The end result: Wide ranges of server responsiveness

All of these factors mean that responsiveness will vary widely depending on what users are doing and how many users there are on your site. *Many operational teams have no idea how many users are on a website or what they're doing.* We've found that fewer than half of the web operators we asked knew even basic statistics such as how many hits per second their site received, even though this is a strong predictor of website performance.

Instead, they looked only at the health of the underlying platforms. It's a myopia that can prove catastrophic.

Consider a blog, for example. Retrieving a blog page is relatively speedy: the content is ready, and it's a matter of a simple request to the database. Your blogging software may even have cached a copy of the requested page.

Compare that to the act of saving a blog. The blog software has to store the new post to the database, then it has to tell other web servers that the new blog is available, and update RSS feed information. It has to modify the list of tags and build a new cached copy of the website. That's a lot of work, and it can take several seconds.

If you're operating a blog and you see a sudden drop in performance, you need to determine whether it's because a few authors are submitting stories or because many visitors are suddenly looking for content. Without the context of what users are doing, you can't diagnose the slowdown.

Client Issues

As if network latency and server delays weren't enough, there's still the client to consider. As web applications become increasingly dependent on edge processing and rich browser frameworks like Flex, Silverlight, Google Gears, and Ajax, the client plays a more important role. More things can go wrong on the browser, where problems are hardest to see.

Desktop workload

Client computers have a finite amount of processing power. This is particularly true for subnotebooks designed to sip lightly from batteries—their power efficiency comes at the expense of CPU.

The desktop may cause delays because of slow network processing or because the user is doing other things that consume cycles the browser needs. The richer your pages, the more the desktop will slow down. This happens in two distinct ways:

- *Processing delays* prevent the browser from retrieving information from memory, launching plug-ins, or performing calculations.
- *Presentation delays* slow down visualization and screen display.

The browser

The modern browser is an operating system in its own right (*www.google.com/google books/chrome/small_00.html*). It's handling multiple tasks (sites, each in their own tab) and managing local resources (Flash, Java, and Greasemonkey plug-ins and add-ons) even as it executes code.

Just like operating systems, browsers are extensible. They offer new ways of visualizing and navigating the Web, each of which poses its own performance and availability risks. New ways of accessing the Web, such as the Cooliris application shown in Figure 8-32, further push the boundaries of desktop processing capacity.

The more we ask the browser to do, the slower it becomes. More and more of a page's latency happens when it reaches the client. Fortunately, desktop performance has increased along with end user demands (thanks to the continued doubling of processor capacity that is Moore's Law)—a typical page from a modern website would have brought a PC from only a few years ago to a grinding halt.

Figure 8-32. While not really a web browser, the Flash-based Cooliris application displays hundreds of images and uses the HTTP protocol

Browser compatibility

One of the first ways a browser can break things isn't actually the browser's fault at all. There are hundreds of plug-ins for the Web, each available in various versions. Adobe Acrobat, Sun Java, Macromedia Flash, and Microsoft Silverlight are just a few examples. Many sites try to run other plug-ins, such as ActiveX objects. If the browser can't correctly launch one of these plug-ins, either because the user won't allow it, because the browser or operating system doesn't support it, or because the plug-in won't load properly, then (as Figure 8-33 demonstrates), your site is broken for that segment of the market.

Even without plug-ins, many websites don't support certain browsers, as shown in Figure 8-34.

Client compatibility can only be addressed with good testing and an understanding of your audience's technical environment. You can learn a lot from web analytics by observing browser types, and you should always consider the impact of plug-ins—many of which won't be included in web monitoring tests—on performance.

Stylesheets and page layout

When HTML was first defined, it included both content and formatting. If you wanted to make a heading red and 22 pixels high, you said:

```
<H1 color=red size=22px>I upgraded your RAM</H1>
```

This was a problem. For a designer, changing the color of headings meant modifying the HTML on every page of the site without touching the content. It was also redundant—every heading needed the `color=red` and `size=22px` attached to it, which increased file size.

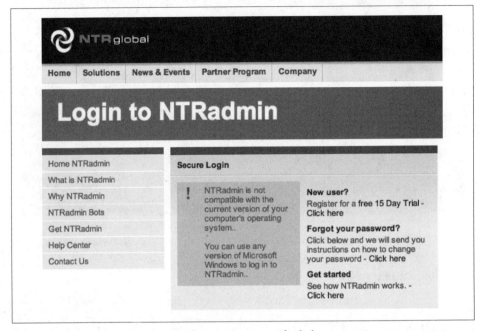

Figure 8-33. A web-based application that requires a specific desktop operating system to run

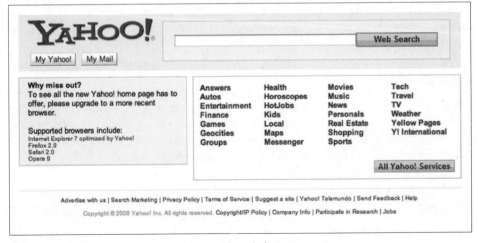

Figure 8-34. Yahoo.com message to visitors with early browser versions

To address this, we can separate formatting (`color=red`) from content and meaning (a heading called, `<H1>`) As a result, to define headings as red, the site will now have two files: one concerned with meaning and one concerned with formatting.

The HTML file includes the content and its semantic meaning:

```
<H1>I upgraded your RAM</H1>
```

A separate document, the Cascading Style Sheet (CSS), tells the browser how to format headings in general:

```
H1{font-size:22px;color:#333;}
```

This way, to change the color, the designer only has to change the formatting once, in one place—the stylesheet. Browsers then use the new formatting every time they render a page linked to the stylesheet.

This not only simplifies the process of changing styles and formatting, it also means that developers don't need to understand layout. It reduces the chances of errors in the editing process. And for large pages, a single note in a stylesheet is more efficient than formatting throughout the page.

Stylesheets aren't all good, however

- They're yet another object for browsers to retrieve, resulting in more turns.
- Stylesheets can contain references to other objects, such as background images, that complicate referrers and make it harder for monitoring tools to determine which components belong to which pages.
- There's additional browser compatibility to check, as not all browsers render stylesheets the same way.

Processing the page

Once a browser has loaded the page, it needs to assemble the contents and display them to the user. Depending on the amount of work involved, this can include executing JavaScript, launching plug-ins, and rendering images. The only way to capture this client-side delay is to be there when it happens, using JavaScript or a desktop agent to time the loading and rendering of the page.

Prefetching

Humans take time to read pages. During that time, the browser and the network are relatively idle. Why not put that idleness to good use?

One approach to improving performance is prefetching: the browser tries to guess what the user will want next, and gets to work loading it. On a page with only five hyperlinks, the browser can preload the five container objects that might come next and have them ready for the user.

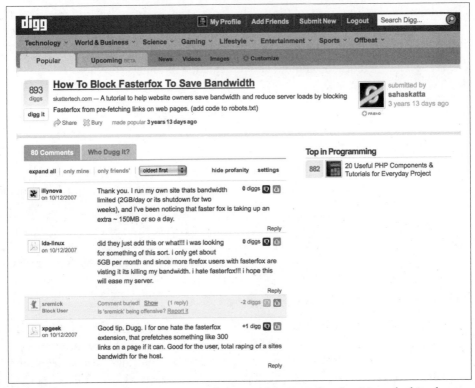

Figure 8-35. Digg criticism of the aggressive behavior of the Fasterfox plug-in's prefetching features

While this seems like a great idea for the user, it's a *tragedy of the commons*—a situation in which individuals acting in self-interest can ultimately destroy a shared resource, even though it's not in their communal interest to do so (*http://en.wikipedia.org/wiki/ Tragedy_of_the_commons*). In our case, the prefetcher receives all the benefit of prefetching, but the damage to performance is shared by everyone. One user's experience is better, but the server is flooded with five times more requests than users will actually see. The site slows down for everyone, including the greedy prefetcher.

Prefetching is considered an aggressive practice when it's initiated by the client, and plug-ins like Fasterfox have been widely criticized for supporting it, as shown in Figure 8-35.

There have been other examples of prefetching built into browsers and plug-ins. Google's Web Accelerator prefetches content, but it does so from Google's regional caches rather than the root server in most cases. Because Google's servers handle the brunt of the load, it receives fewer complaints than client-based prefetching.

Prefetching is useful in the right conditions. It's common to preload the next slide in a slide deck or the next picture in a photo gallery website. There's a very high likelihood that the fetched content will be the next thing the user requests, and it's the site's designer, not the browser, that initiates the prefetching. Good candidates for prefetching are situations in which end users are very likely to go to a single next item, and where that item will take time to load. This is a common Web 2.0 design pattern—when it's done by the site operator, not a browser plug-in.

From a performance and availability standpoint, prefetching makes page monitoring more difficult:

- As with CSS formatting, it's difficult to determine which objects belong to which containers because many "false" container downloads aren't followed by their component objects.

- It can skew page counts and per-visitor traffic estimates if you're collecting analytics through logfiles or passive capture rather than through JavaScript.

- Without detailed information about how prefetching is configured on the browser, it's hard to form an accurate estimate of page render time.

- Extra load on the site increases both network and host latency, and may exacerbate problems that only happen under load.

Other Factors

As the landscape of the Web changes, there are new factors that can undermine performance and availability or make user experience harder to monitor. Here are just a few.

Browser Add-ons Are the New Clients

Browsers can be extended in two main ways using what Mozilla refers to as add-ons. The first category of add-on—known as extensions—affects how a page is loaded and displayed. One extension, Greasemonkey (*https://addons.mozilla.org/en-US/firefox/search?q=greasemonkey&cat=all*), is a scripting platform that can modify pages as they're loaded, letting visitors customize how a site behaves within the browser.

Extensions may separate you from the user experience. Visitors may be playing a game such as WebWars or PMOG instead of interacting with your content. Browsers may be running the NoScript extension to avoid running scripts from untrusted sites. Extensions may also interfere with your JavaScript instrumentation of a page or break it entirely. However, extensions are still running within the browser, unlike the other form of add-on, known as a plug-in.

The Web has evolved in ways its creators couldn't foresee, and today it's the platform for broadcast media, gaming, human interaction, and more. Plug-ins are part of what makes this possible. They're standalone software applications like Java, Quicktime, Windows Media Player, Adobe Acrobat, and Flash. Your browser delegates work, such as playing an audio file, showing a video, or displaying a document to them.

Figure 8-36. Beatport.com is a B2C e-commerce application fully driven by Adobe Flash technology

Plug-ins can cause delay. For example, reading an Acrobat PDF means loading the Acrobat reader, and running a Java program means launching the browser's Java Virtual Machine. It takes time for the browser to initiate these components, and this may introduce additional delay that's hard to capture. Plug-ins take time to run their code or display their content, which further slows down the user experience in ways that are hard to measure.

For the web analyst, plug-ins represent a far bigger concern than just additional delay. Unlike extensions, which run within the browser and have limited access to the user's computer, plug-ins can do whatever they want. They may even communicate independently of the browser (for example, this is how plug-ins like Quicktime know that an upgrade is available).

Entire sites can be built in Flash using only a single container page, such as Beatport (shown in Figure 8-36). Flash is also the basis for much of the video on the Internet and many online games, some so processor-intensive that they only run on high-end machines.

Despite the broad adoption of Flash and JavaScript, there are relatively few standards for monitoring and measuring them. As a result, application developers who want to measure performance and availability must often handcode monitoring scripts and embed instructions in web pages to collect timings and send them back to the servers.

Timing User Experience with Browser Events

As all of these complications should make clear, with the advent of Web 2.0, we've been rethinking what makes pages slow. The focus has shifted from host and network time to client-side interactions and third-party mashups, from efficient page design to better scripting.

What's in a DOM?

The Document Object Model, or DOM, is one of the least understood and poorly explained concepts in the Web today. We'll try to fix that.

We can describe a house as a hierarchy of elements. We might have several categories of information about the house—rooms, occupants, and furniture, for example. Rooms would include house.rooms.bedroom, house.rooms.bathroom, and house.rooms.kitchen. Similarly, occupants might include house.occupants.mom and house.occupants.dad.

We could further extend these hierarchies: house.occupants.dad.pants.color might refer to the color of Dad's pants. Then, if we wanted to change the color of Dad's pants (in our megalomaniacal house DOM) we could set house.occupants.dad.pants.color=paisley.

Browsers have a similar hierarchy of elements that describe everything they're up to. The DOM starts with the browser itself, then the windows and tabs. When you load a page, the browser builds a hierarchy of all the things on that page, such as forms, titles, and headings.

Each of these elements has a value. For example, a form field may be referred to as window.document.form.element.text (or the specific names of those elements.) JavaScript running in that page can do several things with this field by referencing it with its DOM name:

- It can determine the value of an element by referencing it with its name. For example, it can determine what a user has typed into a field.
- It can change the element's value. For example, it can trim the spaces from a text entry in a field.
- It can tell the browser to let the script know when something happens to an element; e.g., it can do something special when the user presses Enter in a field.

These are relatively trivial examples, analogous to changing Dad's pants color in the house example. The real power of the DOM is that it lets JavaScript modify the entire web page—akin to rearchitecting the whole house, adding rooms, and moving furniture around.

On a browser, the DOM is also where details are stored, such as cookies associated with a site. By manipulating cookies with JavaScript, sites can store information on the browser until a visitor returns. The DOM is also where web analytics scripts get much

---

of their data about things like browser version and window size. For more information about DOM, see *www.w3.org/DOM/*.

JavaScript is an event-driven language. It relies on specific events, such as the click of a mouse, the scroll of a window, or the loading of an image to trigger actions. Dozens of events can occur for any element of a page. One of the most fundamental events for a page element is its arrival. When a browser loads a component object, it generates an event to tell JavaScript that the component is now loaded (for more information about DOM events, see *http://en.wikipedia.org/wiki/DOM_events*).

Traditionally, browsers pronounced a page finished when all its component objects were retrieved. You can see this in the status bar of most pages, which changes to "Ready" (or simply goes blank) to indicate that the page has been fully retrieved. This milestone, known as the OnLoad event, was used by many monitoring approaches to measure page load time from the client's perspective.

However, this event isn't as reliable an indicator of page load as it once was. Modern web designers thwart the OnLoad event. Some optimize pages carefully, placing JavaScript high on the page so it can run and so users can interact with the website before the page has finished loading. In these cases, the OnLoad time overstates delay, since performance measured from the OnLoad event is worse than it actually was for the user.

Other sites only start executing JavaScript once the OnLoad event occurs. In these cases, the OnLoad time is optimistic—the user's experience may be worse than what the OnLoad-based measurement says it is if the user needs the script in order to use the page.

There are several initiatives underway to correct this and come up with an accurate framework for client-side instrumentation. Most notably, Episodes, proposed by Steve Souders of Google, tries to standardize event timing and separate the task of instrumentation from the task of capture and reporting (*http://stevesouders.com/episodes/*). We'll return to the concepts behind Episodes when we look at RUM.

Nonstandard Web Traffic

The Web is a conversational, back-and-forth protocol. Web visits are initiated by your browser; servers don't connect to you indiscriminately. However, there are times when a server wants to push data out to a client—a stock ticker, for example, lets clients sign up for a particular stock, and pushes stock price changes out to those clients.

There are several ways to support this server push:

- The Real Time Media Protocol, supported by Adobe's Flash client, transmits a stream of data from server to client.

- A *perpetual GET* approach starts with a client that requests an object by HTTP and a server that returns a never-ending response. The client parses the response as it arrives.

- Web Sockets, a capability in the HTML 5 specification, creates a raw two-way connection between clients and servers. COMET is another similar approach to server push.

- In *long polling*, a browser requests an object, and the server intentionally takes a long time to respond. During this time, if the server has data to send, it can respond immediately; otherwise, at the end of the long response window, it can send back a null response and the browser can issue another long-poll request.

These asynchronous models, in which the client and the server can each act independently of one another, reflect a change in how we use the Web. With the growth of real-time chat and interactive media, we need to find new ways to measure such connections with metrics like message volume, number of streams, jitter, and bandwidth consumption.

RSS feeds and podcasts

Subscribers to web pages receive periodic updates about new content through either the RSS protocol or related web feeds like Atom. All of these feeds, collectively known as RSS (although this isn't entirely accurate), have one purpose: to update subscribers interested in a topic when new information is available.

RSS feeds are similar to traditional web traffic, but they have some important differences:

- They seldom carry a lot of content. The purest RSS feeds are simply structured text with a link to the content and a brief description. While some RSS readers load embedded images, this isn't a popular feature.

- They may offer only an excerpt of the content they're describing in order to encourage readers to visit the website in question to see more.

- They may include advertising; Google's FeedBurner service, shown in Figure 8-37, allows users to embed AdWords ads in content.

- It's hard to track abandonment. A user may receive hundreds of updates via RSS, but only read some of them. While this can be the first step in a conversion if the user loads the content and visits your site, disengagement is common: users simply don't read all the RSS feeds to which they subscribe.

- RSS feeds may be aggregated. A single web portal may request one RSS feed update on behalf of hundreds of subscribers. Some of these portals will tell you the number of subscribers in the user agent itself.

Feed Readers and Aggregators

NAME	SUBSCRIBERS
Google Feedfetcher ▼	73
Firefox Live Bookmarks ▼	9
NewsGator Online ▼	6
Bloglines ▼	5
Netvibes ▼	4
Mac OS X RSS Reader ▼	3
Jakarta Commons Generic Client ▼	2
My Yahoo ▼	2

Figure 8-37. FeedBurner, showing feed aggregators consuming an RSS feed on behalf of their subscribers

For RSS feeds, we care more about availability and freshness. Since RSS feeds are updated in the background, the user's experience won't suffer if the feeds take a bit longer to get there. When we post new content, however, we want it to be available quickly so readers know about it.

Peer-to-peer clients

Video content consumes a tremendous amount of bandwidth. Today's approach of embedding video within a web page using a plug-in may work for viral videos, but it won't scale to Nielsen-level audiences and real-time distribution. We know of one large enterprise that saw its web traffic increase sixfold during the U.S. Open because of employees watching the game, which ultimately forced the company to block access to that site.

One way to address such scaling problems is to enlist the help of the clients. By turning a browser into a broadcaster that can share video content with those nearby, we can reduce the load on the servers and decrease the number of video streams running across the core of the network.

Peer-to-peer technologies, and the P4P working groups, are focusing on such approaches to scaling. We mention them here because they're going to require new ways of monitoring and new metrics to track end user experience. These will include streaming information such as the number of subscribers and the number of peers participating in the stream, as well as more traditional video information like frame rate, jitter, and delay.

 Pando Networks offers a good explanation of the P2P scaling challenges the video industry faces, as well as links to the P4P working group at *http://www.pandonetworks.com/p4p*.

A Table of EUEM Problems

Here's a summary of just some of the problems we've covered and how they affect successful page delivery. As Table 8-1 shows, there are so many things that can break in a web application that it's essential to monitor your website thoroughly.

Table 8-1. Some of the ways websites can fail

Error	Symptom	Possible Source
Server not found	Can't resolve IP address	DNS broken
No TCP response	Server fails to respond to the client's initial TCP request, leaving the client to time out	Network
TCP connection reset	Server establishes a TCP connection (SYN ACK) with the client, then resets it	Server overloaded
SSL negotiation failed	HTTP works, but HTTPS doesn't	SSL not enabled
No HTTP response	Despite a TCP session, HTTP GET receives no response from server	HTTP port wouldn't respond (service hung)
Bad redirect	Server sends client elsewhere, but new destination doesn't exist	Destination not working
Never-ending redirect	Server redirects client to itself	Misconfigured server or load balancer
400 error	HTTP 400 in the browser	Bad request from client (i.e., content doesn't exist)
500 error	HTTP 5xx in the browser	Server error
No content	Request gets a 200 OK, but no content follows	Server error
Response truncated	Server begins to respond, then stops	Server disconnected during response
Incompatible MIME type	Server sends image data, but browser displays it as text only	MIME type configurations on server incorrect
Broken content	User receives incorrect content	Corrupt data or application error
Client code error	Browser reports an error when executing code on page or launching plug-in	Bad JavaScript or old browser version
Components missing	Some elements of the page fail to load	Third-party site

Measuring by Hand: Developer Tools

Your job is to understand end user experience once a system is in production. Long before you deploy the application, your developers will have been using tools to estimate its performance. These include sniffers that can decode network traffic and find granular protocol issues, desktop software that measures responsiveness and suggests corrections, and hosted services that test a page from several locations to analyze its load time across the Internet.

If we've done our job explaining the elements of web performance, you'll be able to understand the results of these tools and pinpoint areas for optimization.

Network Problems: Sniffing the Wire

There's an old saying in networking: sniffers don't lie. Seeing the actual TCP/IP conversations that flow across a network connection and reassembling them into the HTTP transactions between a server and a browser is a time-consuming effort. It's also the best way to find out what really happened.

Wireshark (formerly known as Ethereal), shown in Figure 8-38, is the leading free packet sniffer. You can narrow the traffic it captures to a single TCP/IP session and reassemble the HTTP objects from the data stream. You can also use timing information to measure performance extremely precisely.

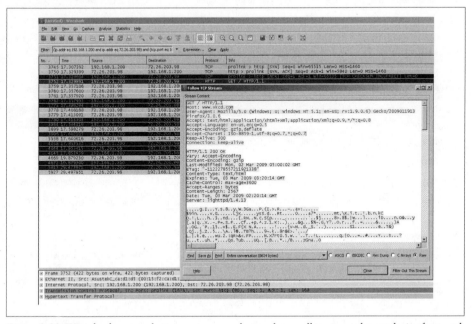

Figure 8-38. Wireshark, a popular open source packet analyzer, allows granular analysis of network traffic

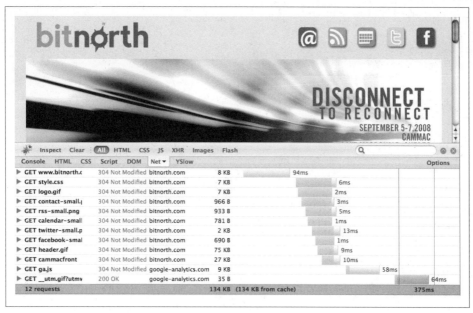

Figure 8-39. An analysis of network latency in Firebug

If you want to understand how an application is working on the desktop itself, however, you need to use desktop tools. Packet sniffers can't show you latency that happens within the browser, and it can be difficult to reassemble modern web pages in an intelligible manner.

Application Problems: Looking at the Desktop

One of the most commonly used measurement tools is Firebug, a Firefox extension that allows developers to break down page load time, examine page components, and browse the DOM (*http://getfirebug.com/*). Figure 8-39 shows a cascade diagram of network latency for a simple page with 12 objects.

The YSlow extension to Firebug, shown in Figure 8-40, takes this analysis one step further, summarizing page performance and suggesting why slowness exists (*http://developer.yahoo.com/yslow/*). Steve Souders summarizes the best practices that are implemented in YSlow in his book, *High Performance Web Sites* (O'Reilly). Google Page Speed (*http://code.google.com/speed/page-speed/*) is another alternative.

Other desktop tools, such as HTTPWatch and Flowspeed for Internet Explorer, are also essential for web developers. All of them let your company fix problems before you go live and train your development team in how to think about performance. If you can report performance issues in ways they understand, using Firebug-style cascade diagrams or YSlow terminology, for example, they're more likely to follow your advice.

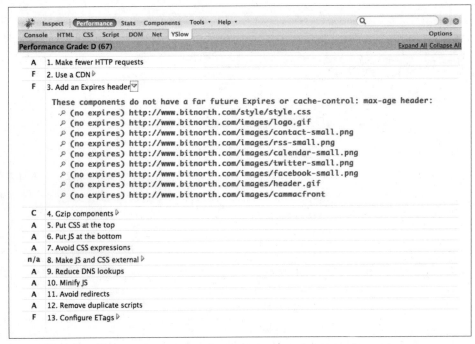

Figure 8-40. The YSlow extension suggests improvements to web page performance

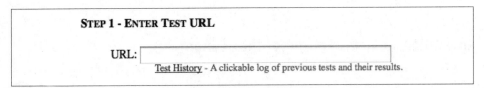

Figure 8-41. Entering a page to test in www.webpagetest.org

Internet Problems: Testing from Elsewhere

Many sites will test web page performance from remote locations. Of these, AOL's PageTest is perhaps the most comprehensive, since it gives you control over many of the wrinkles we've outlined above—such as page caching—to produce more realistic results. WebPagetest.org (available at, strangely enough, *http://www.webpagetest.org*) is a live version of PageTest.

To use the system, first enter the URL you're measuring (Figure 8-41).

Choose the source of the test (Figure 8-42). Since the service is deployed as part of AOL, test servers are based in Virginia, but there is a new system in New Zealand to measure the impact of international traffic.

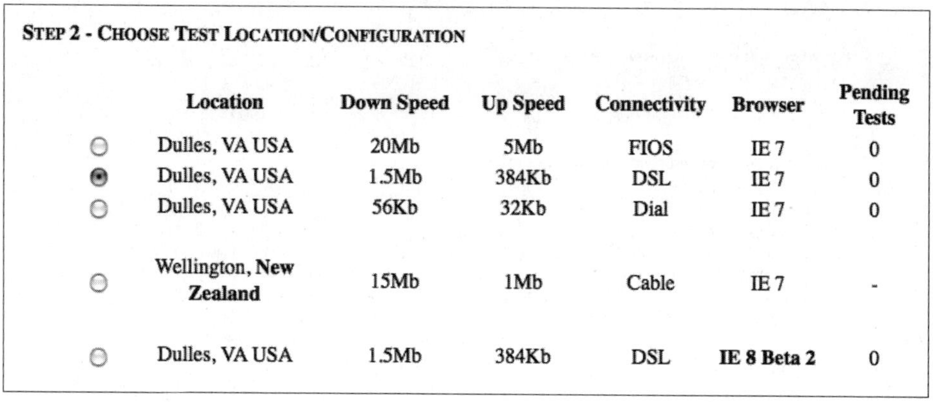

	Location	Down Speed	Up Speed	Connectivity	Browser	Pending Tests
○	Dulles, VA USA	20Mb	5Mb	FIOS	IE 7	0
◉	Dulles, VA USA	1.5Mb	384Kb	DSL	IE 7	0
○	Dulles, VA USA	56Kb	32Kb	Dial	IE 7	0
○	Wellington, New Zealand	15Mb	1Mb	Cable	IE 7	-
○	Dulles, VA USA	1.5Mb	384Kb	DSL	IE 8 Beta 2	0

Figure 8-42. Selecting a testing location

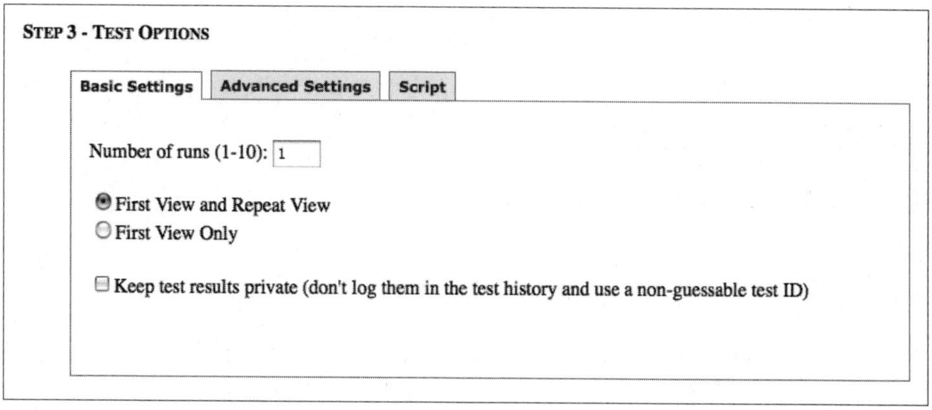

STEP 3 - TEST OPTIONS

| Basic Settings | Advanced Settings | Script |

Number of runs (1-10): [1]

◉ First View and Repeat View
○ First View Only

☐ Keep test results private (don't log them in the test history and use a non-guessable test ID)

Figure 8-43. Configuring the number of test runs

Finally, customize how the test will run (Figure 8-43). You can specify the number of times the test will run, as well as whether to test a repeated view (where the browser cache is already filled).

More advanced options include where to stop measurement (once the entire page has loaded or when the DOM determines the document is complete) and how many parallel browser connections to simulate (Figure 8-44).

The result is an extremely comprehensive analysis of page performance across several runs. The system produces a performance summary like the one shown in Figure 8-45.

It also provides a cascade diagram of the various objects that were loaded within the page, along with many of the metrics and elements we've discussed (Figure 8-46).

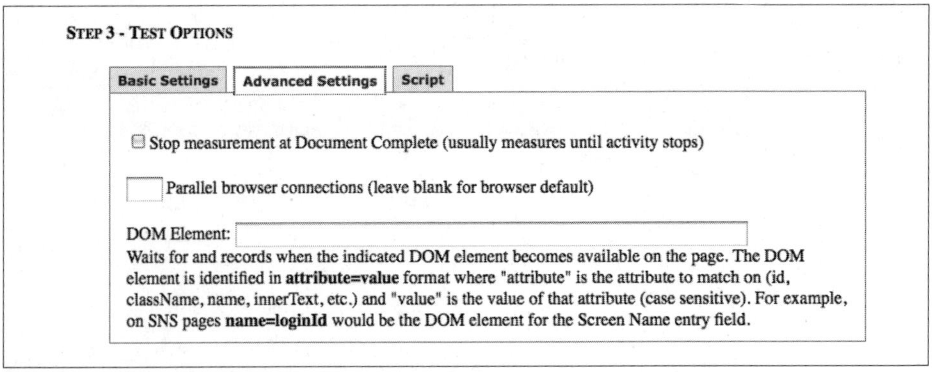

| Basic Settings | Advanced Settings | Script |

☐ Stop measurement at Document Complete (usually measures until activity stops)

☐ Parallel browser connections (leave blank for browser default)

DOM Element: [   ]
Waits for and records when the indicated DOM element becomes available on the page. The DOM element is identified in **attribute=value** format where "attribute" is the attribute to match on (id, className, name, innerText, etc.) and "value" is the value of that attribute (case sensitive). For example, on SNS pages **name=loginId** would be the DOM element for the Screen Name entry field.

Figure 8-44. Adjusting parallelism and defining the end of page retrieval

Optimization Report (punch list of things to fix)

	Load Time	First Byte	Start Render	Document Complete	Fully Loaded	Requests	Bytes In
First View	4.571s	1.438s	1.976s	4.571s	4.880s	25	495 KB
Repeat View	1.877s	0.526s	0.758s	1.877s	2.252s	24	50 KB

Figure 8-45. The high-level report of a PageTest run

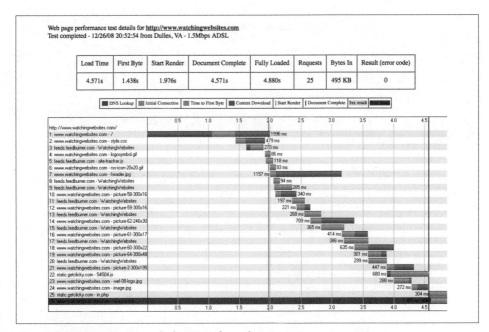

Figure 8-46. A PageTest cascade diagram of page latency

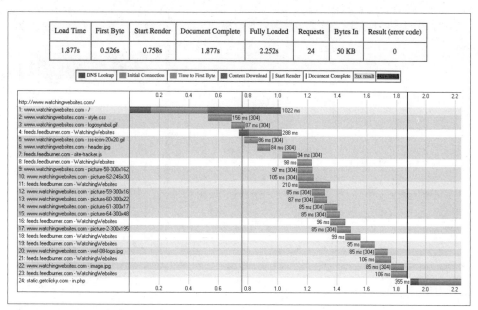

Figure 8-47. A cascade diagram for a PageTest simulation of a visitor whose browser has a full cache

Subsequent test runs reflect the user experience for visitors whose browsers have already cached the page. Get-if-modified requests (HTTP 304) are flagged, since many of these can be improved simply by setting caching parameters correctly so the browser doesn't check to see if it has the most recent version of content that changes infrequently (Figure 8-47).

Perhaps most importantly, the site makes performance recommendations similar to those of YSlow! that point to opportunities for performance improvement (Figure 8-48).

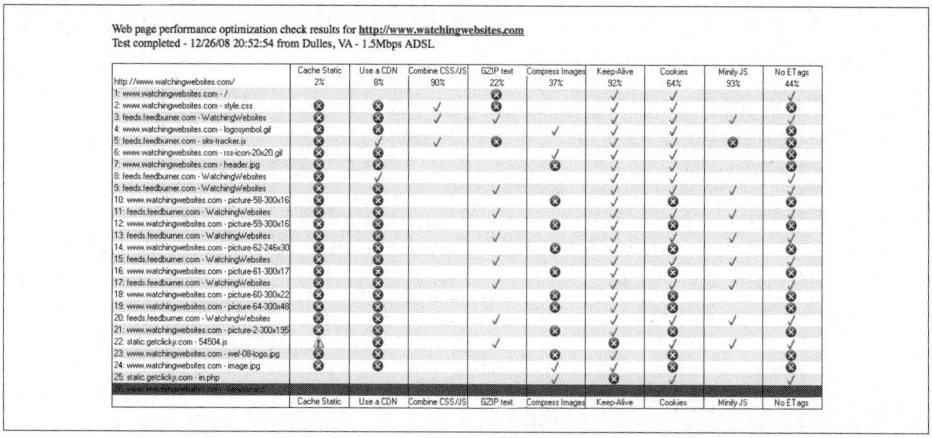

Figure 8-48. A series of optimization recommendations from PageTest

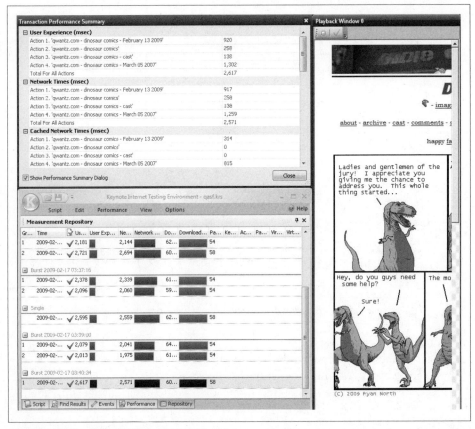

Figure 8-49. Keynote's KITE testing service

Another useful tool for Internet health testing, perhaps more suited to real-time operations than developer testing, is the Keynote Internet Testing Environment (KITE), a free service and Windows-based desktop application that can verify your site's health at various intervals, as shown in Figure 8-49.

KITE can drill down to individual objects on the tested pages to determine where performance and availability issues reside (Figure 8-50).

Getting developers and frontline operators using the same tools is a huge step toward breaking down organizational silos that exist in web monitoring. Operational teams responsible for real-time troubleshooting should train developers on Internet testing services. In return, developers should show web operators the desktop tools they use to examine web applications.

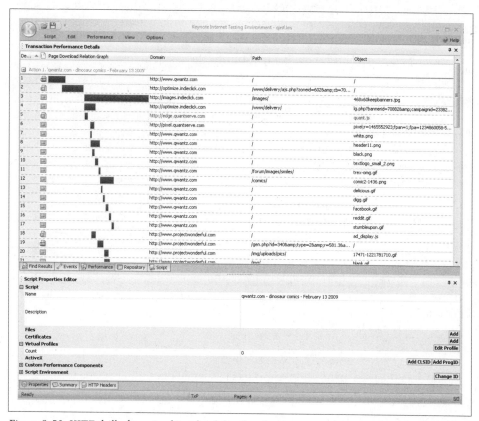

Figure 8-50. KITE drills down to object-level details, allowing you to determine your biggest sources of latency

Places and Tasks in User Experience

Recall the places-and-tasks model we outlined in Chapter 5. It's not just a good way to understand what users did—it's also useful for thinking about their experiences on your website.

Place Performance: Updating the Container

If the page is a place, we can measure the initial loading as outlined earlier. At some point, the page is loaded, and the user can interact with it. Something triggers a change to the place:

- Minor input, such as dragging an item into a shopping cart or upvoting a story.
- Client-side scripts that periodically refresh a display.

- Client-side prepushing (the opposite of prefetching), in which the browser sends content to the server ahead of time. This is a common pattern in blogging (saving a draft of a post) or webmail (uploading an attachment in the background).
- Server events that push new content out to the client across a nailed-up HTTP connection or a web socket.

These updates take time to load and may have errors, but they're only a small fraction of a full page retrieval. It's not correct to measure these updates as a sequence of events the way we do for pages. Instead, it's more accurate to describe them the way we would describe bandwidth: the delay between the server sending new content and the user seeing it. Place performance should therefore be described in terms of the frequency of updates and the latency with which the page is updated or the display refreshed.

Task Performance: Moving to the Next Step

Traditionally, the focus of monitoring is on tasks—and their abandonment—rather than on places. We expect users to move toward a particular outcome. This may still be within a single page (for example, a Flash shopping cart), but we have to measure the progress toward that outcome using a blend of analytics and performance metrics. These can include:

- Delay in loading the next step in the task
- Abandonment at each step
- The time the user spends thinking (considering an offer, filling out a form)

The start of a task is the moment the visitor clicks on the link to begin that task (for example, the Enroll button).

An organization with a mature monitoring strategy could map out places and tasks on its website, assign performance metrics to each of them (with a particular focus on latency in places and productivity in tasks) and tie this to business outcomes from analytics tools. This would provide a unified view of web health across the entire website that linked user experience to business goals.

Conclusions

Hopefully, you now have a good working knowledge of all the factors that affect web availability and performance. Your job is to understand the end user's experience as a result of these factors. To do this, you need to craft a monitoring strategy that can capture your site's performance as experienced by end users.

You might do this by testing critical pages or key functions at regular intervals—a synthetic testing model. Maybe you'll watch individual transactions and measure HTTP timings—RUM. As sites become more complex, you'll likely add your own application logging to round out the picture.

But as you design an EUEM strategy, remember that every web transaction begins with a set of requests and responses across multiple layers of communications protocols. However complicated a web application becomes, it's still all about turns, latency, and avoiding errors.

Now that we know all of the steps involved in web application delivery, we can look at how to measure performance.

CHAPTER 9

Could They Do It?: Synthetic Monitoring

We measure end user experience using two complementary approaches: synthetic testing, which involves testing a website by simulating visitor requests; and real user monitoring (RUM), which involves watching actual user interactions with the site.

Basic synthetic testing of websites is easy to implement, and there are many free options available to operators of fledgling websites. It should be the first kind of performance and availability monitoring you deploy for any web application. While it can't give you the granularity and accountability that comes from watching actual users, it offers peace of mind and an understanding of the percentage of time your site is online and how long it takes to retrieve specific pages.

In this chapter, we'll look at some of the fundamentals of synthetic testing and how to leverage what you already know from your web analytics data to best configure synthetic tests.

The most basic distinction in synthetic testing is between internal tests run behind your own firewall and external tests run from locations around the Internet. While the bulk of this chapter will focus on synthetic testing done outside your data center by a third party, it's important to understand the basics of internal testing to know what data you already have on hand and don't need to use testing services for.

Monitoring Inside the Network

Internal tests are those you run within your own data center to ensure all of your machines are working properly. You can run simple, simulated transactions to each server to verify that everything's working, as shown in Figure 9-1. Many web operators rely on commercial monitoring software, such as HP Sitescope, or open source tools, like Nagios, to do this.

Because you're running your own test systems over a LAN connection with lots of spare capacity, you can afford to generate large numbers of tests every few seconds. This will

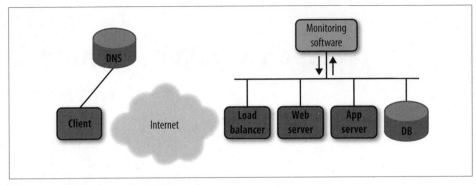

Figure 9-1. Internal testing of website infrastructure components

give you more complete "coverage" of your website, since you'll have smaller intervals between each test during which something can go wrong.

Internal testing is an essential tool for any IT operator. It may take the form of simple "Are you there?" up/down checks run every minute to each machine, or that of more comprehensive HTTP requests that check each machine's response for the correct content.

Using the Load Balancer to Test

For websites whose infrastructure includes a load balancer, a second—and increasingly common—option is to use this device to generate internal tests.

Load balancers provide redundancy by detecting server failures and taking the offending machines out of rotation. To do this, load balancers need to know when a server is broken. They determine this by running their own small tests, as shown in Figure 9-2. Since they're already testing each server, you can use load balancers for monitoring—to a degree. While they're good for up/down alerting, they won't provide long-term performance baselines and trending, although some monitoring tools can extract test results from them and graph them over time using programs like Cacti (*www.cacti.net/*) or MRTG.

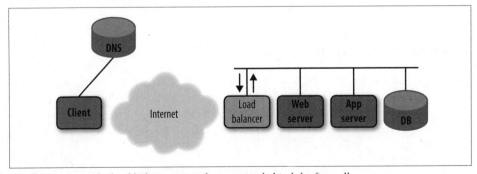

Figure 9-2. Using the load balancer to perform testing behind the firewall

A load balancer can monitor the network, TCP, and HTTP services on the machines across which it is distributing traffic. Any health check sent to a server comes back as "working" or "broken." Because they're constantly sending traffic to servers, load balancers are often the first to know when a bad response comes back. Some go as far as to inject JavaScript into pages as they go past in order to extract performance metrics from user visits.

You should always run internal tests. The load balancer's job is to hide broken servers from the outside world, and as a result, no external test will see a failed server when the load balancer is functioning properly. Internal tests fill in the gaps in your external monitoring, and because you run the tests yourself, you save money by reducing the number of external tests you need to pay for. However, internal device monitoring tools aren't able to properly simulate your visitors' conditions and shouldn't be used as a substitute for external monitoring.

Monitoring from Outside the Network

The Internet can fail in many creative and hard-to-pinpoint ways. Even though these aren't your fault, they're still your problem. To detect these problems, you need external tests that can watch your site from around the world.

A synthetic monitoring solution can:

- Alert you when your site is unreachable or unacceptably slow
- Detect localized outages limited to a region or a carrier
- Identify performance issues specific to a particular segment of your visitors, such as those using a certain web browser or a specific operating system
- Baseline your performance and availability around the world
- Predict whether you'll need to use content delivery networks when entering new markets
- Localize errors and slowdowns to a particular component of your infrastructure
- Generate additional traffic to test the capacity of your system or evaluate a new code release before it goes into production
- Test things you don't control, like mashup content or third-party web services
- Compare your performance to that of your competitors
- Estimate required capacity going forward by simulating load

A Cautionary Tale

A web startup we know had invested heavily in bottom-up device monitoring tools. It had also deployed several web analytics tools, and employees were confident they could

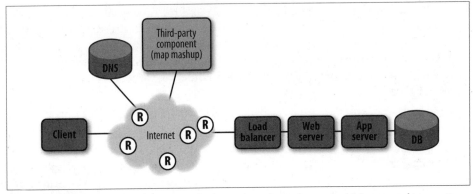

Figure 9-3. Some of the components on which a website depends—and which you need to monitor

finally sleep well at night, knowing that a series of well-designed alerts would notify them immediately if something went wrong.

The company had also been through a dramatic reduction in its operational staff. Management, emboldened by its recent investments in monitoring, decided that the humans weren't important now that the tools were in place.

Early one morning, one of the few remaining IT employees sent a panicked email message to the entire company to let them know that their site wasn't working. The IT team immediately brought up its dashboard, which showed that all servers were operating properly. The dashboard suggested that the infrastructure was, if anything, unusually healthy: web, application, and database servers were faster than normal.

Unfortunately, the company was under siege. A well-orchestrated denial-of-service attack had overloaded some of the core routers linking the company's data center to Europe, where nearly half of its customers were located.

The attack had a significant impact on revenues, costing the company a substantial amount in SLA refunds. That quarter, the company missed its revenue targets, and this ultimately resulted in delayed financing of their startup, at terms that were much less favorable for the founders and managers. Perhaps worse, management lost faith in the team's ability to properly monitor the infrastructure. Even though the problem was resolved in only three hours, the fact that it was discovered almost by accident was unforgivable.

What Can Go Wrong?

Testing your site from outside your own firewall means looking at the many things that can break between a visitor and your website. As you saw in Chapter 8, a web application relies on many components: DNS, routers, CDNs, load balancers, servers, third-party components, client-side scripts, and browser add-ons (Figure 9-3). A failure of any of these components will break or negatively impact the user's experience.

You need to verify the health of all these systems, all the time, from all locations that matter to your business.

Why Use a Service?

It would be expensive for you to set up and maintain dozens of servers in various countries and carriers around the world for the sole purpose of generating requests to your site. While some large organizations can afford their own testing networks, most companies rely on hosted synthetic testing services. There are several reasons for using a third-party service:

Cost
> It's cheaper than building and running your own servers.

No setup time
> These companies offer predefined reports and tools that make it easier to set up tests and report performance.

Viewed as impartial
> As independent organizations, the results of their tests are considered impartial and trusted—something you'll care about when you're using a performance report to resolve a dispute with a customer.

Competitive analysis
> The testing services can show you how your site stacks up against other businesses or even competitors.

Visibility into backbone health
> Because they have many points of presence, hosted services can test connections between their own servers and report on the health of carrier-to-carrier communications. Portals such as Keynote Systems' Internet Health Report (*www.internetpulse.net*) tell you which Tier 1 ISPs are currently experiencing issues. For example, in Figure 9-4, there is increased latency between Verizon and both AT&T and Sprint.

Synthetic testing services are relatively simple to use. Figure 9-5 shows how a basic service works.

1. You configure the test through a web user interface.
2. The service rolls out the test to nodes around the world.
3. The nodes run the test at regular intervals, recording performance.
4. The results are collected in reports you can view or receive by email.
5. If an error occurs, the nodes detect it, and may even capture a copy of the error.
6. Alerts are sent to a mobile device or email client.

Depending on the sophistication of the synthetic testing service, the system may perform additional diagnostics or capture the page error for review, which makes pinpointing and correcting the issue easier.

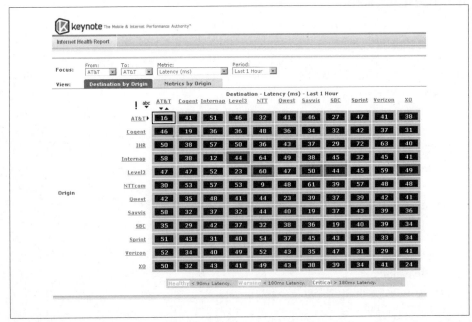

Origin \ Destination - Latency (ms) - Last 1 Hour	AT&T	Cogent	Internap	Level3	NTT	Qwest	Savvis	SBC	Sprint	Verizon	XO
AT&T	16	41	51	46	32	41	46	27	47	41	38
Cogent	46	19	36	36	48	36	34	32	42	37	31
IHR	50	38	57	50	36	43	37	29	72	63	40
Internap	58	38	12	44	64	49	38	45	32	45	41
Level3	47	47	52	23	60	47	50	44	45	59	49
NTTcom	30	53	57	53	9	48	61	39	57	48	48
Qwest	42	35	48	41	44	23	39	37	39	42	41
Savvis	58	32	37	32	44	40	19	37	43	39	36
SBC	35	29	42	37	32	38	36	19	40	39	34
Sprint	51	43	31	40	54	37	45	43	18	33	34
Verizon	52	34	40	49	52	43	35	47	31	29	41
XO	50	32	43	41	49	43	38	39	34	41	24

Healthy < 90ms Latency.   Warning < 180ms Latency.   Critical > 180ms Latency.

Figure 9-4. A Keynote.com Internet Health Report on latency between different monitoring locations

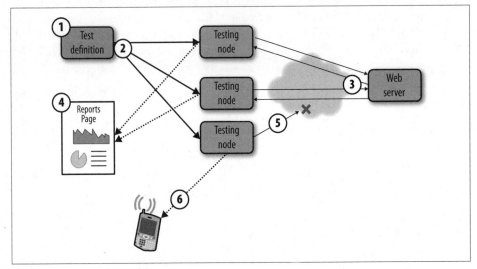

Figure 9-5. The basic steps in a synthetic testing service

Different Tests for Different Tiers

Different synthetic tests probe different layers of the Internet infrastructure we reviewed earlier. Each kind of test provides information on the health of a component or a service, and all are useful for determining where issues lie.

Testing DNS

As you've seen, DNS is an important source of latency for mashups and a common culprit when sites aren't accessible. DNS services are also a common point of attack for hackers, who try to "poison" domain name listings, which can misdirect visitors to other locations. In other words, you need to watch:

- The response time for DNS lookups
- DNS resolution of every site involved in building a page, not just your own
- Whether the DNS lookup returns the correct IP addresses

If you're using a CDN to speed up the delivery of your web pages to the far reaches of the Internet, the CDN may be operating your DNS on your behalf. This isn't an excuse not to test DNS resolution; in fact, you may want to watch more closely to be sure that IP addresses don't change without your approval. However, CDNs use resilient DNS services and Global Server Load Balancing (GSLB), which significantly improve the performance and availability of DNS resolution.

In most cases, you won't test DNS by itself. It will be a part of a synthetic test, and will be shown as the first element of latency in a test's results. Verifying that the content of a page is what you expect it to be will also let you know if your page has been hacked or if users are being redirected elsewhere because of a poisoned DNS.

Getting There from Here: Traceroute

In the previous chapter, we looked at traceroute as a tool for measuring the round-trip time between a client and a server. Traceroutes let you peer into the inner workings of the Internet. But because of the way traceroute collects information—sending several packets to every intervening device—it places a lot of load on the Internet's routers. As the traceroute manpage explains:

```
This program  is intended for use in network testing,
measurement and management.  It should be used primarily for manual fault
isolation.  Because of the load it could impose on the network, it is un
wise to use traceroute during normal operations or from automated scripts.
```

Don't run traceroutes via automated scripts. Instead, use them as a diagnostic tool to tell where something has gone wrong across the Internet devices between a client and a website.

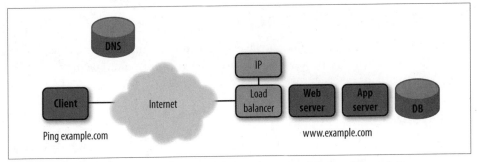

Figure 9-6. Ping testing measures IP functionality on the external device of the website

Recall that it's possible to use TCP, UDP, or ICMP traceroutes; you should use TCP traceroutes because they'll be treated in the same way that HTTP or HTTPS traffic would, while many parts of the Internet treat ICMP traffic differently. Also, don't expect traceroutes to always look clean; some devices won't return information about themselves, or will block downstream devices entirely.

Several synthetic testing services will perform automated traceroutes when they detect an outage, and include this information with the alert they send.

Testing Network Connectivity: Ping

Once you've confirmed that DNS can resolve the IP address of a site, and that the route across the Internet is clear, the most basic test you can run is to send a single packet to a website and receive a response. This is known as a *ping*, or ICMP ECHO. The device that receives the ping responds in kind (Figure 9-6). In most web applications, the responding device is a firewall or load balancer.

A transcript of a ping looks like this:

```
macbook:~ sean$ ping failblog.org
PING failblog.org (72.233.69.8): 56 data bytes
64 bytes from 72.233.69.8: icmp_seq=0 ttl=50 time=57.457 ms
64 bytes from 72.233.69.8: icmp_seq=1 ttl=50 time=58.432 ms
64 bytes from 72.233.69.8: icmp_seq=2 ttl=50 time=56.762 ms
64 bytes from 72.233.69.8: icmp_seq=3 ttl=50 time=56.780 ms
64 bytes from 72.233.69.8: icmp_seq=4 ttl=50 time=58.273 ms
64 bytes from 72.233.69.8: icmp_seq=5 ttl=50 time=58.555 ms
^C
--- failblog.org ping statistics ---
6 packets transmitted, 6 packets received, 0% packet loss
round-trip min/avg/max/stddev = 56.762/57.710/58.555/0.751 ms
```

The ping contains the number of bytes received, the address of the host, and a sequence number, which can be used to identify any lost packets. The hop count is in the direction of the response (one-way) while the time measurement is the round-trip time (back and forth).

On a less reliable network, ping helps us to understand the quality of the connection. Here's a ping run from a wireless network on a train:

```
macbook:~ sean$ ping vangogh.cs.berkeley.edu
PING vangogh.cs.berkeley.edu (128.32.112.208): 56 data bytes
64 bytes from 128.32.112.208: icmp_seq=0 ttl=41 time=296.634 ms
64 bytes from 128.32.112.208: icmp_seq=1 ttl=41 time=431.799 ms
64 bytes from 128.32.112.208: icmp_seq=2 ttl=41 time=352.870 ms
64 bytes from 128.32.112.208: icmp_seq=3 ttl=41 time=479.129 ms
64 bytes from 128.32.112.208: icmp_seq=4 ttl=41 time=503.164 ms
64 bytes from 128.32.112.208: icmp_seq=5 ttl=41 time=291.246 ms
64 bytes from 128.32.112.208: icmp_seq=7 ttl=41 time=777.717 ms
64 bytes from 128.32.112.208: icmp_seq=8 ttl=41 time=391.574 ms
64 bytes from 128.32.112.208: icmp_seq=9 ttl=41 time=722.543 ms
64 bytes from 128.32.112.208: icmp_seq=10 ttl=41 time=1770.265 ms
64 bytes from 128.32.112.208: icmp_seq=12 ttl=41 time=588.587 ms
64 bytes from 128.32.112.208: icmp_seq=13 ttl=41 time=1114.075 ms
64 bytes from 128.32.112.208: icmp_seq=15 ttl=41 time=2086.454 ms
64 bytes from 128.32.112.208: icmp_seq=16 ttl=41 time=1809.736 ms
^C
--- vangogh.cs.berkeley.edu ping statistics ---
19 packets transmitted, 14 packets received, 26% packet loss
round-trip min/avg/max/stddev = 291.246/829.699/2086.454/595.160 ms
```

A ping test ends with a summary of the results. In this example, over a quarter of all packets that were sent were lost entirely, and the average packet took 829.699 milliseconds. This kind of latency and packet loss effectively renders the Internet unusable.

Ping is a general-purpose tool used for testing reachability, packet loss, and latency for any kind of application, including voice, video, file transfers, and so on. It's the most basic test of Internet reachability. Ping tests have some important limitations, however.

The Internet may treat pings differently from web traffic:

- In some cases, firewalls and load balancers simply won't respond to pings, or will block them entirely.

- Some networks may prioritize pings differently, so the results you get won't be representative of HTTP traffic.

- Because a ping only tests whether the server's Internet connection is working, and does not check the actual web service, it won't detect web problems such as missing content, a locked database, or an Apache service that isn't responding correctly.

Nevertheless, ping tests are easy to run and are the backbone of up/down monitoring on the Internet. Some synthetic testing portals offer hosted services that ping an IP address at regular intervals. This may be your only monitoring strategy for nonweb devices that you need to monitor.

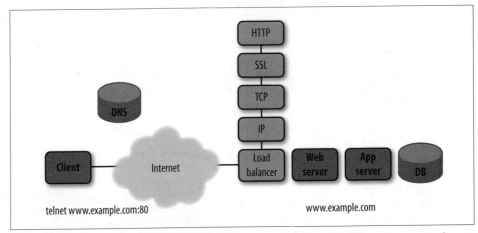

Figure 9-7. HTTP GETs test whether the web service is able to handle a request for a static object

Asking for a Single Object: HTTP GETs

Knowing that your web server is reachable doesn't mean users are getting the right content. Your DNS, network, and Internet-facing devices may be fine, but your website can still be down. The only way to tell whether the web service is working properly is to ask it for something and see what happens.

The simplest synthetic web test asks a server for content (using the HTTP GET method), times the response, and checks for an HTTP 200 status code confirming that the request was handled. In doing so, it tests not only the server's network layer (IP), but also the TCP layer (which manages end-to-end sessions) and any encryption (if present), as shown in Figure 9-7.

The HTTP GET is the workhorse of synthetic monitoring. It retrieves a single object, and in doing so verifies that many systems are functioning correctly. By timing the various milestones in the request and response, a testing service can blame poor performance on the correct culprit, whether it's DNS, network latency, or a slow server. Similarly, it can determine whether an error is caused by a bad network, a broken server, or missing content.

Beyond Status Codes: Examining the Response

Having sent a request and received an HTTP 200 OK in return, you might be tempted to pronounce the test a success. After all, you asked for content and received it. Some synthetic testing scripts stop here—particularly the free ones.

However, if you look deeper into the resulting response you'll see that there's a problem:

```
<div id="bd"><h1>Sorry, the page you requested was not found.</h1>
<p>Please check the URL for proper spelling and capitalization. If
```

```
you're having trouble locating a destination on Yahoo!, try visiting the
<strong><a href="http://us.rd.yahoo.com/default/*http://www.yahoo.com">Yahoo!
home page</a></strong> or look through a list of
<strong><a href="http://us.rd.yahoo.com/default/*http://docs.
yahoo.com/docs/family/more/">Yahoo!'s online services</a>
</strong>. Also, you may find what you're looking for if you try searching below.</p>
```

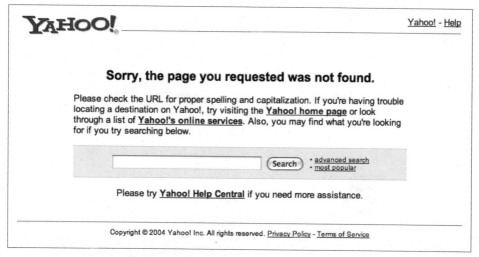

Figure 9-8. An apology page response for missing content that was served with an HTTP 200 OK status code

Despite the HTTP 200 OK response, the page you asked for didn't exist. Instead, you received a polite apology shown in Figure 9-8. In the early days of the Web, the ubiquitous "HTTP 404 Not Found" made it clear that you had asked for something nonexistent. Today, however, apology pages are commonplace, and they can hide errors.

Parsing Dynamic Content

Dynamic content further complicates the detection of errors. When you return to a portal, it probably welcomes you back by name. Your version of that page is different from someone else's version. Similarly, for a media site, news changes constantly. Even when part of the site is broken, much of the page will still render properly, and only one frame or section will indicate a problem (as shown in Figure 9-9). In other words, you'll never retrieve exactly the same page twice. So how does a synthetic test know that it's retrieved a valid copy of *index.html* when every copy is different?

To properly understand the performance and availability of a web application, you need to monitor dynamic content despite the fact that it's changing, as shown in Figure 9-10.

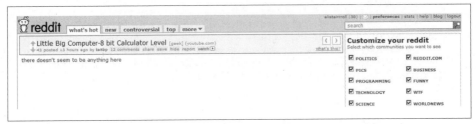

Figure 9-9. An error in a dynamic website, surrounded by properly loaded components

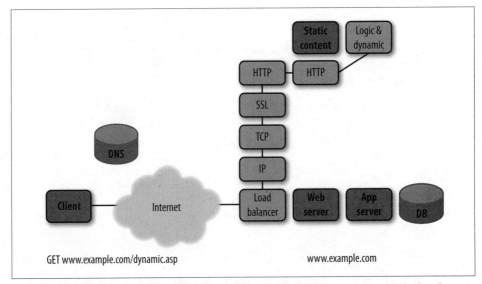

Figure 9-10. Checking for more than just an HTTP status code ensures you measure whether dynamic applications are working

The right way to check for HTTP content is to do the following:

- Check that the site returns an HTTP status code you want (generally a 200 OK).
- Check for content that *should* be there. Not all of a dynamic page changes every time. Often, changes are limited to a particular DIV tag, or certain cells of the table. The synthetic test needs to look for specific strings that are always present to confirm that the page is working. In Figure 9-10, this might be the text "Customize your reddit." You need to configure known keywords for each page you want to test and then make sure the application team doesn't remove them, which would lead to false alarms.
- Check for content that *shouldn't* be there. In Figure 9-10, this would involve making sure you don't see the string "There doesn't seem to be anything here." This is easiest to implement—you just need to be sure you're not getting a specific set of text—but it's error-prone because it won't detect errors that fail to give you the string you're looking for (as in cases of site vandalism.)

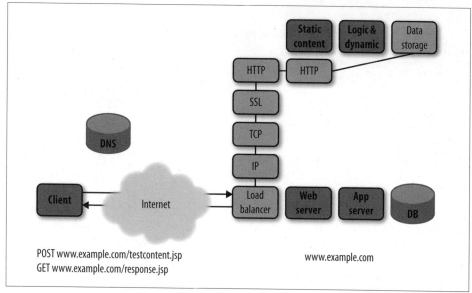

Figure 9-11. To test backend systems, a request must move data to or from data storage

While an HTTP GET for a static page will verify that the network, load balancer, and HTTP service are functioning correctly, it won't include the processing delay that comes from building dynamic pages. To properly measure performance, it's important to simulate what visitors see, even the parts that make the server work. You therefore need to test dynamic pages to measure real site performance.

The time it takes the server to send back the first byte of a response (known as *time-to-first-byte* in many testing reports) is roughly equivalent to the time the server takes to prepare the response, plus one network round trip. Because of this, if you have measurements of static and dynamic pages side by side, you can measure the difference and determine just how long the server is taking to process dynamic content.

Checking data persistence: Database access and backend services

Simply checking a dynamic page may not reveal the delay that backend data services are causing. To test the performance and availability of backend systems, you need to request content that forces the application to communicate with the database, as shown in Figure 9-11.

One way to isolate the database tier so you can test it is to have your developers build a page that exercises the backend of the application, writing data to a record and retrieving it before displaying a message indicating success. This might be a specific URL that, when requested, responds with considerable detail about the health of backend systems.

Here's an example of this kind of page:

```
macbook:~ alistair$ telnet www.bitcurrent.com 80
Trying 67.205.65.12...
Connected to bitcurrent.com.
Escape character is '^]'.
GET backendtest.cgi
<HTML><HEAD></HEAD>
<BODY>
Authentication check: <B>Passed</B> - 60ms
Database read: <B>Passed</B> - 4ms
Database write: <B>Passed</B> - 12ms
Partner feed: <B>NO RESPONSE</B> - N/A
</BODY>
<HTML
```

This is a trivially small response from a server, but it includes tests for authentication, database read and write, and access to a third-party server. You can set up monitoring services to look for the specific confirmation messages and to alert when an error (such as NO RESPONSE) occurs. If your monitoring service captures the page when an error occurs, you'll also have data from the responses at hand.

You can measure the performance of the database tier by comparing the performance of this page to that of a dynamic page that doesn't involve database access, which will help you to anticipate the need for more database capacity. In fact, this approach of creating a custom page that checks a particular backend element, then configuring a synthetic test, can be applied to any service on which your application depends, such as credit card transaction processing, currency conversion, hosted search, and so on. Just be sure to secure the test pages with proper authentication so that the backend service doesn't become a point of attack for hackers.

Beyond a Simple GET: Compound Testing

We've looked at testing various "depths" in the tiers of the web application, from DNS all the way back to the application server and database. Single-object HTTP requests still leave many questions unanswered, though. How long do whole pages take to load? Can users complete multistep transactions? Are page components functioning correctly?

Getting the Whole Picture: Page Testing

The majority of synthetic tests retrieve a page, then all of its components. A standard web page lists all of the components it contains; the testing system parses this page, then retrieves each component. The number of concurrent connections between the client and the server affects the total time the page takes to load, and the result is a "cascade" of objects that ultimately drive the total page load time, as Figure 9-12 shows.

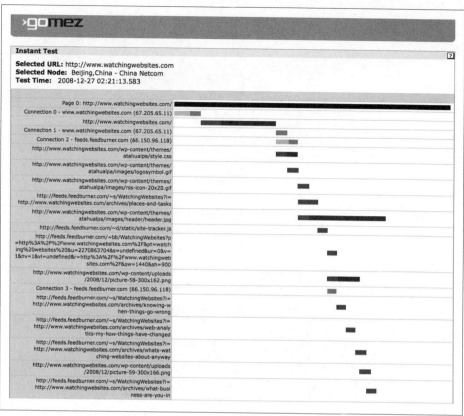

Figure 9-12. A cascade diagram from Gomez showing watchingwebsites.com and all its component objects being loaded from Beijing, China

In the modern Web, a site is only as good as its weakest component. Small components—JavaScript elements, Flash plug-ins, map overlays, analytics, survey scripts, and so on—can limit the speed with which the page loads, or even affect whether it loads at all, as shown in Figure 9-13.

If you're using a compound monitoring service, you'll be testing these components each time you check your own site. Synthetic testing services perform the test from several locations and networks using several browser types, at regular intervals, and report the data in aggregate charts like the one in Figure 9-14.

What if you want to measure a multistep process? It's time to record and replay a transaction.

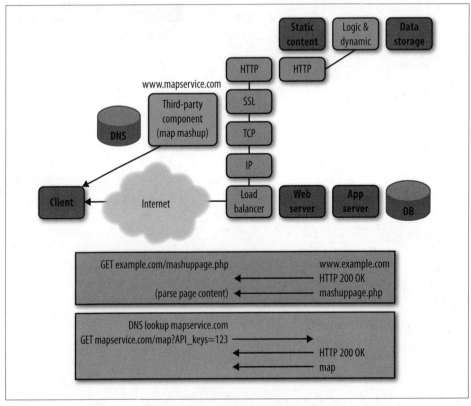

Figure 9-13. A complex page retrieval that includes data from a third-party site such as a mapping service

Monitoring a Process: Transaction Testing

Many synthetic testing services allow you to record a series of transactions and then repeat the sequence at regular intervals. The result is an aggregation of the total time it takes to complete a transaction. Figure 9-15 shows an example of this kind of data for Salesforce.com (*http://salesforce.com*).

This is particularly useful if various steps in a user's experience put load on different systems. Imagine, for example, a purchase on a music site: the visitor searches for a song, posts a review about the band, adds an album to her cart, provides her payment information, and confirms the purchase. By testing a transaction, you see which components are slowest or most error-prone. You can also segment the results to see if certain visitors are having a particularly bad experience, as shown in Figure 9-16.

Figure 9-14. An aggregate performance report resulting from many tests of a website

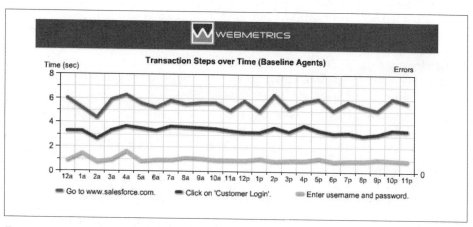

Figure 9-15. A Webmetrics multistep transaction performance report showing the performance of three transaction steps in Salesforce.com

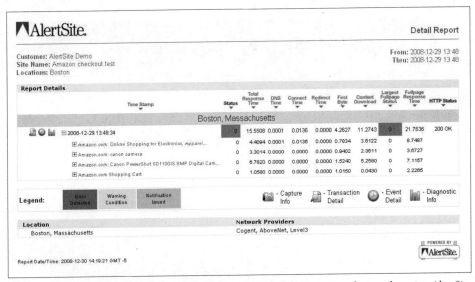

Figure 9-16. Segmenting transaction data by a geographic region or network provider using AlertSite

There are some things you can't monitor with synthetic testing, however. If one step of a process commits to an action, such as charging a credit card for a pair of shoes, the synthetic tests will have all-too-real consequences.

When Bad Things Happen to Good Tests

One publicly traded e-commerce vendor we spoke with described a test that tried to buy a PalmPilot stylus in order to verify its shopping cart process. The company ran the test from dozens of locations, simulating several browsers, every five minutes. Unfortunately, the development team hadn't properly disconnected the test account from the purchasing system. Before this was discovered, the test had purchased hundreds of thousands of styluses, having a material impact on the firm's revenues, requiring the finance team to notify securities regulators of the error.

In another case at a large ISP, the testing team set up a user with the name "test test." The test account hadn't been properly flagged within the provisioning software, resulting in a thousand modems being sent to a new customer named "test."

The moral of the story is that test URLs and test accounts that have been "neutered"— disconnected from the actual transaction engine—are essential for testing, but the creation and maintenance of these accounts must be part of the development process.

One way around this is to work with developers to set up a "dummy" user account and neuter its ability to execute real transactions. However, the synthetic test can't actually verify that the real process works, since it's not using a real credit card. This is one reason it's essential to have both synthetic testing and RUM as part of your EUEM strategy.

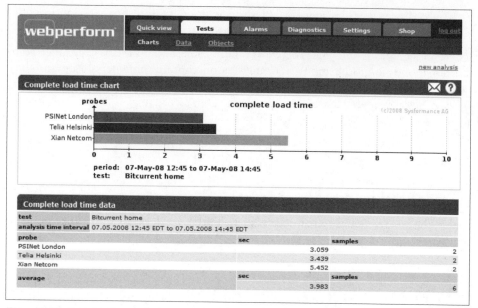

Figure 9-17. Page-level granularity report in Gomez's free Webperform testing service

Data Collection: Pages or Objects?

You've seen how you can probe deep within a site using synthetic testing, and you are now familiar with the way in which synthetic testing systems aggregate HTTP requests into pages and sessions. Different solutions offer different levels of detail in what they report.

Page detail

Some services, such as the one shown in Figure 9-17, report performance only at the page level. They may simply report total page load time, or they may break latency up into its component parts, including DNS latency, time to first byte, connection time, content time, and redirect time.

Object detail

Other services break the page down into its component parts, showing you the latency and problems within the individual pages. This information is more useful to operations and engineering teams, can sometimes lead to quicker answers and resolutions, but also costs more.

Error recording

Some testing services keep a copy of pages that had problems so you can see the error that actually occurred. In addition to showing you firsthand what the error looked like, error recording makes it easier to troubleshoot your synthetic testing setup and identify problems that are causing false alarms, such as mistyped URLs.

There are significant differences between vendors here, and it's the basis of much competition in the monitoring industry. Some vendors capture container objects and their components separately, then reassemble them after the fact. Others capture screenshots of the pages as rendered in a browser, and may even record images of pages leading up to an error.

In other words, you get what you pay for. Ask lots of questions about how errors are captured and reported.

Is That a Real Browser or Just a Script?

One of the main differences between synthetic testing vendors is whether they simulate a browser or run a real browser.

Browser simulation

The simplest way to run an HTTP test is to do so through a script. When you open a command line and use `telnet` to connect to port 80, you're simulating a browser, and you get an answer. Emulating a browser by sending HTTP requests and parsing the responses is an efficient way to perform many tests by simply writing scripts.

Many lower-end testing services rely on this *browser simulation* approach, shown in Figure 9-18.

1. The test service creates simple instructions containing lists of sites and pages.
2. The script interpreter builds properly formed HTTP requests that it sends to the server.
3. The test service examines the resulting responses.
4. The service pronounces the test finished.

While browser simulation is straightforward and consumes few resources on the testing machine, it has important limitations that a second approach—using actual browsers—doesn't face.

Browser puppetry

The other main way to run tests is by manipulating an actual copy of a web browser, illustrated in Figure 9-19, which we call *browser puppetry*.

1. Instead of sending `GET index.html`, the script tells the browser, "Click on the fourth button."
2. The browser then performs the action, which results in a message to the server.
3. The next page is loaded.

4. The script can then examine the browser's DOM to determine whether the request worked.

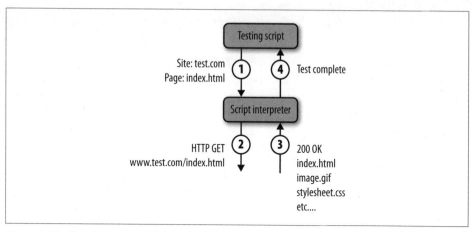

Figure 9-18. How browser simulation works

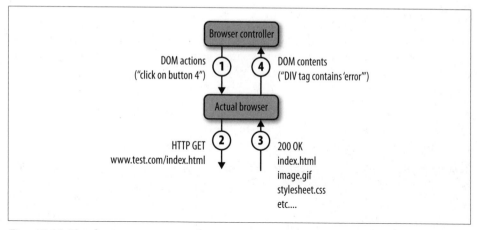

Figure 9-19. How browser puppetry works

This is much more burdensome for the testing service (it must run thousands of actual browsers in memory on its testing platform), but the benefits are significant:

- In a puppetry model, the service can examine all elements of the user experience, including things like text in the status bar, or cookies, by looking at the DOM, rather than just parsing the server's response.

- The puppet browser will include all the headers and cookies in a request that a browser would normally send. Scripted tests require that you manually supply these headers, and also require logic to maintain cookies.

- The puppet browser will automatically handle caching properly, whereas simulation will likely load all objects each time a test is run.

- Forms and complex navigation are much harder to emulate with simulation than they are with a puppet browser, where the test script simply has to say, "Put value A into form B."

The most important advantage of browser puppetry, however, comes from client-side content. In many modern web pages, JavaScript within the page dynamically generates URLs after the page is loaded. Consider, for example, a simple script that randomly loads one of five different pictures. Each time someone visits the site, the container page loads, then JavaScript decides which image to retrieve and builds the URL for that picture.

There's no way to know which image to retrieve without actually running the JavaScript, so this is very difficult for a simulated browser to test. With browser puppetry, the controller simply tells the browser, "Go and get this page." The browser does so, runs the JavaScript, and picks one of the five images. For dynamic, rich websites, puppetry is the right choice.

Configuring Synthetic Tests

Setting up synthetic testing is relatively easy, provided that you have a working knowledge of your web application. To test dynamic features, you'll also need a "dummy" user account that can place orders, leave messages, and log in and out without having an impact on the real world.

Also consider how you name your test accounts and your test scripts. The name "test-script5" will be much less meaningful than "westcoast-retail-orderpage" when you're configuring alerts or trying to tie an outage to a business impact.

Testing services generally bill per test. Four main factors affect how much your bill will be at the end of the month:

Test count
    The number of tests you want to run

Test interval
    How often you want to run them

Client variety
    The number of different clients you want to mimic in terms of browser type, network connection, desktop operating system, and so on

Geographic distributon
    The number of locations you want to test from, both in terms of geography and in terms of the Internet backbones and data centers

You'll also pay more for some of the advanced features outlined earlier, such as browser puppetry, multipage transactions, error capture, and object drill-down.

Test Count: How Much Is Too Much?

You should test every revenue-generating or outcome-producing component of your site, and perform comparative tests of a static page, a dynamic page, and a database-access page. If you rely on third-party components, you may want to test them, too.

That sounds like a lot of tests, which is bad news. Adding tests doesn't just cost more money, it also places additional load on your infrastructure that may impact real users. So you need to be judicious about your monitoring strategy. Here's how.

It's not necessary to run every test from every location. If you already know your site's performance and availability from 20 locations, there's no incremental value to testing the database from each of those 20 locations unless the database somehow varies by region. The trick is to identify *which tests can check functionality from just one location* and *which have to be run from many sources* to give you good coverage.

- *Functional tests* examine a key process, such as posting a comment on a blog. They only need to be run from a few locations. You will have many functional tests— sometimes one test for every page on your site—verified from few locations. They're the things that don't vary by visitor.
- *Coverage tests* examine aspects of your website that vary by visitor segment, such as web performance or a functioning DNS within a particular ISP. You will have few coverage tests—sometimes just a single page such as the landing page—verified from many networks and carriers.

Resist the temptation to run a wide range of functional tests from many locations. You'll make your site slower and your monitoring bill bigger without improving detection or adding to diagnostic information.

Test Interval: How Frequently Should You Test?

Now that you've separated your functional tests from your coverage tests, you need to decide how often to run them.

Availability testing checks to see whether something is working, while performance testing collects measurements for use in trending and capacity planning. Because of this, availability testing needs to run often—every five minutes or less—while baselining tests can run less often.

Problem detection: Availability testing

Operational data is tactical, intended for use by your networking and system administration teams to detect and repair problems, and by your marketing and development teams to verify whether a recent change works as soon as it's released. In the former case, it will be tied to alerting, and in the latter, it will be tied to a particular change, such as a software release or a marketing campaign.

The goal of availability testing is to identify issues and help to localize problems so that you can fix them more quickly. You're likely to change tests whose primary goal is availability testing more often, according to business decisions and release calendars. As a result, you won't be doing long-term trending of this data—a test may only exist in the system for a few weeks before it is replaced with a new one.

Baselining and planning: Performance testing

Performance tests won't change as frequently. You'll keep test results as baselines for months or even years, comparing one holiday season to another, for example. Data from performance-focused tests is aggregated and displayed through dashboards to business decision makers. It tends to focus on availability percentages rather than on individual incidents, and on performance against a service target rather than sudden slowdowns.

You may also use baselining data like this for capacity-planning purposes, helping you to see a gradual slowdown in performance that may require additional servers, and for setting thresholds and alerts in more tactical, operational tests.

Client Variety: How Should I Mimic My Users?

Some desktops are underpowered, which means they take a long time to load and execute JavaScript. Some browsers use older versions of the HTTP protocol or connect across sluggish dial-up links. In other words, not all clients are equal. If you're hoping to understand end user experience, you have to make sure your tests mimic the conditions of your target audience so that the data you're capturing is representative of your end users' experience.

Fortunately, web analytics provides a rich resource for understanding your visitors' environments, letting you define tests that mimic their browsing experience as well as possible.

The folks at *www.eurotrip.com* were kind enough to share their analytics data with us, and we will use it liberally in this chapter. The site's primary focus is on casual European backpackers, and while it is an ad-driven media site, it has some collaborative elements in which visitors share their travel experiences.

Browser type

Figure 9-20 shows that over 80 percent of Eurotrip's visitors are using Internet Explorer and Firefox browsers running on Windows.

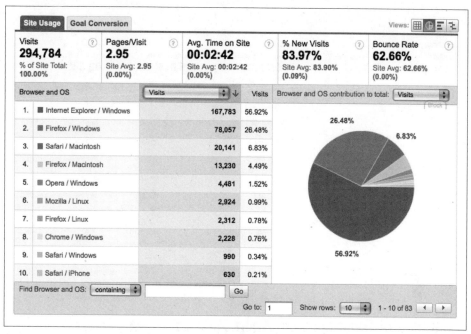

Figure 9-20. Segmentation of Eurotrip.com traffic by browser type

Due to this finding, Eurotrip's tests should probably simulate a Windows-based desktop. If Eurotrip was willing to pay for additional tests, it might also run tests that represent Internet Explorer and Firefox environments.

End user bandwidth

Your analytics tools can estimate what kinds of network connectivity your visitors enjoy. This is a rough estimate at best, because the analytics system can't tell whether delays are related to the core of the network or the edge devices, but it's a good enough approximation for you to use it to define testing.

Some synthetic services will allow you to simulate dial-up connections. Some even have "panels" of thousands of volunteer computers running small agents that can collect performance data from their home networks, businesses, universities, and Internet cafes, giving you an extremely accurate understanding of your site's health from domestic dial-up and residential broadband. This last-mile experiential monitoring can be a good way to determine how an application will function in a new market or new geographic region.

As Figure 9-21 shows, most visitors going to Eurotrip are using DSL or cable modems. With this information, the company can now provision, collocate or request the appropriate carrier types for our synthetic tests.

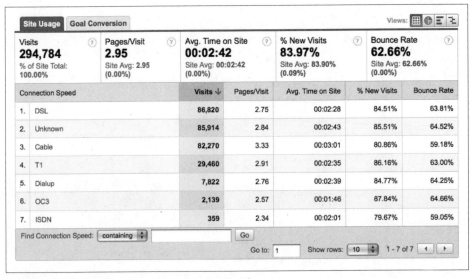

Figure 9-21. *Eurotrip.com users by connection speed*

Geographic Distribution: From Where Should You Test?

Internet latency is often correlated with geographic distance, and you need to test from the places your visitors are based. Knowing what performance is like for visitors from other countries is essential. For one thing, it will help you figure out when an overzealous shark has chewed up a transatlantic backbone on which your users rely. Measuring remote performance is also useful if you're trying to decide whether to deploy a CDN to speed up application performance.

Once again, Eurotrip's web analytics data shows you where its visitors are coming from.

As Figures 9-22 and 9-23 show, most of Eurotrip's visitors are located in the United States, Europe (UK, Germany, the Netherlands, and France), Australia, and Canada. This is consistent with the business goals of the site. If the site were only targeting customers in a particular region, it could ignore testing from other locations and save some money. For example, a used car dealership in San Jose could safely ignore performance from mainland China.

You can get similar data for visits by network or carrier, which may be important if you have customers on one carrier while you're collocated on another. For segmenting performance by carrier, you will want to know whether your tests come from dedicated or shared network connections, and whether they're in a data center or a last-mile location, such as a consumer desktop. You may also need to deploy private testing agents to measure performance from branch offices or important customers if your business needs this data.

Site Usage | Goal Conversion | Views:

Visits	Pages/Visit	Avg. Time on Site	% New Visits	Bounce Rate
294,784	**2.95**	**00:02:42**	**83.97%**	**62.66%**
% of Site Total: 100.00%	Site Avg: 2.95 (0.00%)	Site Avg: 00:02:42 (0.00%)	Site Avg: 83.90% (0.09%)	Site Avg: 62.66% (0.00%)

Detail Level: Country/Territory		Visits ↓	Pages/Visit	Avg. Time on Site	% New Visits	Bounce Rate
1.	United States	116,959	3.27	00:02:58	81.26%	59.52%
2.	United Kingdom	37,172	2.50	00:02:08	88.45%	67.03%
3.	Canada	26,995	4.45	00:03:46	74.42%	54.20%
4.	Australia	11,686	3.55	00:03:59	80.89%	60.03%
5.	Germany	8,109	2.21	00:01:44	91.10%	67.90%
6.	Netherlands	6,052	2.02	00:01:37	91.01%	66.26%
7.	France	5,522	1.96	00:01:26	90.69%	71.35%
8.	Spain	5,360	1.92	00:01:28	91.57%	68.68%
9.	Ireland	4,765	2.05	00:01:28	92.44%	76.79%
10.	Italy	4,759	1.90	00:01:26	89.49%	69.36%

Find Country/Territory: containing | Go

Go to: 1 | Show rows: 10 | 1 - 10 of 199

Figure 9-22. Visits by countries/territories in Google Analytics

Detail Level: City		Visits ↓	Pages/Visit	Avg. Time on Site	% New Visits	Bounce Rate
1.	London	11,578	2.29	00:01:55	87.56%	68.62%
2.	(not set)	8,751	2.62	00:02:39	86.96%	65.52%
3.	New York	3,799	2.82	00:02:12	83.15%	62.83%
4.	London	3,449	2.70	00:02:22	85.79%	63.99%
5.	Sydney	2,965	3.13	00:03:41	82.19%	61.96%
6.	Dublin	2,797	1.78	00:01:12	94.74%	80.51%
7.	Toronto	2,660	3.17	00:03:09	80.98%	61.80%
8.	Melbourne	2,215	3.56	00:03:40	83.16%	62.35%
9.	Vancouver	2,186	3.56	00:03:10	63.91%	47.90%
10.	Amsterdam	2,007	1.92	00:01:38	91.83%	64.42%

Figure 9-23. Top visits by city in Google Analytics

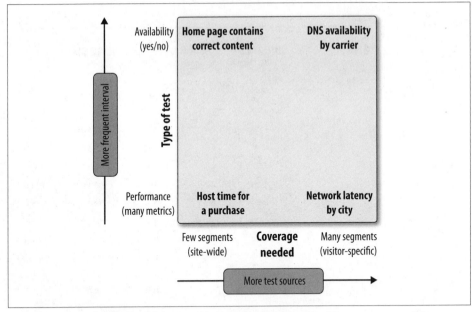

Figure 9-24. Determining test frequency based on the type of test and the coverage needed

Putting It All Together

Knowing which tests are focused on availability versus performance, and which tests need to be segmented for analysis, you can set your monitoring strategy. Some examples of tests across these two dimensions are shown in Figure 9-24.

Now it's time to make your test plan using these two dimensions. For coverage tests, identify the segmentation you need: browser types, bandwidth tiers, geographic regions, networks, and so on. Table 9-1 shows an example of this kind of test plan.

Table 9-1. Determining test frequency based on the type of test and the coverage needed

Test type	Segmentation	Frequency
Functional		Low
	Key pages	Low
	Tiers (i.e. database)	Low
	Third-party components	Medium
	Key transactions	Medium
Coverage		
	Browser types	High
	Bandwidth tiers	High
	Geographic areas	High
	Networks	High

Your plan will allow you to estimate your monthly service costs, as well as the features you need from a synthetic testing vendor.

It should be clear by now that for sites of any size, test management is a full-time job. Tests are created, modified, and retired as the website changes. You need to track test versions and manage which reports are sent to whom, and should treat them as part of the software release cycle.

This is particularly important if those tests rely on specific pass/fail text on a page that may change across releases, or use "dummy accounts" that require maintenance. Tests grow stale quickly, and you'll need to maintain thresholds and alerts as the site's designers change functionality and content.

Setting Up the Tests

Now that you've decided what, where, and how often to test, you need to configure testing. Single-page testing is relatively simple: you provide the URL, the type of test (network, TCP, DNS, and so on), interval, geographic distribution, and any other information that helps the tests to mimic your visitors, as shown in Figure 9-25.

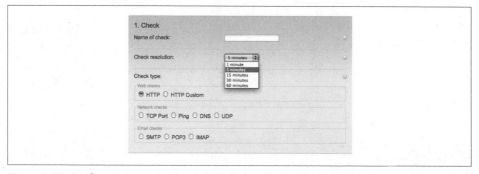

Figure 9-25. Configuring test parameters with the Pingdom testing service

Transactional testing is more complex to configure, and there are many more chances to get things wrong. Fortunately, monitoring service companies often include test recording tools that remove much of the guesswork from setting up tests. Figure 9-26 shows an example of this—you navigate within the target site in a frame, and the recording tool records the steps you take as part of the test.

Once recorded, you can edit the scripts to correct errors made during recording, modify timings, and define what should be used to verify that a page was loaded properly.

Setting Up Alerts

When one of your availability tests fails, it will alert you via email, SMS, or voicemail. Alerts can also be directed to software in the network operations center for escalation

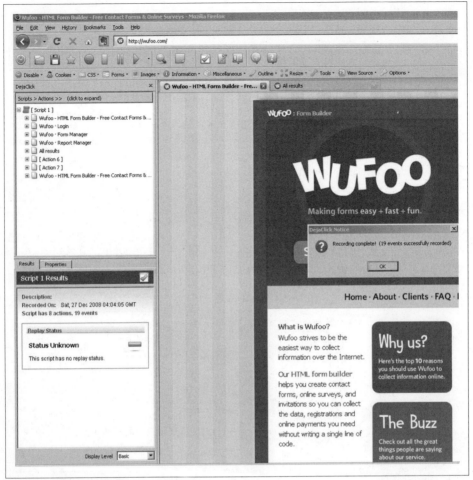

Figure 9-26. Recording a transactional test with Alertsite's test recorder "DejaClick" on the Wufoo.com website

and correlation with device metrics. Some synthetic monitoring vendors also send SNMP traps to enterprise management software tools.

Alerts fall into one of two main categories. *Hard errors* occur when a test encounters a problem, for example, a 404 Not Found message when retrieving a page. By contrast, *threshold violations* occur when a metric, such as host latency, exceeds some threshold for a period of time.

Figure 9-27 shows an alarm configuration screen that defines both hard (availability) and threshold (performance) violations. Setting up a threshold violation alarm requires several details.

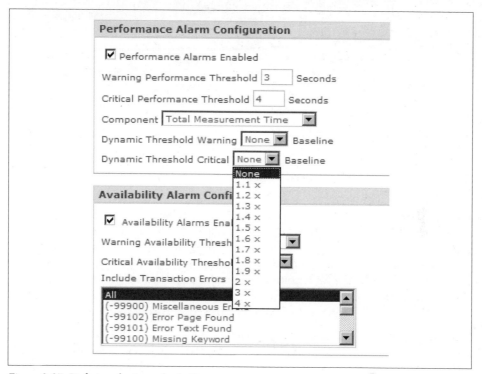

Figure 9-27. Defining alerting rules in Keynote

- The performance threshold. In this example, there are two thresholds, one for "warning" and one for "critical."
- Which metric to look at (in this case, it's the total time to load a page).
- A dynamic threshold violation (this alerts you if the page performance exceeds the baseline for "normal" performance at this time).

For hard error alerts, you simply need to define which kinds of availability problems to notify about. When defining any kind of alert, you may also have to configure rules about suppression (so that the service only tells you when it's broken and when it's recovered, rather than flooding you with error notifications) and verification (so that the service checks several times to be sure the error is real before telling you about it).

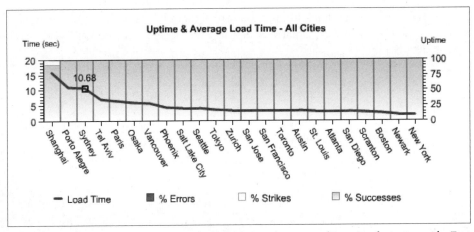

Figure 9-28. Webmetrics transaction performance report by geographic region for a site on the East Coast of North America

A good alerting system lets you sleep at night knowing that if there's an issue with your website (availability or performance), you'll be alerted. That system loses its effectiveness if there are too many false positives, but tuning an alerting system can take time. For example, most sites will exhibit different response times from different geographic locations, even within the same country. Setting a single response time alert threshold across all the geographies from which you're testing from will either cause you to miss critical alerts (if set too high) or get too many false positives (if set too low).

Some services have far more complex alerting systems, including escalation, problem severity options, time zones, maintenance windows during which to ignore downtime, and so on.

Whichever option you choose, your alerts should contain as much business information as possible, such as how many users were affected, the change in website conversions during the outage, or whether the incident happened during a planned adjustment to the site. *The perfect alert contains all of the information you need to localize, diagnose, and repair the problem, and to assess its impact on your business.*

Aggregation and Visualization

We've seen several examples of how synthetic tests report data. Let's look at a few other common representations of end user experience. One way of reporting performance is to segment it by region or carrier to identify the worst-served parts of the Internet, as shown in Figure 9-28.

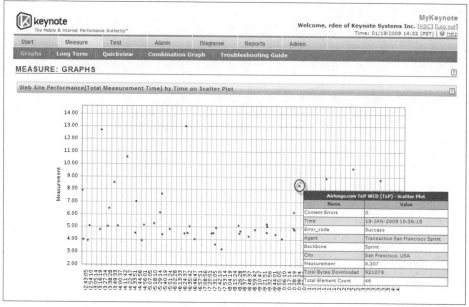

Figure 9-29. A scatterplot of individual test results

Segmented reports show you the slowest or least-available tests along a particular dimension. But what if you want to focus on a particular element of latency? Synthetic tests collect a great deal of data, and you need to look across all of it to properly understand what normal performance is like.

By displaying a metric—such as total transaction time—as a histogram, you can quickly see what kind of performance is most typical. On the other hand, you may not want to aggregate data, but rather see individual tests to better understand a particular period of time. In this case, a scatterplot visualization like the one shown in Figure 9-29 works well.

This kind of report is only available from services that keep the raw test data for considerable lengths of time. Some lower-cost services will aggregate this information into averages to reduce their storage requirements.

Finally, matrix reports are ways of comparing two segments to one another. In the case of Figure 9-30, these segments are locations and websites.

Matrix reports are useful for identifying problem areas at a glance, such as two carriers that aren't forwarding packets to one another well. From a matrix report you can drill into the history of a specific region or site.

As you define and prepare these reports, be sure to share them with managers and members of other monitoring groups, including marketing, QA, and web design, to accustom them to the information that synthetic monitoring can provide.

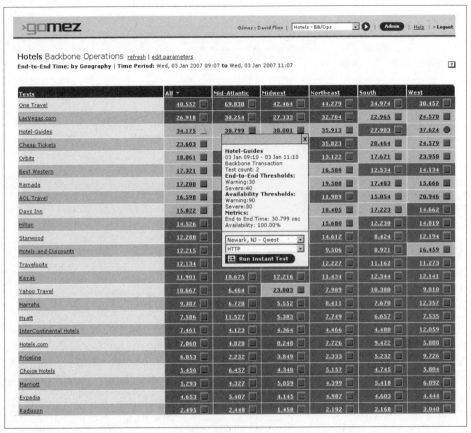

Figure 9-30. Gomez matrix report of websites by region

Advantages, Concerns, and Caveats

Synthetic testing is your first answer to the question, "Could they do it?" It's a reliable, controllable, repeatable measurement that you can use regardless of how much or how little traffic there is on your website. It also has some significant shortcomings.

No Concept of Load

Synthetic testing services don't know how much traffic is on your site. A testing service is blissfully unaware of whether your site is experiencing a deluge of traffic or is so slow that it has idle servers. This means they're missing an important possible cause of web latency, because the more visitors you have, the longer the site takes to respond. Alerts and thresholds on synthetic testing systems can't take into account how busy the site is.

You should, of course, be concerned if your site becomes slow when nobody's using it, but a synthetic testing service won't alert you to that fact as long as the latency is within acceptable limits. Dynamic baselining, in which the service learns what "normal" latency is like at a certain time of day, is somewhat of a proxy for load, assuming your website gets the same loads at the same times of day.

Muddying the Analytics

Synthetic tests generate traffic. If you're running tests on a site, exclude the synthetic tests from overall analytics *before you start testing* or your visitor count will be artificially high, as shown in Figure 9-31.

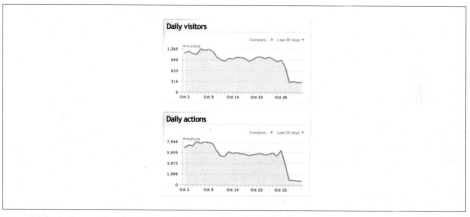

Figure 9-31. The sudden drop in traffic on October 27 is the result of excluding synthetic testing from web analytics measurements

Checking Up on Your Content Delivery Networks

If you're using a CDN to speed up traffic, you should test the performance of retrieving a page from the CDN, but also of retrieving the same page from your own servers. To accomplish this, you'll need a second domain name that bypasses the CDN. By comparing both test results, you'll see how much the CDN is helping with performance and whether you're getting your money's worth.

Rich Internet Applications

Most of this chapter has dealt with monitoring HTML-centric applications. The modern Web is changing that in two important ways. First, more and more applications rely on rich clients written in JavaScript, Flex, and so on. These clients do much of the presentation and display on the end user's computer. Instead of retrieving whole pages, the client retrieves smaller nuggets of information from the server, often structured in XML or JSON. The testing service must be able to emulate these kinds of calls.

This is an area where browser puppetry works better, because there's a real browser executing the tests. Nevertheless, rich clients have destroyed much of the standardization that we enjoyed with traditional HTML-centric, page-by-page web design, forcing us to rethink web monitoring.

In some cases, the user may also install client-side software, such as a streaming video helper or a plug-in that uses a non-HTTP protocol like the Real-Time Streaming Protocol (RTSP) or even UDP instead of TCP. If this is the case, you need to extend your synthetic testing to test these new protocols, and this will limit the synthetic testing vendors you can use.

Video monitoring is a special case that will change which metrics matter to you. Rather than response times, you'll care about startup time (how long video takes to begin) and rebuffering ratio (what percentage of visitors saw a message saying "rebuffering..." while playing the video). You may even measure details like frame rate and effective bit rate that can only be collected within a media player. If you want to measure this synthetically, you'll need "media player puppetry" from your testing provider.

Site Updates Kill Your Tests

Changes to a website are the most common cause of false alarms. Test scripts are inherently "brittle"—when the site changes, the tests stop working properly. It may be a transaction that that no longer follows the same navigation path through the site, a new URL, or a different response that the service interprets as an error.

There are no simple technical solutions to this problem. They're human issues, and you need good processes like change control management to overcome them. Remember, however, that tests using browser puppetry are less likely to falsely report errors when you change a page: if a URL is altered but the button's name remains the same, the test will still work.

Generating Excessive Traffic

Synthetic tests still consume resources on your servers. We've seen bad test plans that caused over 60 percent of all the site's traffic via the testing itself. This is a tremendous waste of money and makes it even harder to identify real users who are having problems amidst the noise.

Data Exportability

Synthetic testing data contains historical information that you may want to use elsewhere. When selecting a synthetic testing platform, you should be aware of how much data the system stores and for how long. This is particularly true if your organization relies heavily on data warehousing or business intelligence systems. Data should be downloadable via XML, CSV, or a real-time feed.

Competitive Benchmarking

Synthetic tests let you keep tabs on your competitors. You can monitor their perform-ance and availability and use it to set your own service-level targets. For SaaS compa-nies, knowing that your competitors are slower or less reliable than you are is an important selling point.

Some synthetic testing companies offer comparative benchmarks that rank members of a particular industry segment against one another.

Tests Don't Reflect Actual User Experience

It's easy to let synthetic traffic lull you into a false sense of security. Often, synthetic tests will be fine, even as users are suffering. This happens for two main reasons:

- Your tests don't simulate the network, region, bandwidth, browser, or desktop properly.
- Users are doing something on the site that you aren't testing for.

This is one of the main reasons that you should always use synthetic testing in con-junction with RUM, which we'll cover in the next chapter.

Synthetic Monitoring Maturity Model

You need synthetic transaction monitoring as early as possible, preferably while you're still building and testing your application. The less real traffic you have on your site, the more valuable synthetic testing is. Once you've got more traffic, you can reduce the number and frequency of your tests and look at actual user performance instead.

Many companies limit the use of synthetic testing to IT only, without considering its broader impact on other departments. As an organization starts to share synthetic data, it shifts from testing machines to testing business processes, as well as using properly structured tests to speed up problem resolution. Eventually, synthetic testing becomes a key performance indicator for marketing campaigns and competitive benchmarking.

Maturity level	Level 1	Level 2	Level 3	Level 4	Level 5
Focus	Technology: make sure things are alive	Local site: make sure people on my site do what I want them to	Visitor acquisi-tion: make sure the Internet sends people to my site	Systematic en-gagement: Make sure my relationship with my visitors and the Internet continues to grow	Web strategy: Make sure my business is aligned with the Internet age
Who?	Operations	Merchandising manager	Campaign manager/SEO	Product manager	CEO/GM

Maturity level		Level 1	Level 2	Level 3	Level 4	Level 5
EUEM	Synthetic	Availability and performance: Checking to see if the site is available from multiple locations, and reporting on performance	Transactions and components: Multistep monitoring of key processes, tests to isolate tiers of infrastructure	Testing the Internet: Monitoring of third-party components and communities on which the application depends	Correlation & competition: Using the relationship between load and performance; comparing yourself to the industry and public benchmarks	Organizational planning: Using performance as the basis for procurement; uptime objectives at the executive level; quantifying outages or slowdowns financially

CHAPTER 10

Could They Do It?: Real User Monitoring

A company once ran a beautiful monster of a marketing campaign.

The campaign was an attempt to drive traffic to its e-commerce site. Shortly after launching the campaign, sales dropped by nearly half. Synthetic tests suggested everything was fine. Web analytics reported an increase in visits, but a huge drop in conversions across every visitor segment. It looked like the campaign had appealed to a large number of visitors who came to the site but didn't buy anything.

Management was understandably annoyed. The official response amounted to, "Don't ever do that again, and fire the guy who did it the first time."

Fortunately, one of the company's web operators was testing out new ways of monitoring end user performance at this time. He noticed something strange: a sudden spike in traffic, followed by the meltdown of much of the payment infrastructure on which the site depended. This payment system wasn't part of the synthetic tests the company was running.

As it turned out, the company had hit upon an incredibly successful promotion that nearly killed the system. So many people were trying to buy that the checkout page took over 20 seconds, and often didn't load at all. Nearly all of the visitors abandoned their purchases. Once the company responded by adding servers, upgrading the payment system, and fixing some performance bottlenecks, they tripled monthly revenues.

It's one thing to know your site is working. When your synthetic tests confirm that visitors were able to retrieve a page quickly and without errors, you can be sure it's available. While you know it's working for your tests, however, there's something you don't know: *is it broken for anyone, anywhere?*

Just because a test was successful doesn't mean users aren't experiencing problems:

- The visitor may be on a different browser or client than the test system.
- The visitor may be accessing a portion of the site you're not testing, or following a navigational path you haven't anticipated.

- The visitor's network connection may be different from that used by the test for a number of reasons, including latency, packet loss, firewall issues, geographic distance, or the use of a proxy.

- The outage may have been so brief that it occurred in the interval between two tests.

- The visitor's data—such as what he put in his shopping cart, the length of his name, the length of a storage cookie, or the number of times he hit the Back button—may cause the site to behave erratically or to break.

- Problems may be intermittent, with synthetic testing hitting a working component while some real users connect to a failed one. This is particularly true in a load-balanced environment: if one-third of your servers are broken, a third of your visitors will have a problem, but there's a two-thirds chance that a synthetic test will get a correct response to its HTTP request.

In other words, there are plenty of ways your site can be working and still be broken. As one seasoned IT manager put it, "Everything could be blinking green in the data center with no critical events on the monitoring tools, but the user experience was terrible: broken, slow, and significantly impacting the business." To find and fix problems that impact actual visitors, you need to watch those visitors as they interact with your website.

Real user monitoring (RUM) is a collection of technologies that capture, analyze, and report a site's performance and availability from this perspective. RUM may involve sniffing the network connection, adding JavaScript to pages, installing agents on end user machines, or some combination thereof.

Three Ways to Watch Visitors

Three of the technologies we cover in this book overlap considerably. We've already looked at web analytics, which records visitor requests and shows you what visitors did. And we've seen WIA, which shows how those visitors interacted with pages and forms. RUM collects performance and availability information from user traffic.

Because these three technologies all capture information about a visitor's session, it's common for vendors to combine analytics, WIA, and RUM features in a single product. So, for example, you might purchase a RUM product that allows you to diagnose a usability issue. While looking at POST parameters for visitor sessions you notice that the form field for "quantity" often contains five-digit zip codes. This is a usability problem, discovered by a RUM solution.

There's overlap in the industry, which inevitably leads to vendors jockeying for position and overreaching their claims in order to compete with one another. Ultimately, any solution that gives you better visibility into end users is a good thing, and hopefully by understanding the technology behind these tools, you can ask smarter questions and choose a solution that's right for your business.

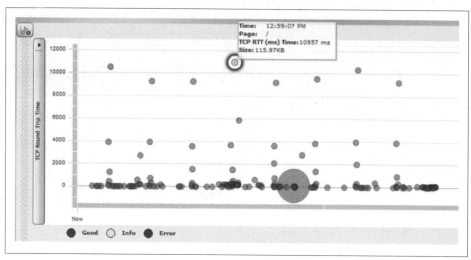

Figure 10-1. A scatterplot of page requests over time in Coradiant's TrueSight, showing relative TCP round-trip time

RUM and Synthetic Testing Side by Side

For this book, we're using a simple distinction between synthetic testing and RUM. If you collect data every time someone visits your site, it's RUM. This means that if you have 10 times the visitors, you'll collect 10 times the data. On the other hand, with the synthetic testing approaches we saw in the previous chapter, the amount of data that you collect has nothing to do with how busy the site is. A 10-minute test interval will give you six tests an hour, whether you have one or a thousand visitors that hour.

Here's a concrete example of RUM alongside synthetic data. Figure 10-1 shows page requests to an actual website across an hour. Each dot in the figure is an HTTP GET. The higher the dot, the greater the TCP round-trip time; the bigger the dot, the larger the request.

While requests happen throughout the hour for which the data was collected, there are distinct stacks of dots at regular intervals. These columns of requests, which occurred at five-minute intervals, are actually synthetic tests from the Alertsite synthetic testing service, coming from Australia, Florida, and New York.

Figure 10-2 highlights these five-minute intervals. The tests from each of the three locations have "bands" of latency—tests from Australia had the highest round-trip time, as we'd expect. Notice that there would have been no data on Australia without synthetic data. Also notice that the only way to discover the excessively large request (the big dot) was to watch actual visitors—there's no way for a synthetic test to detect this. Finally, notice the dozens of requests that happen in those five minutes—an eternity of Internet time.

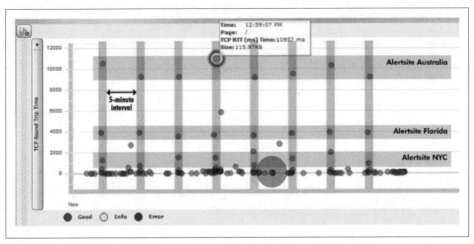

Figure 10-2. The same scatterplot in Figure 10-1, with synthetic tests identified

Synthetic tests give you an idea of what users *might* experience, but RUM tells you what actually happened—whether users could accomplish the things they tried to do. In this respect, RUM is the natural complement to web analytics. However, you cannot use RUM on its own: it's useless if users don't visit the site, because there are no visits to analyze.

How We Use RUM

We've already covered why you need to look at end user experience (Chapter 8), but here are some specific uses of RUM technology:

- Using performance records to prove you met service-level targets with customers
- Supporting users and resolving disputes based on a record of what actually happened
- Speeding up problem resolution with "first-cause" analysis to localize the issue
- Helping to configure and verify synthetic tests
- Defining testing scripts from actual user visits

Proving That You Met SLA Targets

When a service provider and its customers argue, it's usually because they don't have all the facts. Instead, they resort to anecdotes and recrimination, each providing incomplete evidence to support the view that they're right and the other is wrong.

This is especially true for SaaS websites. When you're the service provider, you need to be gentle. If you're in the wrong, you'll be issuing a refund and apologizing soon. If the problem is the customer's fault, you have the opportunity to fix it and help her save face.

You need to know what actually happened, which is where RUM excels. If you have reports on what the end user experience was like, you can tell subscribers precisely where delays occurred.

To make the most of dispute resolution, your RUM solution must segment traffic by the application being used, by subscriber (usually the company that's paying the bill), and by individual user. It must also generate reports by elements of latency and by type of error. By distributing this data to sales and support teams on a regular basis, you'll be well equipped to prove that you did what you said you would.

Supporting Users and Resolving Disputes

While dispute resolution normally happens with aggregate data, call centers work with individuals. If your call center has access to RUM information, call center personnel can bring up a user's session and see what went wrong.

If the problem is on the user's side, you can add the issue to a FAQ, modify trouble-shooting processes to help customers serve themselves in the future, or let the design team know where users are getting stuck, all of which will reduce call center volumes. On the other hand, if the user has encountered a legitimate problem that must be fixed, the session records will be invaluable in convincing engineering that there's an error and helping QA to test for the problem in future releases.

"First-Cause" Analysis

RUM data will seldom diagnose a problem completely—there are simply too many components in a web transaction to pinpoint the root cause by looking at web traffic. Rather, RUM will tell you where in your delivery infrastructure to look. In this respect, it is a "first-cause" rather than a root-cause approach.

Consider, for example, RUM data broken down by type of request: a static image, a static page, a dynamic page, and a page that writes to the database. If there's a sudden performance problem with dynamic pages, it's likely that the application server is to blame. If that data is then segmented by server address and URL, you know where to start looking.

This kind of segmentation is how IT operations teams solve problems naturally. When you adopt a RUM solution, you need to make it an integral part of your problem res-olution workflow and escalation procedures, with the new data in order to reap all of the benefits.

There are, however, emerging end-to-end passive analysis tools, like ExtraHop, that monitor not only HTTP, but other protocols as well, and can drill down into many of the backend systems behind the web tier to help with troubleshooting.

Helping to Configure Synthetic Tests

When you're developing your synthetic test strategy, you need to know what an acceptable response time is. In the previous chapter, we looked at how you can use data from web analytics to help ensure that your synthetic tests are watching the right parts of your site. RUM data can help you determine what the acceptable results should be for those tests.

Imagine, for example, that you want to test the login process. You've got a synthetic test for *http://www.example.com/login.jsp* that you'd like to run. You can take the RUM data for the 95th percentile of all logins, add a standard deviation, and you will have a good estimate of a "normal" performance range for that page.

Chances are, however, that you'll deploy a synthetic testing solution well before you deploy RUM, so a more likely use of RUM is to validate that your synthetic tests are working properly and that your test results match what real users are seeing.

As Content for QA in New Tests

Session records provide the raw material for new tests. Because they record every HTTP transaction, you can feed them into load testing systems to test capacity. What's more, if you share a copy of a problematic visit with the release management group, the release team can add the offending test to its regression testing process to make sure the issue is addressed in subsequent releases.

Capturing End User Experience

Armed with thousands of measurements of individual page requests, you can answer two important questions: could visitors use the website? What were their experiences like?

The first question involves problem diagnosis and dispute resolution. When someone calls with a problem, you can quickly see what happened. You can even detect and resolve a problem before she calls, because your RUM tool has seen the error occur.

The second question involves segmentation and reporting. You can look across all requests of a certain type—a specific browser, a geographic region, a URL, and so on—and analyze the performance and availability of the website for capacity planning or to understand how the site's performance affects business outcomes.

Both questions are vital to your web business. To answer them, you first need to collect all those page measurements. The work you'll need to do depends heavily on which RUM approach you use.

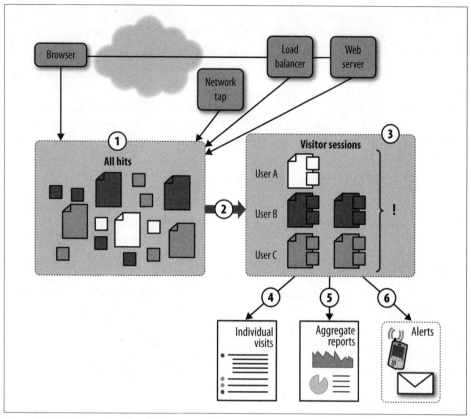

Figure 10-3. The basic steps in all RUM solutions

How RUM Works

RUM solutions vary widely, but always involve the following basic steps, as shown in Figure 10-3.

1. **Capture**. The monitoring system captures page and object hits from several sources—JavaScript on a browser, a passive network tap, a load balancer, or a server and its logfiles.

2. **Sessionization**. The system takes data about these hits and reassembles it into a record of the pages and components of individual visits, along with timing information.

3. **Problem detection**. Objects, pages, and visits are examined for interesting occurrences—errors, periods of slowness, problems with navigation, and so on.

4. **Individual visit reporting**. You can review individual visits re-created from captured data. Some solutions replay the screens as the visitors saw them; others just present a summary.

5. **Reporting and segmentation**. You can look at aggregate data, such as the availability of a particular page or the performance on a specific browser.

6. **Alerting**. Any urgent issues detected by the system may trigger alerting mechanisms.

These six steps seem fairly basic, but as with most monitoring technologies, the devil's in the details. How much information the RUM solution records, and what you can do with it, depends on what information that solution collects, how it decides what's "interesting," and how it stores and retrieves user data.

Server-Side Capture: Putting the Pieces Together

Collecting performance at the server's side of the connection takes a bottom-up approach. First, your RUM tool gathers the individual HTTP objects, then organizes them into pages and visits, then calculates performance metrics. This is the approach many server agents, logfiles, and passive analyzers use, including those from Computer Associates, Coradiant, HP, and Tealeaf.

Often, these objects have something in common that lets the RUM tool group them together. It may be a cookie unique to that visit (known as a *session cookie*) or some other piece of information, such as an IP address or a browser. The RUM tool uses this information to assemble all of the objects into a user's visit.

Within those objects, some are containers (such as those ending in *.html*) and some are components (such as those ending in *.gif*). This allows the RUM solution to identify where pages begin and end, though in practice, it's hard to do this well and there are many rules and tweaks that a RUM tool uses to reassemble a user session properly.

Once a RUM tool has grouped a visit into pages, it uses the timing of the various objects to determine page load time.

Client-Side Capture: Recording Milestones

If you're collecting performance at the browser's side of the connection using JavaScript, you're recording milestones that occur as a page loads in the browser. JavaScript embedded in the page records key events, such as the start of a page or the moment that all the objects it contains have been retrieved from the server. The JavaScript then sends the timing of these milestones and some metadata about the page to a collector.

With this model, you don't need to worry about reassembling the individual objects that make up a page. In fact, you may ignore them entirely. You also won't need to work out which sessions belong to which visitors—after all, you're running on one visitor's browser, so you only see one visit and can mark page requests on behalf of the service to associate them with one another.

Gomez and Keynote both offer this model, and many large websites (such as Netflix, Yahoo!, Google, and Whitepages.com) have deployed homegrown client-side collection. The developers of the Netflix solution have made their work available as an open source project called Jiffy (for more information on how Netflix is instrumenting page performance, see *http://looksgoodworkswell.blogspot.com/2008/06/measuring-user-experience-performance.html*).

You may not have to choose between server-side and client-side collection. Some vendors offer a hybrid approach that collects user experience data at both the client and the server for a more complete perspective on performance.

What We Record About a Page

There are dozens of facts about a web page that you will want to record. All of them are useful for diagnostics, and many of them are good dimensions along which to segment data within reports. These facts include performance metrics, headers and DOM information, error conditions, page content, correlation data, and external metadata.

Performance metrics

You can track the following timing metrics for every page:

TCP round-trip time
 The round-trip time between client and server. This is the time it takes a packet to travel across the Internet from the client to the server and back.

Encryption time
 The time to negotiate encryption (SSL, for example) if needed.

Request time
 The time for the client to send the server a request.

First byte
 The time for the server to respond to the request with a status code and the requested object (known as the first byte or host time).

Redirect time
 The time taken to redirect a browser to another object, if applicable.

Network time
 The time it takes to deliver the object to the client.

TCP retransmissions
 The number of TCP segments that were lost (and had to be retransmitted) during the delivery.

TCP out-of-order segments
 The number of TCP segments that arrived out of order and had to be reordered by the client.

Page render time
 The time it takes for the browser to process the returned object—generally, the time taken to render a page.

Application-specific milestones
> The start of a rich media component, the moment a visitor drags a map, or other timing events specific to a particular application that are recorded as part of the page's performance.

End-to-end time
> The total time taken to request, receive, and display a page.

Visitor think time
> The time the visitor takes once the page is loaded, before beginning the next action.

Some of this data may not be available, depending on how you're capturing user experience. Page render time, for example, is something only client-side RUM can measure. Similarly, low-level networking statistics like TCP retransmissions aren't visible to JavaScript running on the browser.

Headers and DOM information

Every web page—indeed, every HTTP object—includes request and response information in the form of headers sent between the browser and the web server. Because HTTP is an extensible protocol, the number of possible headers is unlimited. Most RUM solutions will collect the following data about the request:

Browser type (user agent)
> The browser requesting the object.

URL
> The requested object.

POST or URI parameters
> Any data the browser sent to the server.

Referrer
> The referring URL that triggered the browser's request.

Cookies
> Any cookies the browser sent that it acquired on previous visits.

The server's response contains additional information about the object being delivered, which the RUM solution can capture:

MIME type
> The kind of object that the server is returning.

Object size
> The size of the requested object.

HTTP status code
> The server's response (200 OK, 404, etc.).

Last-modified date
> The time it takes for the browser to process the returned object—generally, the time taken to render a page.

Compression type
> Whether the object is compressed, and if so, how.

MIME type
> What type of object is being delivered. This helps the browser to display the object.

Response content
> The object itself, or specific strings within the object.

Cachability information
> Details on whether the object can be cached, and if so, for how long.

New cookies
> Any cookies the server wants to set on the browser.

Metadata is important. If the server says an object is big, but the actual object is much smaller, it's a sign that something was lost. Similarly, an unusual MIME type may mean that content can't be displayed on some clients. As a result, RUM tools often capture this kind of data for segmentation ("What percentage of requests are compressed?") or to help with diagnostics.

Server-side RUM tools will collect this data from actual requests, while client-side tools will assemble it from the browser's DOM, so the metadata you can collect will depend on how you're capturing page performance.

Error conditions

RUM tools are on the lookout for everything from low-level network errors to bad HTTP status codes to specific content in a page. Most watch for a built-in list of error conditions. Again, the RUM approach you choose will determine which errors you can detect. In particular, client-side RUM can't detect errors that happen before the client-side JavaScript has loaded (because the monitoring tool isn't capturing data yet), so it's used less often than server-side RUM for error detection.

Page content

Many pages your website serves contain additional context about the content of the page. Perhaps it's the price of a purchase, the type of subscriber (for example, "premium"), or maybe the visitor's account number.

Some RUM tools extract this kind of information from pages and associate it with a visit. With a server-side RUM tool, you specify which HTML content it should extract as it goes by, using an interface like the one shown in Figure 10-4. Every time page content matches those specifications, the RUM tool captures the content and stores it with the page.

One common piece of data to extract from page content is the title of the page. Most HTML objects have a `<Title>` tag that provides the name of the page in plain language. This name is often different from and more readable than the page's actual URL. Instead of talking about "page4.aspx," you're now discussing "the checkout page." If you

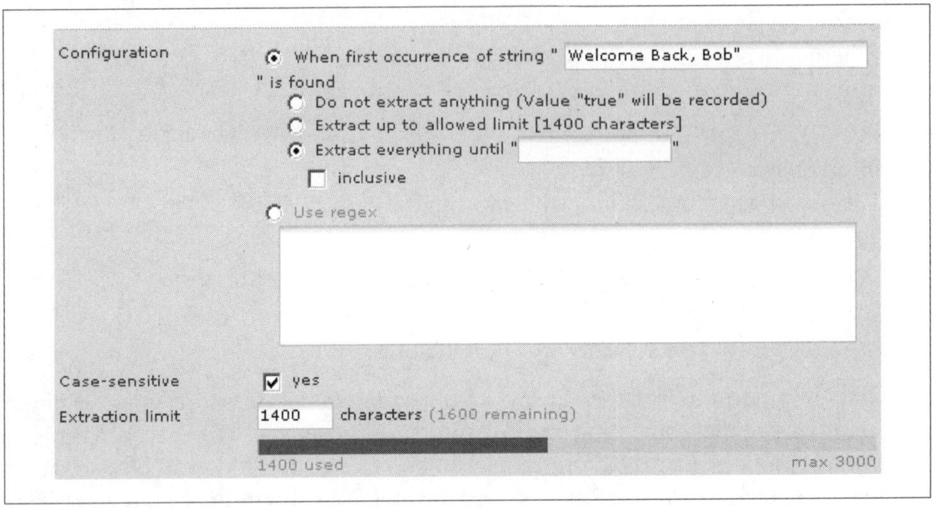

Figure 10-4. Configuring content extraction in Coradiant's TrueSight

capture additional page metadata, such as the total amount spent, you can make the visit record even more relevant and tangible.

Page content is also useful for segmentation, allowing you to ask questions like, "How much did sales go down for users whose transactions took longer than five seconds?"

If you're using client-side RUM collection, metadata collection happens differently. The JavaScript used for performance instrumentation can also extract information from the DOM, such as the page title or some text in the page, and include that data in its report on the page's retrieval, just as an analytics tool records page tags.

Some tools can even capture the entire page rather than just specific strings. This lets you search through pages after the fact, and is useful for problem diagnosis, particularly when problems are related to content or to user input (Figure 10-5).

Capturing all page content also lets you search across all visits for specific occurrences of a problem. We know of a case in which a disgruntled employee vandalized a site's pages on his last day on the job. When an outraged customer first reported the problem to the call center, the company had to determine how many others had seen the changed content. Nobody knew to look for the string in question until it had happened, but being able to search through historical data to find all instances of the string of expletives the employee had left let the company defuse the situation.

Correlational data

If you're planning to combine your RUM data with other data sources, you may need to extract strings that will help you correlate those sources.

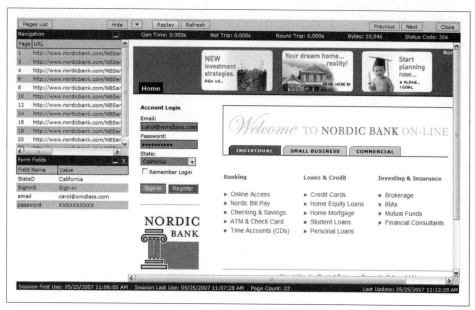

Figure 10-5. Replaying an entire user visit from a complete HTML capture with Tealeaf

- *Timing information* is the most common form of correlation. It lets you line up RUM with synthetic tests and other forms of analytics, such as conversion rates. For this, the RUM tool needs to synchronize its clock with that of the other data collectors. You can then merge records of user visits by timestamp to understand how user experience affects conversions.

- *Visit-specific information* (such as session cookies) is even more useful, because it lets you associate an individual visit with other systems, such as a record of an individual visit in a WIA tool or an individual customer survey in a VOC tool. This also lets you segment individual conversions (in analytics) by the actual experience that visitor had on the site (in RUM).

- *Personally identifiable data*, such as an account number or full name, can help you track customers as they move from a website to a call center or even a retail outlet. With this data, you can bring up a visitor's online session when he calls support or sales, and offer better service by seeing what happened, just as you do with WIA tools.

Actually joining RUM data to other information in a data warehouse is quite another matter, which we'll address in Chapter 17, but it's wise to collect correlation information ahead of time in case you need it in the future.

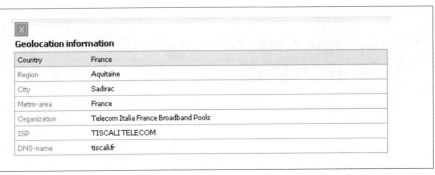

Figure 10-6. Additional visit metadata based on a visitor's IP address

External metadata

Some RUM packages look up the visitor's IP address in a database and include the municipality, country, and carrier in the session log, as shown in Figure 10-6. While not always accurate, this gives you some insight into where users are coming from.

Many of the databases used for geographic lookup also return the owner of the IP address, which will either be a company or a service provider. Service provider information helps to diagnose networking issues such as peering point congestion, since segmenting by service provider can reveal a problem that's happening in one carrier but not others.

Figure 10-7 shows how the various elements we've just seen can be extracted from a page in a visit.

In this figure, the delivery of a page (and its components) results in a record of the page request that includes:

- Performance data, such as host time, network time, and SSL latency based on the timing of specific events.
- Metadata from within the browser's DOM about the visitor's environment and the HTTP request/response.
- External metadata, like IP address and world time.
- Specific data that can be used to correlate the request with others, such as timing, address, and session cookies.
- Geographic and carrier information based on the IP address of the request.
- Data extracted from the content of the page, including the name of the page ("Checkout"), the identity of the visitor ("Bob Smith"), and the value of the page ("$158.40").

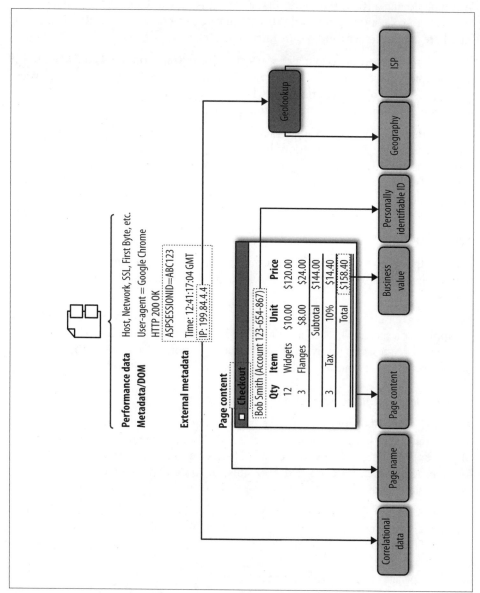

Figure 10-7. Extracting performance information and metadata from a page request

Deciding How to Collect RUM Data

Having information on every user's visit is tremendously useful, both for troubleshooting individual incidents and for determining whether your website is living up to its promises.

When deploying a RUM tool, your first decision is how to collect all this data. Your approach to collection significantly affects what you can do with your RUM solution, and there are advantages and disadvantages to each approach.

We've seen that there are two major ways of collecting data: on the server side and on the client side. Server-side collection approaches include server logging, reverse proxies, and inline sniffers or passive analyzers. Client-side collection approaches include desktop agents and JavaScript instrumentation in the browser.

Much of this collection technology resembles the monitoring approaches we looked at for a web analytics implementation, but it's focused more on performance measurement. Consequently, passive analysis approaches are more common in RUM than they are in web analytics because they can collect network timing information and detect failed requests.

Server Logging

Web server logs give you only basic information about visitor performance. You'll have the timestamps at which each object was requested, and basic data like what was requested.

You may have more advanced logging, either through specialized software or within the application server. This can tell you about key milestones of an object request, such as when the request was received, when the server responded, and when the server was finished sending the object.

How server logging captures user sessions

A logging agent sits between the server's operating system and the application container. Some open source RUM tools, such as Glassbox, track each call to a Java Virtual Machine (JVM) and can provide a detailed hierarchical analysis of an application call to see exactly which function had problems, or to determine which database table caused a slowdown.

Server-side RUM tools like Symphoniq, which also rely on client-side JavaScript, can correlate end user experience with platform health, which allows them to span the gap between IT operations and end user experience.

How server logging captures timing information

Logging agents on application servers can time requests to the application from the network, as well as database requests or other backend transactions. Most server agents are more focused on the performance of the application tier (breaking down delay by component or query) than they are on reassembling a user's session and overall experience.

Server logging pros and cons

Logging is essential for application administrators, and forensic analysts may require logging to detect fraud or reproduce incidents. But it's not a popular indicator of end user experience unless it's combined with a client-side monitoring approach.

Here are some of the advantages of using server logging:

- It runs on the server.
- It can capture server health data (CPU, number of threads, memory, storage, etc.).
- It can get granular information on transactions within the application container/ JVM.
- It can include backend transaction latency (such as database calls).

However, server logging has some important limitations:

- It consumes server resources.
- Aggregating logfiles across servers is always problematic.
- A single visitor may hit multiple servers, making records incomplete.
- Servers have limited visibility into WAN health since they're behind load balancers.
- You can't see the ultimate rendering of the page that the end user sees.
- It doesn't see CDN performance.
- It can't measure mashups.
- When the server's down, so are the logfiles that could otherwise tell you what broke.

Reverse Proxies

Reverse proxy servers are located between the web server and the client and can be used to monitor end user experience. While this approach has fallen out of favor in recent years because it adds a point of failure to infrastructure, many load balancers behave much like reverse proxies and may have a role to play in performance monitoring.

How reverse proxies capture user sessions

A reverse proxy server terminates client requests and forwards them on to servers. Similarly, it terminates server responses and sends them to the client. It may respond to some requests, such as those for static images, on its own to offload work from the servers. Because it's terminating connections, it is also the endpoint for SSL encryption, so it may have access to data in plain text that is encrypted on the wire.

The result of reverse proxy data collection is a log of HTTP requests that resembles that of a web server, although some proxy servers offer more granular information that yields better visualization and analysis.

How reverse proxies capture timing information

Reverse proxy servers that record timings track milestones in a connection. The incoming request from the client, the status code response from the server, the first byte of the object the server sends, and the end of object delivery are used to calculate the performance of a page.

Because a reverse proxy is between the client and the server, it can measure the network health and performance of *both* ends of a connection. In other words, it may have two sets of TCP round-trip time information, one representing the Internet connection to the client and one representing the LAN connection to the server.

Reverse proxy pros and cons

Reverse proxies are servers in the middle of a connection. Unless they have to be there, they're probably another point of failure and delay for you to worry about. If you have a load balancer with logging capabilities, however, this may be an option you can use.

Reverse proxy collection provides the following advantages:

- It sits in the middle of the connection, so it sees both perspectives.
- If the proxy is already terminating SSL, it may simplify the monitoring of encrypted traffic.
- It may already be in place as a load balancer.
- It may be able to inject JavaScript, simplifying client-side instrumentation.

Some of the disadvantages of using a reverse proxy include:

- It introduces an additional single point of failure.
- It may introduce delay.
- It can't see the ultimate rendering of the page to the end user.
- It doesn't see CDN performance.
- It can't measure mashups.
- It may be a point of attack or vulnerability, and represents one more server to worry about.
- It's difficult to diagnose problems when the proxy is the cause.

Inline (Sniffers and Passive Analysis)

While reverse proxies actually intercept and retransmit packets across a network, there's another way to sit between the browser and the web server that doesn't interfere with the packets themselves: sniffing. This approach uses either a dedicated device (a *tap*, shown in Figure 10-8) or a spare port on a network switch (known as a *SPAN port* or *mirror port*) that makes a copy of every packet that passes through it.

Collectively, these approaches are known as *inline capture* or *passive analysis*.

Figure 10-8. A network tap that makes a copy of all traffic flowing through it.

Should I Use Taps or Switches to Capture?

While network taps and mirror/SPAN ports both work well, there are times when you'll want to use one versus the other.

Connecting a tap requires that you disconnect the network cable. This results in downtime, even if only for a moment, which means you'll need to install monitoring during a maintenance window. Taps can also be expensive, consume a power socket on your rack, and in some cases may require configuration. However, if a network tap fails, it simply turns into an expensive piece of wire along which traffic can still flow. Once implemented, taps require little or no maintenance.

On the other hand, mirror ports use up a port on a switch that might be put to better use. If you paid a lot for your switch, or if it's full, you may not be able to use a network port for monitoring. You can implement a mirror port without disconnecting anything, so this approach is easier to set up without disrupting operations, and mirror ports don't use up another power outlet. Under heavy load, older switches won't copy all traffic to the mirror port, however; fortunately, most modern switches can make a copy of traffic at wire speeds without breaking a sweat.

The biggest issue with mirror ports is that they often change or go down due to networking teams and upgrades. It's hard for a team of analysts that doesn't own the routers to know what's going on when a mirror port suddenly disappears. Consequently, we recommend using mirror ports if you need to test something for a few weeks, and using network taps if you're implementing a more permanent solution. It's always a good idea to have a network tap in front of your load balancers just in case.

Sniffing traffic is a common practice for networking professionals. They rely on sniffers to detect virus traffic, even when there's no trace of an infection on a server. They use them to pinpoint Ethernet issues or to figure out which applications are running on a LAN. In fact, they use them anywhere they need to know what's really going on across a network connection.

In recent years, they're using enhanced versions of sniffing technology to measure end user experience. This approach, called *inline RUM*, is sometimes referred to as passive analysis because the monitoring doesn't generate any additional traffic on the network (as opposed to a synthetic monitoring approach, which is "active" because it generates traffic).

How inline devices capture user sessions

A tap or a mirror port copies traffic indiscriminately—every packet on the active network that's tapped is copied onto the monitoring connection. This means any inline RUM solution has to be good at blocking out nonweb traffic and at reassembling individual packets into web pages at wire speeds.

How inline devices capture timing information

To capture per-session timing, the device watches for important milestones—the start of a page request, the delivery of the last object, and so on—and uses timers and TCP/IP sequence number information to calculate latency. Because the device can see when packets arrive at and leave the server, it can measure extremely precisely. In fact, some inline RUM tools aimed at the financial industry such as SeaNet (*www.seanet-tech.com*) can report trade notification timings to the microsecond.

Inline device pros and cons

Inline RUM is precise and powerful, and sees problems even when the servers are broken or the pages aren't loaded in a browser. It can also be expensive—we're talking about network equipment, after all—and you probably won't be able to use it if you don't own your own hardware, because you're hosted by someone else or running your application in a cloud environment. It also doesn't see the end user's ultimate experience, because it's not on the browser itself.

Inline devices provide the following benefits:

- They don't lie: what you see on the wire is what happened.
- They are transparent, so there is no load on clients or servers.
- They do not present a point of failure in the network.
- You can upgrade and modify them without a maintenance window.
- They see performance even when the web page doesn't load.
- They work for any HTTP request, even when JavaScript isn't executing (mobile devices, RSS feeds, RESTful APIs, etc.).

Of course, inline capture devices have some important shortcomings:

- They are more expensive than other options.
- They require physical deployment in a network you control.
- They can't see the ultimate rendering of the page to the end user.
- They don't see CDN performance.
- They can't measure mashups.
- They have a hard time reassembling pages when the page contains RIA components such as AJAX or Flash, which may make additional HTTP requests at any time.

- They capture huge amounts of data, so storage may be an issue.
- They require a copy of the SSL key when sniffing encrypted traffic.
- You must ensure that security and compliance officers are OK with deployment because you're collecting data that is potentially sensitive.

Agent-Based Capture

One way to collect end user experience data is to put an agent on the user's desktop. This agent can see every aspect of application use, not just for the web application, but also for other applications. Want to know if the user's playing Minesweeper while she's on your site? Client agents will tell you. They've got access to the client's operating system, too, so agents know how healthy the network is and how much CPU resources are being used.

Unfortunately, you probably can't use them.

How agents capture user sessions

Agents are software applications installed on client desktops. They're used almost exclusively in enterprise applications, where they're part of company-wide management platforms that handle everything from antivirus updates to backup systems. They sit between the operating system and the applications, watching traffic between those applications and the operating system's resources.

Aternity, for example, makes desktop agents that track traffic flows to and from applications and that summarize the data and look for exceptions before sending performance metrics back to a management console.

How agents capture timing information

Agents see messages flowing in and out of applications. They can watch for specific strings or for operating system events (such as a window opening or a mouse click). They can also watch for network events like a new DNS lookup or an outbound HTTP request. Agents keep track of the timing of these events, as well as key operating system metrics.

Agent pros and cons

Agents see everything, but you need to own the desktop to install them. If you are able to make use of agents, you can take advantage of the following:

- They provide the best visibility into what the user is really doing.
- They can see system health information (CPU, memory).
- Much of the instrumentation work is done by the client, so this approach scales well as the number of users grows.

Agents have the following disadvantages:

- To use them, you will require access to the end user's desktop, so they are a non-starter for most Internet-facing web applications.
- They cannot see the network outside the end user LAN segment, so IP addressing, packet loss, etc., may be incorrect.
- They require different software for different operating systems (Linux, Windows, OS X, etc.).
- They slow down the client.
- Agents must be maintained by IT.

JavaScript

JavaScript changed the web analytics industry, and now it's transforming RUM. JavaScript-based monitoring sees what the user sees. This means it has a better view than any other web monitoring technology into the final assembly of the page, which may include client-side logic, plug-ins, and so on. It's the only way to capture the performance of mashups and third-party content.

What's more, JavaScript code can access everything the browser knows about the session and the user. This includes data such as cookies stored from previous visits or data on the number and size of browser windows. You can use this information to augment user performance with business and visitor context.

How JavaScript captures user sessions

JavaScript RUM begins with page instrumentation, just as web analytics does. You insert a snippet of JavaScript into your web pages or use an inline device like a load balancer to inject the snippet into pages as they're served. Either way, the visitor downloads a monitoring script that runs on the client.

The script records milestones of page arrival, and then sends performance metrics to a collector—a third-party service, a server, or the inline device that injected the script initially. To do this, the script requests a small image and appends the message it wants to send to the collector as a series of parameters to the URL. This is similar to JavaScript used for web analytics; in this case, however, the message's parameters contain performance and availability information.

Imagine that you're using performance monitoring service Example.com. Your JavaScript watches the page load, and at the end it determines that there were eight objects on the page and that it took 3.5 seconds (3,500 milliseconds) to load.

It then sends a request similar to the following:

```
http://www.example.com/beacon.gif?loadtime=3500&objectcount=8
```

The monitoring script doesn't care about a response—the use of a tiny image is intended to make the response as small as possible. The RUM system now knows that a page loaded, that it had eight objects, and that it and took 3.5 seconds.

The rest of the work, such as reporting, aggregation, and data storage, happens on the RUM service or appliance that received the request for the small object.

How JavaScript captures timing information

Recall from our earlier discussion in Chapter 5 that JavaScript is an event-driven language. To instrument a page as it loads, a monitoring script starts a timer and marks off the moments when important events occur. The first important event is the moment the page loads, and to capture this, the first part of the script appears right at the top of the page. This is the "first byte" time.

As the page loads, the browser generates other events, such as the onLoad event, which signifies that all objects have loaded. Simply by knowing the time the page started and ended, we can determine a useful performance measurement—how long the page took to deliver, otherwise known as *network time*.

Most JavaScript measurement happens in a similar fashion. Using the system's time (known as *epochtime*), measurements are determined by calculating the elapsed time between two events.

There's a problem, however. JavaScript is page-specific. When you load a new page, you load new JavaScript. There's no way to start a timer on page A (when the user clicks a link) and then stop the timer on page B (when the page loads), because everything related to page A ends when page B is loaded in its place.

There are good security reasons for this. If JavaScript didn't work this way, someone who'd instrumented site A with analytics could watch everything users did for the rest of their online time, even after leaving site A. This feature of JavaScript provides security and privacy to web users at the expense of being able to monitor their page performance.

Fortunately, developers have a way around the problem that doesn't undermine security. When a user is about to leave page A in a visit, the browser fires an event (onBeforeUnload) telling JavaScript that it's about to get rid of the current page and load a new one. JavaScript stores the current epochtime in a cookie, which is then available for the newly loaded JavaScript on page B.

JavaScript uses a cookie to store the time at which the user clicked the link. The script on page A effectively passes that start time to the script on page B, where it can be used to calculate the elapsed time—how long it took the server to receive and respond to the click that launched page B.

Despite its appeal, JavaScript still has many problems. Timing through JavaScript is more complex than for other collection models that time things independently of page loads, because *the JavaScript that monitors performance is itself part of the page being monitored.*

A recent initiative, called Episodes, addresses several of these problems.

JavaScript pros and cons

JavaScript sees everything from a user's perspective—when it's loaded properly—including third-party content and mashups. However, implementing it is usually vendor-specific, making switching services difficult. Furthermore, JavaScript can't see outside the browser's sandbox.

The following are advantages to using JavaScript:

- It sees all objects from all locations, so it's good for mashups and sites coming from CDNs.
- It sees client-side delay, so it knows when scripts or plug-ins are causing problems, and it measures "perceived render time."
- It knows exactly what the components of a page are.
- It can instrument user actions (clicking play on a video, for example) and make them part of the timing.
- It works in cloud computing and managed hosting environments because there's no need for access to servers and no hardware to install.

JavaScript still has some key limitations, however:

- If the JavaScript isn't loaded, you don't get any data, so it's not good for diagnosing problems.
- Power users may skip a page before JavaScript can run, resulting in gaps in monitoring.
- Using JavaScript increases page size and delay.
- It doesn't work for documents (PDF), RSS feeds, some mobile devices, or anywhere that there's no JavaScript being executed.
- Additional coding is required to instrument events beyond those in the DOM.
- It can't see anything outside the browser sandbox (TCP round-trip time, out-of-order segments, public IP address, etc.).
- It can't measure the server delay on the very first page of a visit because it lacks a timer from the previous page—there is no "previous page."
- It must be maintained along with other software within the web page, and is subject to release cycles and QA.
- It may introduce some privacy concerns, similar to web analytics, causing users to block third-party scripts.

JavaScript and Episodes

An effort by Steve Souders of Google (and the author of YSlow while at Yahoo!) may address the just described issues and give us an industry-wide approach to performance monitoring for rich Internet applications.

To understand Episodes, let's first look at the limitations of JavaScript monitoring today.

onLoad is not a reliable indicator of page load time
> Many applications aren't ready for the user at precisely the moment the browser's onLoad event occurs. Some are ready before the onLoad event because they've carefully loaded what users need first. Others have additional code to execute before pages are truly complete and ready for users. In both of these cases, we require a way to report a "truly ready" event.

No standardization of important milestones
> Browsers support a limited number of timing milestones. Modern websites have unique milestones, such as the moment when video starts playing. Coders must write code to generate their own timing data to mark these milestones, so there isn't an easy way to compare the performance of two sites. The result is many proprietary definitions of timing. There's also no consistent way to track timing of the user's click on the preceding page, forcing coders to resort to cookies to store click time.

What is monitored and how it's collected are intertwined
> This is by far the biggest problem with JavaScript-based RUM today, and it's the one Episodes fixes most cleanly.

> In Chapter 6 we saw how stylesheets separate web design from content, making it easier for a designer to change the color of a heading across an entire site with just a single stylesheet change. Stylesheets are an example of specialization in web design: developers can code the application and make it visually appealing, while authors can focus on content.

> A similar problem exists with proprietary RUM approaches. The person who builds the application is not the person who's in charge of monitoring it. The developer knows which important milestones exist in the page—the rendering of a table, the loading of a video, or small messages back to the server. At the same time, the person monitoring the application knows what he wants to watch.

> Unfortunately, to monitor an application with JavaScript today, many developers are forced to design not only what is monitored, but also how it's reported back to a service for analysis. The timing of the page, the metrics to report, and the mechanism for reporting them are all intertwined in much the way content and formatting were with HTML in the early years of the Web. As Steve Souders says (*http://stevesouders.com/episodes/paper.php*), "There are drawbacks to the programmatic scripting approach. It needs to be implemented.... The switching cost

is high. Actually embedding the framework may increase the page size to the point that it has a detrimental effect on performance. And programmatic scripting isn't a viable solution for measuring competitors."

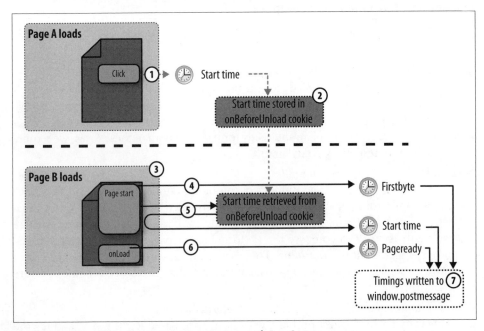

Figure 10-9. How Episodes captures page timings with JavaScript

Episodes does for EUEM what stylesheets did for web design: it provides a model in which the developer defines milestones and measurements, but one in which those measurements can be collected independently by someone in charge of operations and monitoring.

How Episodes works

Figure 10-9 shows how Episodes works, particularly Steve Souders' *episodes.js* reference application.

Monitoring of a page's performance begins when the visitor leaves page A.

1. The monitoring script records the current time (Starttime) when the visitor clicks a link.

2. Starttime is stored in a cookie on the browser.

3. When page B loads, it includes a script near the start of the page.

4. That script records the current time as soon as it runs (which approximates the first byte of the page) in the DOM (in `window.postmessage`), calling it "Firstbyte."

5. The script also retrieves the Starttime left by the previous page from the locally stored cookie.

6. At the end of the page (the `onLoad` event) it records the Pageready timing. It may also record custom events the application developer wants to track (such as the start of a video). By measuring the elapsed time between these milestones, other timings (such as server time and network time) can also be calculated.

7. All of this information is stored in `window.postmessage`, where any other tool can receive it.

A browser plug-in could read the contents of that space and display information on timings. A synthetic testing site could grab those timings through browser puppetry and include them in a report. And a JavaScript-based RUM solution could extract the data as a string and send it back to a RUM service.

Where Episodes really shines, however, is in operational efficiency. So far, the developer has simply recorded important milestones about the page's loading in a common area. If the page changes, the developer can just move the snippets of code that generate Episodes milestones accordingly. If new functions (such as the loading of a video) need to be measured, the developer can publish these new milestones to the common area.

As a result, switching RUM service providers is trivial—just change the script that assembles the milestones and sends them to the RUM service. There's no need to change the way developers mark up the events on the pages. In the same way CSS separates the page's meaning from its formatting, Episodes changes the page's functional timings from the way in which they are collected and reported.

Episodes proposes several standard names and timings, as shown in Table 10-1.

Table 10-1. Episodes names and timings

Metric	What it is	How it's calculated
Starttime	The moment the previous page unloads (approximates a user's click)	Stored in the onBeforeUnload cookie by the preceding page's JavaScript and retrieved by the current page's script when loaded
Firstbyte	The moment the content is received from the server	Measured when the browser executes an Episodes message near the top of the page
Frontend	The time it takes for the browser to get the page prepared once a server response has been received	Pageready – firstbyte
Backend	The time it takes the server to prepare and send the content	Firstbyte – starttime
Pageready	The moment at which the page is ready for the user	Browser's onLoad event
Totaltime	The total time it takes a page to load	Pageready – starttime

You can calculate custom timings from new milestones and these default ones.

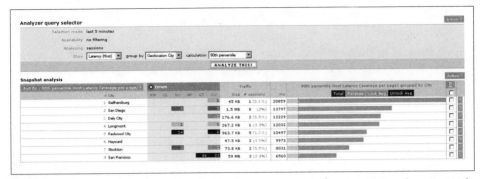

Figure 10-10. An aggregate report of host latency by city across visitor sessions in Coradiant TrueSight

The right answer: Hybrid collection

So what's the right approach for collection?

A combination of inline monitoring and a JavaScript-based programmatic script that's compatible with the Episodes approach is the right choice for RUM. The inline device has the most visibility into what is really happening, even when pages aren't loaded or servers aren't working, and is invaluable for troubleshooting. The JavaScript approach shows client-side activity, as well as mashups, CDN delivery, and third-party content that closely mimics end user experience. Together, they're unbeatable.

RUM Reporting: Individual and Aggregate Views

RUM yields two kinds of data: individual visits and aggregate reports.

Individual visits are great for diagnosing an issue or examining why a particular page was good for a particular user. They're the primary use of RUM in troubleshooting and customer support environments. They usually consist of a list of pages within a visit, along with timing information; in some products, each page can be viewed as a cascade diagram of container and component load times.

While looking at individual visits is useful, however, it's important to recognize that if you worry about a problem affecting only one visitor, you may overlook a more significant, widespread issue.

Aggregate calculations, on the other hand, give you a broad view of the application as a whole, as shown in Figure 10-10. They can, for example, show you a particular metric ("host latency") across a segment of your traffic ("the login page", "users from Boston", or "handled by server 10").

Aggregation also means making Top-N lists, which helps to prioritize your efforts by showing you the slowest, most errored, or busiest elements of your infrastructure. Finally, aggregate data is the basis for baselining, which decides what's normal for a particular region or page, and lets you know when something is unacceptably slow or broken.

Support teams, QA testers, and developers tend to use individual visit views, while aggregate data views are more often used for reporting and defusing SLA disputes.

Analyzing aggregate data has to be done properly. This is the place where statistics matter—you need to look at percentiles and histograms, not just averages, to be sure you're not missing important information or giving yourself a false sense of security.

RUM Concerns and Trends

Watching end user activity does present some concerns and pitfalls to watch out for, from privacy to portability and beyond. We've already considered many of the privacy concerns in the section on WIA, so be sure to check there for details on data collection.

Cookie Encryption and Session Reassembly

Some websites store session attributes in encrypted cookies. Unfortunately, obfuscating personally identifiable information may make it hard to reassemble a user's visit or to identify one user across several visits. Whenever the visitor changes the application state (for example, by adding something to a shopping cart) the entire encrypted cookie changes.

Your development team should separate the things you need to hide (such as an account number) from the things that you don't (such as a session ID). Better yet, store session state on the servers rather than in cookies—it's safer and makes the cookies smaller, improving performance. This is particularly true if your sessionization relies on the information in that cookie.

Privacy

RUM tools may extract content from the page to add business context to a visit record. While this is less risky than collecting an entire page for replay (as we do in some WIA tools), you still need to be careful about what you're capturing.

When you implement your data collection strategy, you should ensure that someone with legal authority has reviewed it. In particular, pay attention to POST parameters, URI parameters, and cookies. You'll need to decide on a basic approach to collection: either capture everything except what's blocked, or block everything that's not explicitly captured.

A permissive capture strategy might, for example, tell the RUM solution to blank out the POST parameter for "password." Unless it's explicitly blocked, it will be stored. Permissive capture means you may accidentally collect data you shouldn't, but it also means that a transcript of the visit will contain everything the visitor submitted, making it easier to understand what went wrong during the visit.

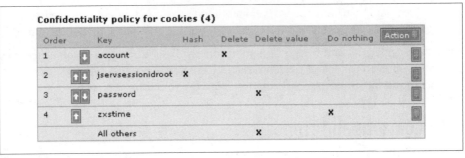

Figure 10-11. Configuring confidentiality policies in Coradiant TrueSight

On the other hand, a restrictive capture strategy will capture *only* what you tell it to. So you might, for example, collect the user's account number, the checkout amount, and the number of items in a shopping cart. While this is the more secure approach (you won't accidentally collect things you shouldn't), it means you can't go back and look for something else later on. Figure 10-11 shows an example of a restrictive capture configuration screen in a RUM tool—everything that isn't explicitly captured has its value deleted from the visit record.

RIA Integration

We've looked at programmatic RUM using client-side JavaScript. More and more applications are written in browser plug-ins (like Flash and Silverlight) or even browser/desktop clients (Adobe AIR and Sun's Java FX, for example.)

The methods described here for sending messages back to a hosted RUM service work just as well for RIAs. The application developer has to create events within the application that are sent back to the service. Episodes is a good model for this because it's easily extensible. As part of their RUM offerings, some solutions provide JavaScript tags or ActionScript libraries that can also capture multimedia data like startup time, rebuffer count, rebuffer ratio, and so on.

Storage Issues

As we've noted, capturing user sessions generates a tremendous amount of information, particularly if those sessions include all of the content on the page itself. If you're planning on running your own RUM, make sure your budget includes storage.

Many server-side RUM tools allow you to extract session logs so that they can be loaded into a business intelligence (BI) tool for further analysis (Figure 10-12).

With a hosted RUM service, it's important to understand the granularity of the offering, specifically whether it can drill down to an individual page or object, as well as the length of time that the stored information is available. Some systems only store user session information for sessions that had problems or were excessively slow.

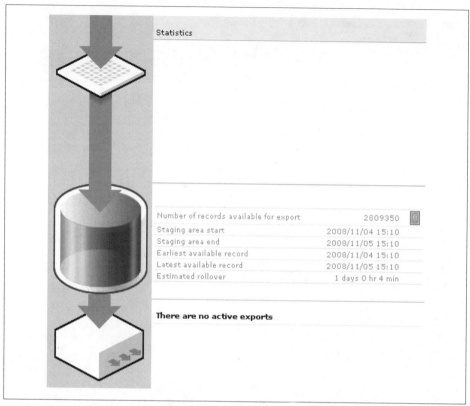

Statistics

Number of records available for export	2809350
Staging area start	2008/11/04 15:10
Staging area end	2008/11/05 15:10
Earliest available record	2008/11/04 15:10
Latest available record	2008/11/05 15:10
Estimated rollover	1 days 0 hr 4 min

There are no active exports

Figure 10-12. Bulk data export in Coradiant's TrueSight

Exportability and Portability

RUM data must be portable. Whatever technology you deploy, you need to be sure you can take your data and move it around. Often, this will be in the form of a flat logfile (for searching) or a data warehouse (for segmentation and sharing with other departments).

With the advent of new tools for visualization and data exchange, you will often want to provide RUM in real time and in other formats. For example, if you want to stream user events to a dashboard as structured data, you'll want a data feed of some kind, such as the one shown in Figure 10-13.

You may also want to overlay visitor information atop third-party visualization tools such as Google Earth, particularly if you're trying to find a geographic pattern. For example, you may want to demonstrate that visitors who are prolific posters are in fact coming from a single region overseas and are polluting your community pages with blog spam, as is the case in Figure 10-14.

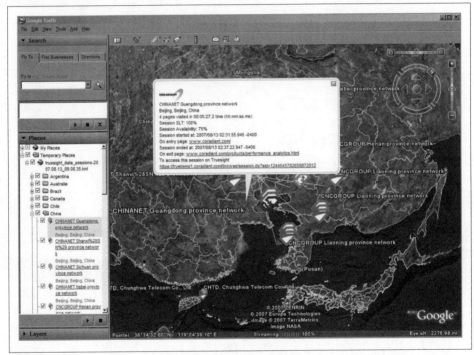

```
Mozilla Firefox                                                                              _|□|X
File  Edit  View  History  Bookmarks  Tools  Help

#version 1.1
#x-record-type,x-wp,x-page-id,x-session-id,x-server-id,x-page-name,cs(Host),cs-uri-stem,cs-uri-query,cs(Referrer),x-redirect-
"PAGE","2,54,61,62,82,84","1285466347040054565","1282372006803678942","10.20.1.228","redirect","fraserext.fraserworks.org","/
"PAGE","2,61,62","1285466379770306323","1282372006803679881","10.20.1.228","low+frequency+page","fraserext.fraserworks.org","
"PAGE","2,45","1285466366128332224","1282372006803679726","10.20.1.231","ListNextPrev","raytheonhc.myasp.com","/BasePages/Lis
"PAGE","2,45","1285466379860483857","1282372006803679726","10.20.1.231","ListNextPrev","raytheonhc.myasp.com","/BasePages/Lis
"PAGE","2,44","1285466380088024857","1282372006803679887","10.20.1.232","low+frequency+page","www.employmentment.harris.com","
"PAGE","2,44","1285466380153036568","1282372006803679887","10.20.1.232","header","www.employmentment.harris.com","/myasp.com/
"PAGE","2,61","1285466380166668065","1282372006803679888","10.20.1.228","low+frequency+page","mirant.myasp.com","/myasp.com/j
"PAGE","2,45","1285466366233189826","1282372006803679726","10.20.1.231","Log","raytheonhc.myasp.com","/Logging/Log.aspx","par
"PAGE","2,61,62","1285466380436152094","1282372006803679881","10.20.1.228","low+frequency+page","fraserext.fraserworks.org","
"PAGE","2,44","1285466380475997980","1282372006803679887","10.20.1.232","low+frequency+page","www.employmentment.harris.com","
"PAGE","2,44","1285466380495920928","1282372006803679887","10.20.1.232","low+frequency+page","www.employmentment.harris.com","
"PAGE","2,61","1285466380629090086","1282372006803679888","10.20.1.228","low+frequency+page","mirant.myasp.com","/myasp.com/j
"PAGE","2,61","1285466380790570798","1282372006803679888","10.20.1.228","low+frequency+page","mirant.myasp.com","/myasp.com/i
"PAGE","2,61","1285466380806299440","1282372006803679888","10.20.1.228","low+frequency+page","mirant.myasp.com","/myasp.com/i
```

Figure 10-13. Raw data of individual object requests from a streaming API

Figure 10-14. User visits showing performance and availability, visualized in Google Earth

These kinds of export and visualization are especially important for gaining executive sponsorship and buy-in, since they present a complex pattern intuitively. When selecting a RUM solution, be sure you have access to real-time and exported data feeds.

Data Warehousing

Since we're on the topic of data warehousing, let's look at some of the characteristics your RUM solution needs to have if it is to work well with other analytical tools.

- It must support regular exports so that the BI tool can extract data from it and put it into the warehouse at regular intervals. The BI tool must also be able to "recover" data it missed because of an outage.

- It must mark session, page, and object records with universally unique identifiers. In this way, the BI tool can tell which objects belong to which pages and which pages belong to which sessions. Without a way of understanding this relationship, the BI tool won't be capable of drilling down from a visit to its pages and components.

- If the data includes custom fields (such as "password" or "shopping cart value"), the exported data must include headers that allow the BI tool to import the data cleanly, even when you create new fields or remove old ones.

We'll look at consolidating many sources of monitoring data at the end of the book, in Chapter 17.

Network Topologies and the Opacity of the Load Balancer

A load balancer terminates the connection with clients and reestablishes its own, more efficient connection to each server. In doing so, it presents a single IP address to the Internet, even though each server has its own address. This means that the server's identity is opaque to monitoring tools that are deployed in front of the load balancer, including inline monitoring devices and client-side monitoring.

To overcome this issue, some load balancers can insert a server identifier into the HTTP header that the RUM tool can read. This allows you to segment traffic by server even though the server's IP address is hidden. We strongly suggest this approach, as it will allow you to narrow a problem down to a specific server much more quickly. You can use a similar technique to have the application server insert a server identifier, further enhancing your ability to troubleshoot problems.

Real User Monitoring Maturity Model

We've seen the maturity model for web analytics; now let's look at the model for web monitoring. There are two parallel types of monitoring: synthetic testing and RUM. As the organization matures, its focus shifts from bottom-up, technical monitoring to top-down, user-centric monitoring. It also moves from simple page analysis to the automatic baselining and alerting of transactions, and to tying the performance and availability of the site back to analytics data about business outcomes.

PART IV

Online Communities, Internal Communities, and Competitors

By now, you have a comprehensive view of visitors' experience with your site. More and more, however, visitors aren't interacting with you on your website. They're doing so elsewhere, in chat rooms, wikis, and social networks. Part IV of the book looks at how to monitor online communities—and how those same techniques can be used to track intranets, internal communities, and even competitors. Part IV contains the following chapters:

- Chapter 11, *What Did They Say?: Online Communities*
- Chapter 12, *Why Care About Communities?*
- Chapter 13, *The Anatomy of a Conversation*
- Chapter 14, *Tracking and Responding*
- Chapter 15, *Internally Focused Communities*
- Chapter 16, *What Are They Plotting?: Watching Your Competitors*

What Did They Say?: Online Communities

Today's companies engage with customers more directly than ever before. There are three big reasons for this change: *new forms of interaction* and *consumer technology* that ultimately lead to *more vocal markets*. This isn't simply better communication—it's fundamentally new kinds of discourse with customers and markets spurred by what Clay Shirky calls "the power of organizing without organizations."

New Ways to Interact

Corporate communications was once a formal affair, sent on letterhead and approved by marketing. No more. Today's companies are leaky sieves through which unregulated and unapproved messages flow between customers and employees at all levels of the organization. Social media redefines what we think of as public relations, demanding new tools and new approaches.

Career marketers may be tempted to stifle such communications in order to better control their messages and the way the brands they represent are perceived. They'll fail for one simple reason: customers like to connect. Companies that engage directly with customers for marketing, support, and research get strong competitive advantages— lower costs, stronger loyalty, better conversion rates, and improved product designs.

Consumer Technology

One reason that PR is being redefined is that we're all just a click away from one another. Employees have Blackberries and iPhones that connect directly to social media. Even when they're not at work, these tools keep workers linked to their organizations. This has resulted in the creation of many informal relationships between an organization and its market. For example, Comcast's Twitter-based support program began from an employee's casual use of the microblogging tool. Search, too, makes it easy for people

Figure 11-1. When communities strike back: the Motrin Moms viral video

to keep track of what their employers are up to, breaking down the traditional walls that kept formal marketing teams blissfully unaware of the consequences of their work.

Employees and customers can now connect on external systems, outside of the control of the company itself, often with a degree of anonymity and candor that circumvents employment agreements, leaving their employers naked and without recourse.

And they do so anytime, anywhere.

Vocal Markets

Direct engagement is born of necessity as consumers find their voice. Thirty years ago, disgruntled buyers could do little more than scratch out their righteous indignation in a letter to the local paper, throwing themselves on the mercy of the editorial court. Today, social media is the great amplifier. Consider a controversial advertising campaign for Motrin, launched in late 2008. Twitter users voiced their objection to the ads' portrayal of mothers, using microblogging and YouTube videos to get their point across (*http://adage.com/digital/article?article_id=132622*), as shown in Figure 11-1. Motrin killed the campaign and apologized (*www.mathewingram.com/work/2008/11/16/flash-flood-mom-bloggers-and-motrin/*).

In a connected world, an angry mob forms in an instant, undermining millions of dollars and thousands of hours of work. If you're not trying to engage customers directly, you can be blind to—or blindsided by—an online movement that fundamentally changes your business.

A happy community, on the other hand, rewards you with free marketing and positive word of mouth. Businesses that understand this fundamental shift are connecting directly with their markets through communities and social media, abandoning formal messages and rigid hierarchies in favor of dynamic interactions at all levels of the organization. Marketing is a dialogue, sometimes in the extreme: Skittles went so far as to replace its website with a Twitter feed showing mentions of its candy, effectively turning its brand over to its customers.

How did we get here?

Where Communities Come from

Online communities have been around for decades, but it took recent advances in usability, widespread Internet adoption, and the proliferation of social media to make online communities and social networks something organizations could no longer afford to ignore.

Digital Interactions

The Internet on which the Web runs is a communications platform. Long before the Web, nerds with common interests found ways to talk amongst themselves. In fact, people using computers started to connect with each other during the early 1960s. Even before the Internet was introduced, military organizations and large corporations were implementing ambitious networks to share and store information.

By the 1970s, bulk email platforms kept groups in touch using *listservs*. These special-interest mailing lists are still widely used today in the form of Google Groups, Yahoo! Groups, and MSN Groups. Some community platforms also emerged in academic environments, running on university minicomputers.

Bulletin board systems (BBSs) brought computer-based communities out of academia and into the public psyche. In 1978, Ward Christensen—snowed in and bored during a storm in Chicago—decided to write the first BBS, known simply as CBBS.

Online Communities: A Historian's Perspective

Computer networks have existed through the 1950s and onward, although they vary in terms of what the computers let you do and how you can interact—the whole reason TCP/IP took off with such velocity is that it is so flexible and can shoot through other networks. But there was a whole range of networks out there long, long before the 1970s.

One of the first ones that really pushed itself as being a "community" was PLATO (*http://en.wikipedia.org/wiki/PLATO_system*). But let's go back further. Home com-

---

puters with modems? In that case, one might want to look at the BBSs of the 1970s, like CBBS. I made a documentary on those.

Some of the BBSs out there grew quite quickly once people started hooking on. Nothing would count past 1982 for a "first online community," although the introduction of FidoNet in 1984 would probably be a big pointer for when we had all sorts of networked folks interacting.

We're talking probably a few thousand folks linked into pre-1982 communication.

—Jason Scott,
owner of *www.textfiles.com*,
Internet historian,
and director of the acclaimed *BBS Documentary*

In the world of amateur technologists, BBSs were an instant hit. Fueled by the popularity and affordability of the personal computer, thousands of hobbyists began building and operating their own. They all featured the same basic functions:

- A *message base* that allowed users to communicate with each other locally or regionally (through inter-BBS protocols such as FidoNet)
- A *file sharing system* that allowed users to upload and download files
- Turn-based and real-time "door" *games* that allowed users to compete against each other
- Real-time *chat*, allowing users to converse with one another

Early BBSs often included other functions such as voting, BBS lists, polls, and art galleries (Figure 11-2) that helped personalize them and increased visitor stickiness. BBSs became popular enough that people traveled from various countries to meet each other in person. They foreshadowed the modern Web—in fact, nearly everything that's happened on the Web has happened before in the BBS world on a far smaller and geekier scale.

A dominant feature of BBSs was a message base like the one shown in Figure 11-3. This was a forum that allowed users to interact with one another, typically organized around a topic or special-interest group. Some of these areas were private, hidden to other users if they didn't have the required credentials.

Some BBS communities are still strong today—the Whole Earth 'Lectronic Link (WELL) was founded in 1985, but made the transition to the Web and is still around (as part of Salon.com). Until relatively recently, however, online communities were the domain of technologists, activists, and fringe groups. Widespread adoption of computers with Internet access wasn't enough for communities to become truly mainstream. They also had to be easy to use.

Figure 11-2. The main menu of a BBS, yesterday's equivalent of index.html

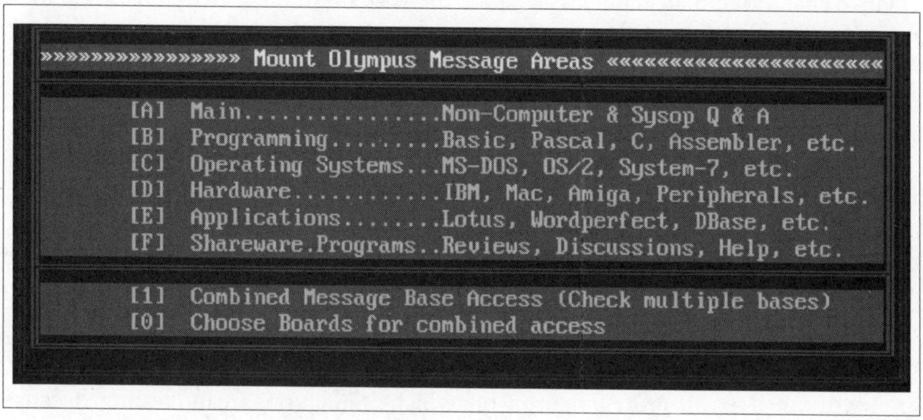

Figure 11-3. Mount Olympus BBS's Message Base

Making It Easy for Everyone

To use an early BBS, visitors had to type cryptic modem codes into terminals, often dialing for hours to try to get through to perpetually busy phone lines. Then they stared at text-only screens and slow-moving cursors. It wasn't an easy way to connect with others, and it attracted a certain kind of user.

Despite the efforts of communities in the 1980s that built client-side graphical mouse-driven interfaces, these systems were generally proprietary, platform-specific approaches that discouraged widespread adoption, for example, Hi_Res BBS, Magic BBS, and COCONET (*http://www.bbsdocumentary.com/software/IBM/DOS/COCONET/*). Accounts were also tied to the individual BBS—if you frequented 10 different BBSs, you had 10 different email messages to check, which made it hard to centralize your online identity.

Today, users don't need special hardware or lengthy codes—they just need a web browser. Online communities of many different types welcome everyone. But at their core, communities haven't changed much at all: they exist for interaction. In the end, *online communities are about the conversation.*

Online Communities on the Web

The first version of the Web wasn't about connecting people to one another; it was about connecting people to information. It was a universal system for information retrieval, where everything was interrelated by hyperlinks. It followed ideas established in text-based linked directories such as Gopher.

At that time, there were other protocols and applications for communication, from email to chat rooms to instant messaging, so the Web was more focused on information retrieval. The Web's first destinations were directories and search engines, such as AltaVista, shown in its early form in Figure 11-4.

Deciding What Mattered

Directories weren't enough. Once the world had tasted the Web, and found it good, it needed tools to make it better. As the number of destinations on the Internet grew, users needed ways to find out what was interesting. Search engines based on relevance algorithms (such as Google's PageRank) provided better, more up-to-date results than static directories.

Eventually, communities emerged that could suggest good content. As we now know, groups of web users are great at figuring out which destinations are interesting, because they can quickly flag bad sites and throw the weight of their numbers behind good ones. Slashdot, Reddit, and Digg, shown in Figure 11-5, are examples of dominant aggregation communities today.

Community recommendation reflected a shift in how the Internet worked. Instead of users deciding what they were interested in and "pulling" it from the Web, these sites "pushed" content toward users by telling them what their community thought they'd find interesting. This is an important change that has affected how most online applications are designed today, built around lifestreams, friendfeeds, alerts, and status updates.

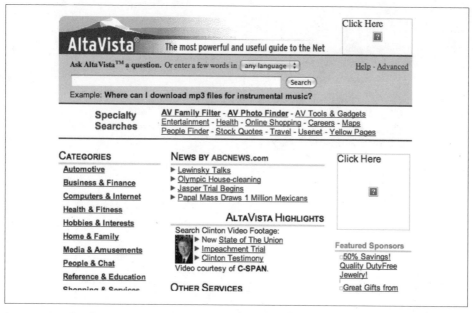

Figure 11-4. The Altavista.com search engine of many years ago emphasizes a directory of information organized by editors

Figure 11-5. The front page of digg.com

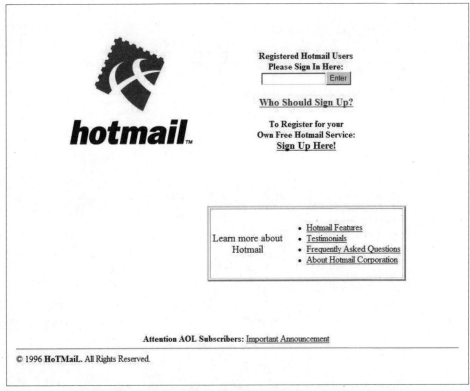

Figure 11-6. Hotmail.com in 1996, courtesy of cidfadon's Flickr photostream (special thanks to @maczter for finding this for us)

Email for Everyone, Everywhere

While relevance-based search and recommendation communities were making it easier to find the right content, people began using the Web to communicate. By 1996, enough people had web browsers that Hotmail, shown in Figure 11-6, was able to sign up millions of customers for its online messaging service in just a few months. No longer did users have one email address per ISP subscription; now, everyone in a household could have his or her own.

Personal email required an address book. This meant that mail service providers had a map of who each user knew, which would later become important for building out a "social graph" of their relationship to others.

Instant Gratification

Email messages were great, but only for specific kinds of interaction: relatively long, carefully constructed messages that weren't responded to immediately. Several of the large online portals added instant messaging (IM) functionality as part of their services.

IM gained immediate popularity. Email address books and IM contacts merged. Trillian, originally a client for Internet Relay Chat, was one of the first IM clients that spanned multiple IM networks, and today's IM clients, such as the one shown in Figure 11-7, are independent of a specific service provider—it's possible for a person on one network to talk to users on others.

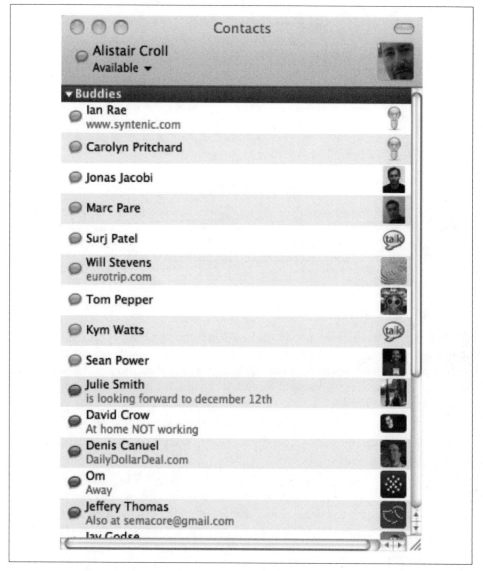

Figure 11-7. The Adium client for Mac OSX

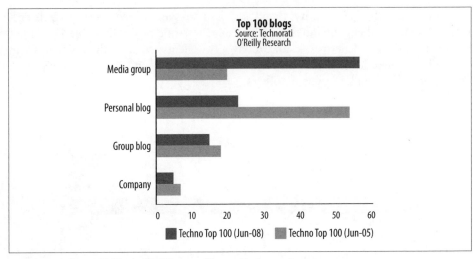

Figure 11-8. Top 100 blogs of 2008, according to Technorati

Power in Numbers

By March 2008, instant messaging platforms already accounted for very large user bases. According to Comscore (via TechCrunch), in terms of monthly users, MSN had 235 million, QZone managed 100 million, Yahoo! hosted 97.6 million, AIM accounted for 27.3 million, ICQ for 30 million, and GTalk handled 4.9 million.

Everyone's a Publisher

The Internet also made everyone a publisher. While creating and maintaining a website manually was hard work—few mainstream users had the time to hand-edit the HTML of their personal websites to keep the world apprised of their daily goings-on—blogs (first known as "weblogs") lowered the bar for online authors, and millions of users launched personal websites.

Within a few short years, blogging became a dominant force online. Whereas in 2005 most of the top 100 bloggers were individuals, as Figure 11-8 shows, today they are media organizations.

Still, blogging wasn't for everyone. Social networks were quick to provide simpler online presence. MySpace made it easy for the average web citizen to have a single page online, and appealed to a younger audience than ever. Friendster and Facebook, shown in Figure 11-9, focused on the relationships between their users. Everyone invited their friends, and a few dominant players emerged as a result of the strong network effects these sites enjoyed. By filling out how we knew our friends, we were creating what we now call a "social network." For business relationships, sites like LinkedIn and Spoke focused on referrals and employment history.

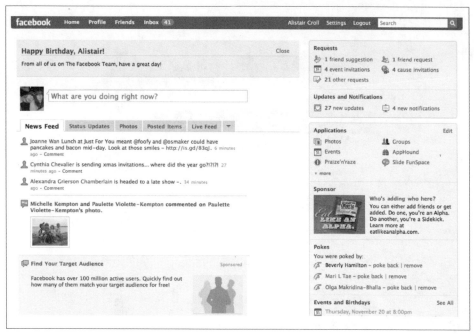

Figure 11-9. Inside Facebook.com

 A map of a person's social network is also sometimes called a *social graph*. The term social graph refers to graph theory, and entered the Internet vernacular when Mark Zuckerberg of Facebook used it to describe the company's map of social interrelationships.

Another key element of these social sites was the ability to send small updates to our communities, keeping them apprised of what we were doing, and maintaining a degree of "ambient awareness" within our network. For a deeper look at ambient awareness, refer to *http://www.nytimes.com/2008/09/07/magazine/07awareness-t.html*. (You may need to register to access the article, but it's well worth the time and effort to do so.)

Microblogging Tells the World What We're Thinking

Meanwhile, IM was changing. Most IM applications allowed users to set their status so contacts would know if they were available for a chat. But users started to repurpose the status field so their community of friends would know what they were doing in a lighter-weight version of social network sites' status updates, as Figure 11-10 shows.

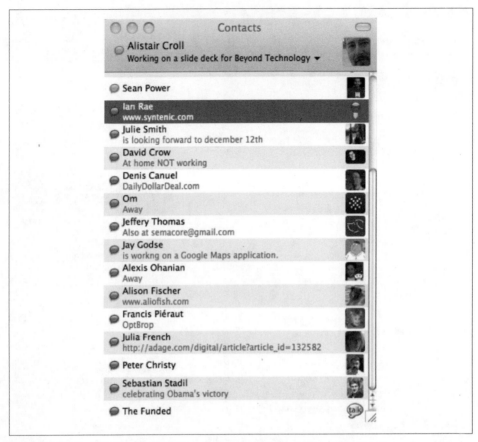

Figure 11-10. An IM client displaying many status messages

Twitter, Identi.ca, FriendFeed, and others took this repurposing of the status field to its logical conclusion in the form of microblogging, sending small updates to a community of followers. Just as with blogs, microblogging is still personal and casual, and the topics covered are often far ranging, as shown in Figure 11-11.

However personal and fleeting the bulk of its messages may be, microblogging's immediacy is hard to beat. From real-time forest fire updates that helped fire departments to manage wildfires, to eyewitness accounts of terrorist attacks, microblogging is positioned to be the protocol for real-time human interaction. It's also the best way to tap into the collective sentiment of the Web at a particular moment in time: If Google tracks what the Web knows, Twitter search tells you what its citizens are interested in right now.

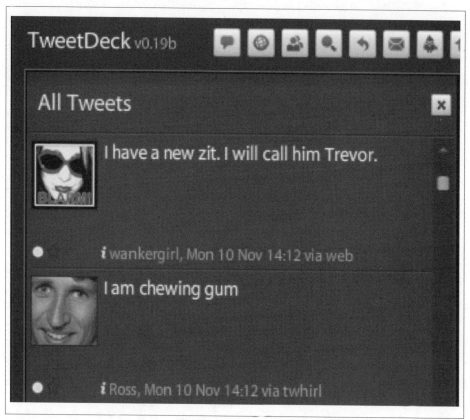

Figure 11-11. Conversations on twitter today

If you doubt the rise of microblogging, remember that blogs were in the same situation just a few years ago, as Figure 11-12 shows. Personal blogs quickly gave way to influential media sources that are today supplanting traditional print publications. Microblogging may have as significant an impact on cable news stations as blogs are having on print journalism, with news desks already creating "Twitter correspondents" to track and spread breaking stories and celebrities rushing to establish a presence and follow one another.

It would be wrong to write off microblogging as just short-form blogging. Twitter differs from social sites in several ways, and one of the most important is the openness of the social network. Unlike Facebook, where others need your permission to receive your updates, on Twitter you can follow anyone without his or her approval, as Figure 11-13 shows.

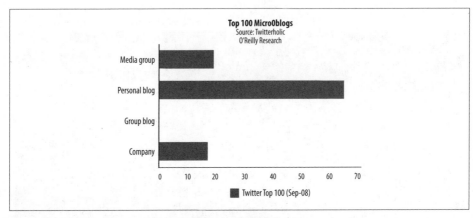

Figure 11-12. Top 100 microblogs, according to Twitterholic

Figure 11-13. A social graph on Twitter

This pattern of *asymmetric following* leads to interesting behaviors, many of which have been developed by the community itself: naming others with @name, grouping similar content with #hashtags, and reTweeting (RT) to amplify content the community feels deserves broader attention (for a brief Twitter primer, see *http://www.watchingwebsites .com/archives/twitter-survival-guide*). Microblogging is also home to short-lived bursts of information, often pointing to other content rather than being content sources themselves.

All of these community platforms—news aggregators, social networks, microblogging, blogs, and comment threads—are also shifting toward rich media and spreading out to mobile devices, creating a vast range of community places in a wide variety of forms. Ultimately, the Internet today is a spectrum of conversations, each taking place atop different platforms, as shown in Figure 11-14.

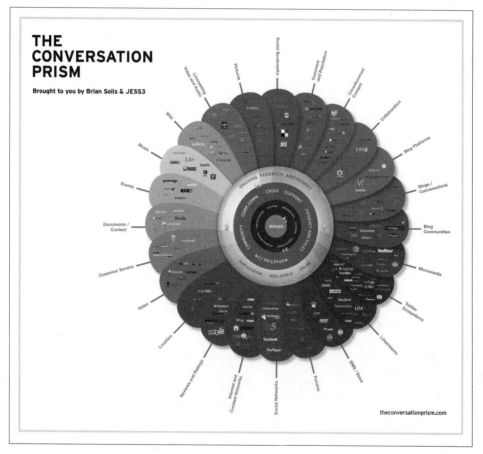

Figure 11-14. The conversation prism by Brian Solis and JESS3

Like it or not, it's your job to care about them all.

Community monitoring is the remaining piece of a complete web monitoring strategy. In the next few chapters, we'll show you why communities matter to you, how to figure out who's talking about you, where they're talking, what they're saying, and how to track them and join the conversation.

CHAPTER 12

Why Care About Communities?

Simply put, you should care about communities because they want you to. A 2008 study by Cone, LLC, found that most online consumers want companies to have a presence online, and that many want them to be interactive with their customers, and use social networks to help them with problems (Figure 12-1).

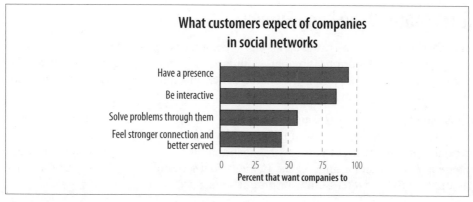

Figure 12-1. Most users of social networks want companies to join them

Many of the benefits of communities are associated with marketing communications—a new kind of public relations. That's only natural. Marketing communications is about getting messages out to the world, and as conversations, social networks fit that bill well. But communities do much more than simply deliver messages.

They are the start of a long funnel
> Audience engagement starts long before someone visits your site. It begins in the communities and online venues where your market first hears of you.

They amplify messages more efficiently
> If the community adopts your message, it will amplify that message across the Web in a new and more effective form of promotion that people trust.

They help customers to help themselves
> Community-based customer self-service not only lowers your own costs, it actually provides better service than you can offer by yourself—and customers prefer it.

They make your business more agile
> By short-circuiting traditional market research, you can go straight to the source and design better products or services faster than ever.

They provide the first signs of online liability
> Others may be doing things that have a legal impact on you, such as distributing your intellectual property illegally or slandering you. If you run your own platform, you may also be liable for what's said on it. Staying in touch with your community is an excellent way to discover these problems early on.

They improve referrals
> A community dramatically expands your social network and your reach, whether you're looking for customers, suppliers, or employees.

Before diving into what makes up these conversations and how to join them, let's take a closer look at each of these.

The Mouth of the Long Funnel

As we've seen, web analytics is all about outcomes. Visitors arrive at your site, and—assuming you've done your job right—a number of them progress inexorably toward a goal you've set. This is often represented as tiers in a funnel, starting when the visitor arrives and tapering toward the outcome you intend.

This "short funnel" is increasingly outdated. It no longer accurately describes the way companies engage with a web audience. Today's long funnel starts the moment a potential visitor hears about you—with a mention on a blog or a comment on Twitter. Sometimes this is not a single event, but instead a series of interactions over time that gradually builds up interest in your product or service.

Tying what happens out on the Web at large back to outcomes on your own site is difficult. In fact, it's the major challenge for the web analytics industry in coming years. There's no consistent way to link several impressions across multiple sites you don't control, and since privacy concerns prevent you from identifying individual users, you can't know which specific events on a social network triggered activity on your site.

This is changing. Social networks like Facebook, as well as identity frameworks like OpenID and cross-site blogging platforms like Disqus, all help web users leave trails—and build personal identities—as they navigate many websites. This kind of tracking information can let you know every time that someone interacts with your brand or message online, and may one day eventually tie a person's actions elsewhere on the Web back to an outcome on your website or in the real world. Unique URLs targeted at a specific message or a segment of the social network can also be used to track who clicks what, tying a link in a microblog to its outcome.

We may not need to rely on such tracking methods. Consumers are increasingly willing to disclose tremendous amounts of personal information in return for web functionality like bookmarking and sharing. Twitter profiles, for example, are visible to the world and often contain personal details as well as links to blogs and other personal records. While profiles like these will help tie community interactions to outcomes, using them may pose ethical and legal risks to your organization resulting from privacy and confidentiality issues.

There are other challenges in tracking the long funnel. Many of the social networks on which discussions start and opinions form want to make money by charging marketers for insights into what communities are thinking. At the same time, analytics firms want to be the primary source of insight into web activity. There's an inevitable power struggle brewing between analytics companies and social network platforms.

Whatever happens, monitoring the mouth of the long funnel is becoming an increasingly vital component of a complete web monitoring strategy, whether it's done through analytics tools that instrument online communities, clever manipulation of short URLs, or services provided by online communities directly.

A New Kind of PR

With so many free web platforms now available, it's easy to craft a message and build grassroots support, provided you can generate genuine attention from an audience. Getting genuine attention is the hard part.

In an industry once dominated by one-to-many broadcasting, online marketing analytics brought accountability to advertising. Viral marketing approaches made it easy to spread messages. And now community marketing—done right—makes it possible to overcome the natural skepticism of a media-saturated audience and communicate with it directly. Figure 12-2 shows how information flows in each of these models.

Broadcast Marketing Communications

Broadcast marketing was the twentieth century's dominant form of promotion. It cost money to design and produce a message, and still more money to print and ship it, or to license some spectrum and transmit it. Broadcasting a single message to a broad audience offered the economies of scale needed to market cost-effectively. Where the industrial revolution gave us mass production, broadcast marketing gave us mass media.

Computerization and lower production costs meant that advertisers could target their messages to specific audiences according to publication, TV program, or geography—with computers, the incremental cost of a personal message was nearly zero. Marketers

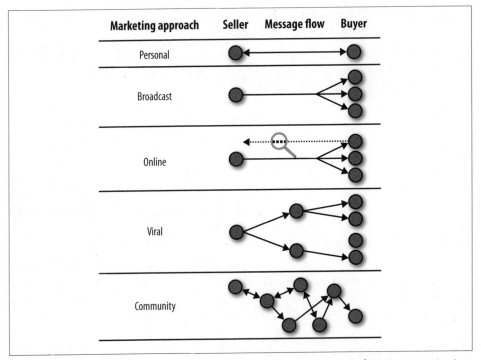

Figure 12-2. How messages flow between buyers and sellers in various marketing communications approaches

used databases and form letters to try to optimize which prospects received which messages. But while they were tailored to some degree, those messages were tightly controlled.

In the traditional marketing communication model shown in Figure 12-3, the marketing team creates a message that combines the interests of the product team, legal and ethical restrictions, and the company's brand and overall positioning. The message is then distributed through mass media, public relations, and other channels. The audience is segmented, but largely unqualified. In broadcast marketing, that's OK: volume trumps precision. The more people you reach, the larger the response.

Unfortunately, broadcast audiences are nearing saturation. The average American sees a tremendous amount of advertising, as shown in Figure 12-4: 245 ad exposures daily, of which 108 come from television, 34 from radio, and 112 from print (these are AMIC.com estimates—some of the most conservative). And that's just the ads they notice—some estimates claim that they're exposed to thousands a day.

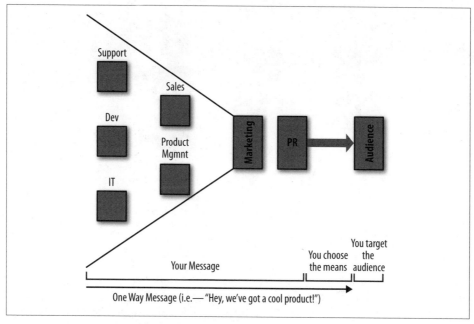

Figure 12-3. Traditional marketing communication model

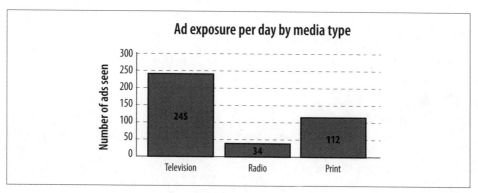

Figure 12-4. The amount of ad exposure Americans receive per day by media type

It's not just saturation that's killing traditional marketing. People are also ignoring the messages. They tune out promotions, skipping them with their Tivo. TV producers have switched to product placement as the world skips past paid advertising. *30 Rock* runs an episode mocking demand for product placement, and in doing so, showcases a product. Tina Fey looks straight at the camera, winks, and says, "Can we have our money now?" Even the great Stephen Colbert tells his Nation to eat Doritos as part of the show itself.

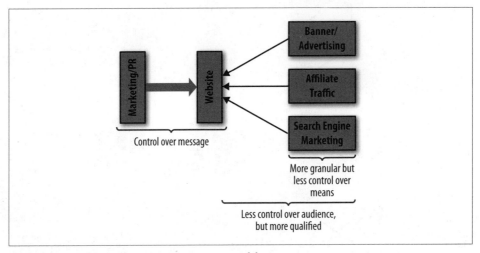

Figure 12-5. Online marketing communication model

Broadcast audiences are both weary and wary. Ad revenues are in free-fall, and newspapers are foundering. The broadcast era is waning, not only because people are saturated and bored with its impersonal content, but because the precise tracking that's possible with online campaigns makes broadcast media look like a blunt instrument. It's unaccountable, unwieldy, and overpriced, and advertisers are switching to other approaches.

Online Marketing Communications

With online marketing, advertisers see the results of their efforts. Outcomes are tightly tied to campaigns, making online approaches a much more attractive resource for getting a message out.

Online marketing can be better targeted, too. Through careful segmentation along dimensions such as geography, gender, age, or keyword, there's a much higher chance that a message will resonate with its recipient, turning a prospect into a convert. And those messages can be adjusted over time, mid-campaign, to maximize their effectiveness.

In the online marketing model, advertisers still control the message, although it might take the form of teaser ads, search results, affiliate referrals, or paid content. They can also choose their target segments, but they have much less control over the distribution of the message—they've outsourced that part of the work to the other online properties instead. Figure 12-5 shows how online marketing works.

Online marketing is still paid marketing. Ad dollars are tracked for effectiveness to provide accountability, and to let advertisers optimize their spending. It also has two fundamental problems:

- First, *a message only goes as far as it's paid to*. It's not amplified by the audience. To reach more people efficiently, advertisers need a communications model that scales well, growing by itself rather than costing them more money to reach more people.
- Second, the *audience still largely ignores the message* because it's untrusted. Ads are noise, and with the notable exception of relevance-based advertising, like paid search, people mostly ignore them.

One technique, viral marketing, addresses the first challenge by letting the audience itself amplify the message.

Viral Marketing: Pump Up the Volume

The term *viral marketing* was coined by Jeffrey Rayport in 1996. It describes a form of marketing that's designed to be repeated. Borrowing from concepts within epidemiology, viral marketing "infects" the audience, encouraging those who hear the message to amplify it. Done right, it's a much more efficient way to garner attention, it's cheaper to deliver, and it enjoys *demand-side economies of scale*. That is, the volume of the message grows as more people hear it.

In nature, when an initial carrier is infected by a virus, that carrier then spreads the disease to the immediate environment, and from there, the infection spreads as carriers transition across borders to previously uninfected territories. Figure 12-6 shows an example of viral spread across population groups.

Distribution costs for viral marketing are lower because the audience becomes the medium. Making campaigns viral isn't easy, however. You have to craft the right message and convince the audience to amplify it across many sites and platforms.

The classic example of online viral marketing is Hotmail. In its first 18 months of operation, Hotmail grew to 12 million users. At one point in 1998, the company was signing up 150,000 users *a day*. By comparison, a traditional print publication hopes to grow to 100,000 subscribers over several years. And Hotmail did it all with the seven short words shown in Figure 12-7.

During those 18 months, Hotmail spent less than $500,000. There were other factors that influenced its success, of course: most people didn't already have web-based email; many people were connecting to the Internet for the first time; and the service was free. But an inherently social tool like email proved the perfect growth medium for the virus.

 There is some debate between the Hotmail founders and the main investors, Draper Fisher Jurvetson, as to whose idea it was to implement the email footer. In any case, the story had a happy ending: Microsoft acquired Hotmail in 1997 for $400 million, touting more than 9 million members worldwide at the time.

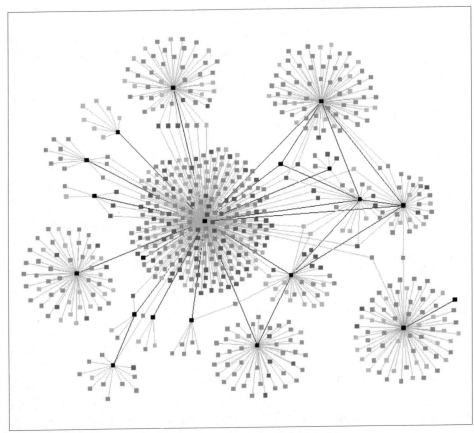

Figure 12-6. A CDC contact tracing map of real airborne infectious disease

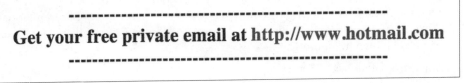

Get your free private email at http://www.hotmail.com

Figure 12-7. The Hotmail footer is perhaps the Internet's most successful viral marketing campaign

The Bass Diffusion Curve

Viral marketing isn't new. Decades before Hotmail's record-breaking growth, Frank Bass published research on message diffusion and introduced marketers to the Bass Diffusion Curve.

Traditional advertising campaigns are a function of two things: how many people hear your message, and how effective that message is in getting them to act. In other words, to get better results, you can either advertise to more people or you can improve the

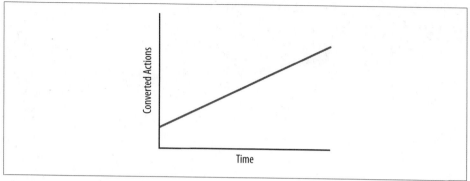

Figure 12-8. Traditional marketing diffusion curve

offer. This implies complete control over message power (the offer) and people who hear it (the reach).

```
Converted buyers = message power × people who hear it
```

Assuming a regular monthly expenditure on advertising (reaching the same number of people each month), the result of this traditional marketing model looks like the straight-line message diffusion curve shown in Figure 12-8. A convincing message reaching a certain number of listeners each month will result in a specific slope to the diffusion line—and the better the marketing message, the steeper the slope.

But Bass realized that there's a second channel for your message—*the buyers themselves*. If your message is sufficiently engaging, your audience *becomes* your marketing medium. This second channel isn't in your control, but it's free. It's word of mouth.

The Bass model adds a second element to the equation:

```
Converted buyers = message power × people who hear it +
  (buyers × WOM power × People who hear it)
```

In other words, in addition to traditional advertising reach, there's word-of-mouth reach. And word of mouth grows as more people become amplifiers who adopt and repeat the message. The more amplifiers you have, the further your message travels until, ultimately, everyone has heard it.

The effect of word-of-mouth diffusion is strikingly different from traditional marketing, and yields the diffusion curve shown in Figure 12-9.

Early on, word-of-mouth diffusion isn't very effective. Word of mouth starts off slowly, but as the number of buyers, users, or visitors builds, it can quickly outpace traditional marketing.

This doesn't happen by itself. Viral marketing requires certain things to work well:

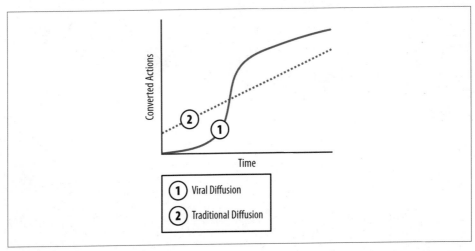

Figure 12-9. Viral marketing diffusion curve

A good story

Some marketing messages are inherently viral. Apple's Steve Jobs calls inherently viral products "insanely great"—something so wonderful that you feel smarter when you tell others about it.

Something of real value

While virality pays for its own distribution, you'll need to work harder to craft your message. The Web is full of chain letters and goofy videos, but they're not viral marketing unless they somehow lead to an outcome—they're just entertainment.

Consider Gmail, Google's web-based email offering. Unlike Hotmail, Gmail launched into a market that already had entrenched web mail platforms, such as Yahoo! and Hotmail. Google tried to differentiate itself by offering a hundred times the storage capacity of competing services, and by using an invitation-only model to drive adoption. By limiting invitations, the company was able to create perceived scarcity, which in turn drove up the perceived value of the service.

Support from community leaders

You can improve viral marketing effectiveness by targeting *supernodes*—extremely connected invividuals—in social networks that have a broader reach and whose amplification has a greater impact because it spans several different groups.

A large end audience

Viral marketing reaches a saturation point as the target audience hears and tires of the message (which is what gives it an S-shaped diffusion curve), so bigger markets are worth targeting. If your market is small or you can reach it efficiently through traditional means, it may be better served with more traditional, personalized advertising.

A platform for distribution

Hyperlinks, videos, and pictures work well because they're easily shared. The easier you make it for someone to tell another person about your message, the better you'll do. Hotmail worked well because the product was itself a platform for spreading the message.

Seth Godin's *Unleashing the Ideavirus* (Hyperion) and *Purple Cow* (Portfolio Hardcover) deal with the issue of how to transmit ideas that others will pick up.

For an excellent overview of crafting messages that are easy to remember and pass on, see Chip and Dan Heath's "Made to Stick" website at *www.madetostick.com/*.

Malcolm Gladwell's *The Tipping Point* (Back Bay Books) is required reading on supernodes and how messages grow rapidly.

It's important to set the right expectations for viral marketing within your organization, since it has fewer concrete, short-term effects.

In the early stages, when you're seeding your message into the community, conversions from word of mouth will be low. Once a message gets into the hands of *social hubs* (networks of people with many casual friends and followers), the likeliness of message transmission greatly increases. If each user you bring in adds more than one user, you have a positive viral coefficient, and growth will follow.

Robert Zubek explains: "The viral coefficient is a measure of how many new users are brought in by each existing user. It's a quick and easy way to measure growth: if the coefficient is 1.0, the site grows linearly, and if it's less than that, it will slow down. And if the coefficent is higher than 1.0, you have superlinear growth of a runaway hit" (*http://robert.zubek.net/blog/2008/01/30/viral-coefficient-calculation*).

You can help prove the value of viral transmission to your organization by tracking word-of-mouth conversions separately from broadcast marketing. Get viral marketing right, and you'll have a huge, cheap channel for your message. Hotmail adoption from June 1996 to June 1997, shown in Figure 12-10, closely follows the curve predicted by Bass' research 30 years earlier.

In Hotmail's case, the message was embedded in emails its users sent, and Hotmail controlled most of what was said as a result. You won't be so lucky—as your message spreads, it will be diluted and misinterpreted. The Web will edit and repurpose it. Reactions and comments to the message might become more important than the message itself. Sometimes, entire communities will emerge around an idea as it becomes a part of online culture. Writer and evolutionary biologist Richard Dawkins calls these cultural ideas *memes*. They're transmitted from mind to mind, evolving and competing with other memes for prominence.

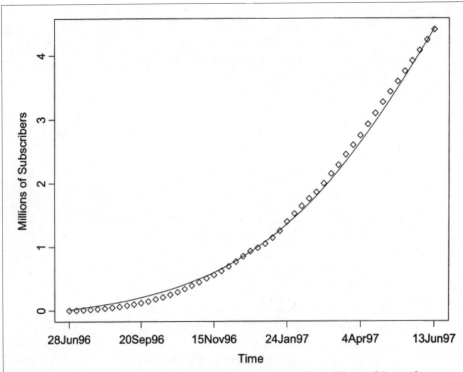

Figure 1: The cumulative number of Hotmail subscribers in millions (diamonds) at each weekly period closely matches the fitted predictions of the estimated Bass diffusion model (the line). The data is taken from Roberts and Mahesh [1999].

Figure 12-10. Diffusion curve plotting Hotmail adoption over one year

In the same way, communities relabel and repurpose viral marketing messages as they enter the Web's zeitgeist, often at the expense of the original intent.

Consider, for example, a failed martial arts backflip that became the "Afro-ninja" Internet meme (*www.youtube.com/watch?v=b_NQCTbvRnM*), garnering millions of viewers. That this video was in fact related to an ad for the video game Tekken 5 was completely lost, and the marketing opportunity was missed entirely, despite its rapid spread to all corners of the Internet. Communities tend to do that. So why would we want to hand over our message to a Web that might change it?

Because we're all skeptics.

We live in an attention economy where being noticed is the best currency. In his book *Free*, *Wired* Editor-in-Chief Chris Anderson cites social scientist Herbert Simon, who in 1971 predicted the rise of this attention economy, saying, "a wealth of information creates a poverty of attention." Paris Hilton and Google both make money from their ability to steer our attention toward a particular product or service. *And nothing gets our attention like the advice of a trusted source.*

Community Marketing: Improving the Signal

Community marketing and viral marketing are closely related, since both rely on the audience to amplify the message. But while viral marketing focuses on message spread, community marketing combines message redistribution with the genuineness of a message crafted by others.

 There are subtle differences in the types of community marketing messages that exist today. David Wilcox at the Social Reporter has a great blog post about this at *http://www.designingforcivilsociety.org/2008/03/we-cant-do-that.html*.

In community marketing, you engage the community through interaction, hoping that you're able to make yourself the subject of discussions. It's more about engaging and creating platforms for interaction like the one in Figure 12-11 than about crafting clever videos. It's also often about listening and joining, rather than crafting the message yourself.

Community engagement helps to promote products or services, particularly for business-to-consumer marketing efforts. It also helps increase brand awareness. But communities are extremely wise—if the community marketing effort isn't tied to genuine value, the community will rebel and the message will backfire, as it did for Belkin. Revelations that an employee had paid for positive product reviews online prompted anonymous letters claiming that the company engaged in even more deceptive practices. These far more damaging claims would likely never have surfaced if the disingenuous behavior of the initial employee hadn't come to light. There's a fine line between using a community to share your news and becoming the news itself.

As online consumers become increasingly savvy, they can tell pitches from proponents with great precision. They flock to trusted peers. For example, sites like Chowhound, TripAdvisor, and Yelp (shown in Figure 12-12) have changed how consumers check out restaurants.

Figure 12-11. AppleInsider fan community

Figure 12-12. Yelp reviews are from real users, which makes them far more credible than those from marketers

Figure 12-13. A message that Oracle does not control

Perhaps because of this, communities terrify traditional marketers. Community discussions peel back the veneer of a product, unearthing all manner of wart and weakness. Communities roam off in unexpected directions, as shown in Figure 12-13. Sometimes, they'll talk about the message you've created; other times, they'll talk about messages that one of your competitors has created about you. Most of the time, they simply won't talk about you at all. When they do, however, you need to be ready to engage them.

Marketing communications in the age of widespread online communities consists of many different groups within an organization having many different dialogues. Gone are the hierarchies, the message approval, and the controlled release. In their place are many flows of information linking your organization to the rest of the Web, as shown in Figure 12-14. It's very different from the controlled, predictable communications of broadcast marketing.

With little control over delivery or message, there are still five things you can do:

- *Create platforms* where communities can flourish, such as groups, hashtags, and forums.
- *Find mentions* of your message and your brand across the Web using search and referring URL tools.
- *Engage with communities* so you can nurture and gently steer conversations.
- *Provide relevant content* so you can grow customer loyalty and help the community to correct misinformation itself.
- *Build products and services that don't suck.*

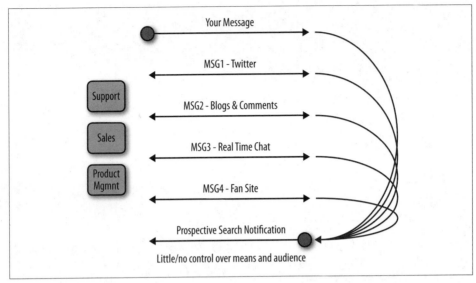

Figure 12-14. Marketing communications in the age of online communities

Many people equate community management with next-generation marketing. It's an obvious association: public relations has always been about getting the right message to as many people as possible, and because online communities are cheaper and more genuine vehicles for that message, they're attractive. It would be wrong, however, to think that marketing is the only reason companies need to engage communities.

Support Communities: Help Those Who Help Themselves

One of the most compelling reasons to create and engage communities is that customers love helping one another. They're good at it. It's also much cheaper for customers to service themselves than it is for you to handle their support calls.

We've known for many years that online service is cheaper than the in-person alternative. A 1997 study conducted by Booz Allen Hamilton showed that a web banking transaction cost a bank just $0.01, compared to $0.27 for an ATM transaction and $1.07 for a transaction carried out in person. Dell Financial Services estimates that it saves $1.41 for every transaction completed online versus a transaction completed by an employee.

It's not just about cost savings. Communities often provide better support more quickly than the manufacturer itself. To harness the power of better, faster answers that cost less, you need to create a platform for discussion and moderate it. Because support communities benefit both end users and companies, they are one of the fastest-growing community initiatives within businesses.

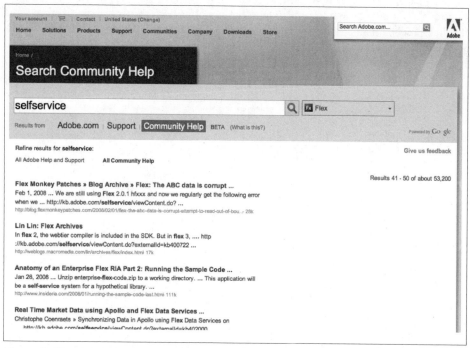

Figure 12-15. Adobe's support community

Consider Adobe's community help site, shown in Figure 12-15, for example. It's run by the company itself on systems that it controls, staffed by subject matter experts, moderators, and editors. In fact, the company often promotes its online community rather than its support sites.

Other support platforms are less tied to a particular company. Mahalo Answers, for example, is a self-organizing community that tries to help its members with a variety of topics. And sites like Stack Overflow, shown in Figure 12-16, build around a particular subject matter or topic—in this case, developers.

Support communities can just as easily appear on Google groups, IRC channels, or anywhere a company's users congregate. Most of the members of these communities will arrive through word of mouth, or by searching for a particular name, brand, or string.

What Makes a Good Support Community?

Users have to be able to explain their problems easily and find the best potential answers. Often, this is a combination of robust search and a means for community members to upvote the most relevant or well-prepared responses. This kind of functionality is one of the reasons support-focused sites like Get Satisfaction, shown in

Figure 12-17, have grown in recent years—they offer better rating, organization, and feedback mechanisms than generic communities such as mailing lists or Usenet.

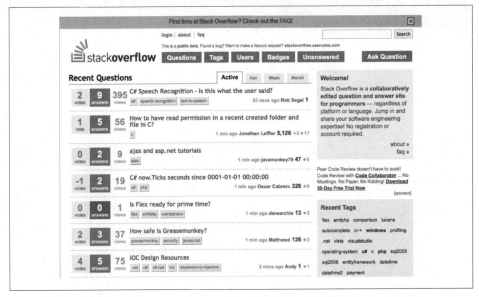

Figure 12-16. Stack Overflow is a developer-oriented question-and-answer community

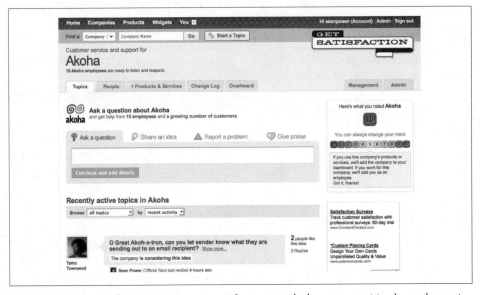

Figure 12-17. Get Satisfaction is a customer satisfaction portal where communities demand attention from companies

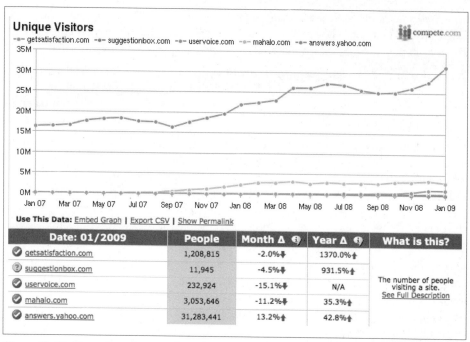

Figure 12-18. *Up-and-coming sites have healthy growth; established sites have huge numbers of visitors*

Community self-service sites are growing quickly. While Yahoo! Answers dominates the self-support field today with millions of unique visitors a month, some of the more targeted, topical support sites we've seen are rapidly gaining market share (Figure 12-18).

Risk Avoidance: Watching What the Internet Thinks

The Web is an open forum, and there's very little control over what's said. This can affect your business in important ways.

First of all, online content may be *libelous or slanderous*. With web communities, everyone in the world has a soapbox. Open conversation is the norm and the Web is a relatively free form of expression, but if people post things that can be proven false, you can often challenge them or have them removed.

Don't be overbearing with your retaliation to online commentary, or your challenge may backfire. Search engine caches, whistle-blower sites (like Wikileaks), and even attentive users taking screenshots will preserve copies of what was said and distribute them on platforms beyond your control. It's always better to deal with the criticism openly, and to let the community draw its own conclusions, as long as you can support your claims with facts.

A second source of risk is *intellectual property violation*. So much of your company's assets are tied up in brand equity, proprietary information, media, and other forms that you may experience significant material losses when others use that information.

While digital piracy might seem the most obvious form of intellectual property risk, other, subtler, ones are also important. For example, some vendors may repurpose your brand or slogan, modify it, and sell it on clothing; you need to defend that brand across all media. Communities provide excellent early warning against this kind of behavior.

Finally, there's *your own liability*. If you set up a platform for a community, you have a moral—and sometimes legal—responsibility to moderate what goes on there. This may depend on the subject matter (healthcare, stock tips, and financial advice are good examples of high-risk topics) or audience (sites aimed at children are particularly restricted). You may be responsible for what others post to your platform, since you're the one operating it, and you're definitely responsible for reviewing complaints about offensive or damaging content and taking appropriate actions.

During the recent U.S. electoral campaign, for example, attackers posing as Obama supporters created posts that included embedded video links to malware-laden video sites. It's not clear who would be responsible for infections delivered in this way, but courts may yet find that companies have an obligation to use recognized best practices to protect visitors from such risks.

Even if you don't have legal reasons to moderate content, you should still maintain your community. Jason Kottke makes a good case that untended communities, like uncared-for neighborhoods, bring out the worst in people (*www.kottke.org/08/12/does-the-broken-windows-theory-hold-online#*).

Business Agility: Iterative Improvements

Companies that don't understand their markets are doomed to failure. You may have the best product in the world, but if you're selling it to the wrong audience—or if they don't actually need it—you'll lose. Communities keep you nimble by showing you what they need and providing immediate feedback on what's working and what's wrong.

A Climate of Faster Change

When business cycles were slower and customers lacked the means to engage companies directly, firms had slow product release cycles. They could take months or even years to design new products, build them, release them into the market, and see how they fared. If the companies were clever, they built instrumentation into their products and reviewed the resulting data when the time came to build new ones. But it was still a slow process, shown in Figure 12-19.

This cycle existed whether the company was releasing a car, a phone service, or a new ad campaign. To know if things had improved, the company looked at the results. This

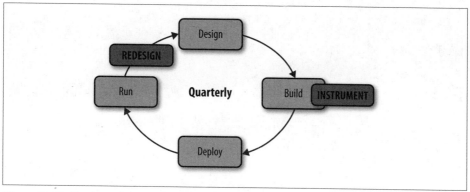

Figure 12-19. Traditional design/feedback cycle

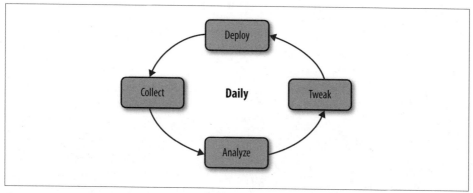

Figure 12-20. Modern design cycle incorporating community feedback in real time

approach also assumed that the problem the product solved was well understood and that the solution to that problem was clear—it just needed to be built.

A long product cycle and periodic review of feedback is simply too slow for competitive, fast-changing business climates, particularly if your company is web-based or offers a service whose policies and offerings can be quickly adjusted. You not only need to tweak the solutions you offer in real time, you also need to adapt your fundamental business model to tackle new problems or take advantage of changing market requirements.

To do this, you need to go straight to the source—your customers—and get the answers from them with a daily cycle of collection, analysis, and iterative tweaking (Figure 12-20). This turns your business into a real-time business that measures and adjusts itself quickly.

Communities provide an excellent way to find out what your users think. You can share feedback with internal teams, such as product managers, and you can repurpose this feedback as marketing collateral. Provided you set expectations properly and manage the occasional disappointment, you can even invite customers to help you define your

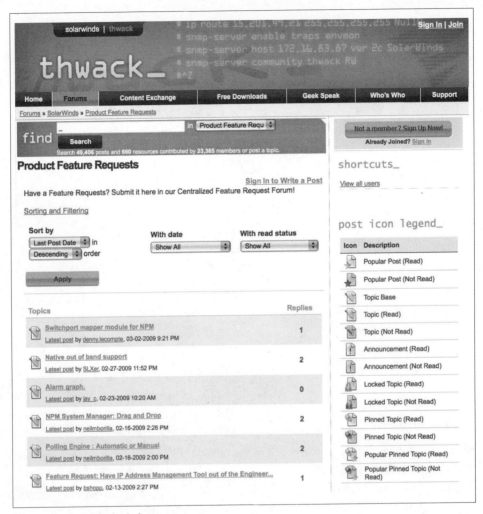

Figure 12-21. SolarWinds feature request

product road map, as network management vendor SolarWinds does on its website, shown in Figure 12-21.

Getting Leads: Referral Communities

Another important business reason for communities is referral. Whether your company is looking for new customers or for new employees, websites are quickly becoming the most efficient tool for establishing new relationships.

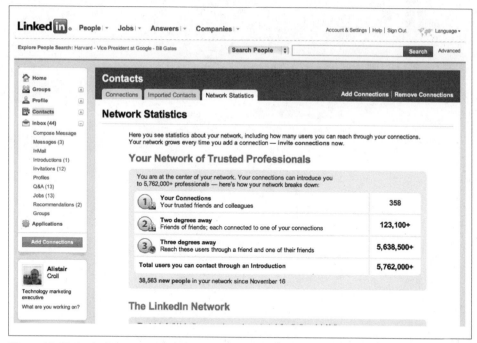

Figure 12-22. LinkedIn is a social community targeted at business users

While sites like LinkedIn, shown in Figure 12-22, are primarily focused on individuals rather than companies, they are important resources for your business and need to be treated like a community, either by training your employees on their use or by monitoring your presence.

Clearly, there are many reasons to engage your community, whether your goals are better marketing, improved support, broader reach, or simply to find out what's going on across the Web in case you need to get involved.

The best way to learn about communities is to join the conversation. And like any conversation, that's a matter of figuring out who's talking, what they're saying, and where they're saying it.

CHAPTER 13

The Anatomy of a Conversation

Communities are all about conversations.

This isn't a new idea, and it isn't ours—*The Cluetrain Manifesto* (*www.cluetrain.com*), talks about this concept extensively. It's just taken a long time for mainstream businesses to embrace it. But embrace it they have, and today's marketing organization is a veritable group hug with communities in the farthest reaches of the Web.

Having looked at some of the reasons that you need to engage in a dialogue with your audience—both on platforms that you run yourself and those run by others—it's time to turn to the conversations themselves.

All conversations consist of three distinct components: who's talking, what they're talking about, and how they're talking. The same is true of communities, and these three components can help you understand the dynamics of community monitoring.

The Participants: Who's Talking?

Some of the community participants are members of your own organization, helping to grow and moderate it, but the vast majority of community members are outsiders.

Internal Community Advocates

Communities thrive on activity. As with conversations, an awkward silence makes everyone uncomfortable. If you're organizing and moderating a community, it's your job to fill in the gaps in the conversation so that the other participants stick around.

Nurturing and engaging communities requires the involvement of employees from the mailroom to the boardroom. If you're running a community by yourself, such as a customer support forum or a documentation wiki, you'll need a lot of internal contributors and advocates. On the other hand, if you're just joining communities that already exist elsewhere, you won't need as much internal support, because the community's already self-sustaining and there's less need to administer or police things.

Even if you're only concerned with communities that you join, it's still wise to keep upper management apprised of what you're up to. This way, you'll surface any concerns they may have about discussing your company in a public forum.

Here are some of the internal advocates you'll need on your side.

Executive sponsor

Online communities rarely succeed without executive buy-in. A senior member of the company has to be convinced that the community is good for business. She needs to know the differences between traditional marketing and community marketing, or the cost benefits of self-service portals.

Many executives will pay lip service to community initiatives—after all, it's political suicide to say you don't want to talk to customers—but far fewer of them will really support the effort. There are simply too many things that could potentially go wrong with an unfettered discussion for most executives, who are notoriously risk-averse.

To get executive sponsorship, you'll have to answer one of the toughest questions of all: *why are we investing in a community?* We've given you some of the reasons, from better marketing to improved support to risk mitigation, but you need to actively communicate and get buy-in for this motivation. Many community managers go straight to community building without knowing why they're doing it. This is a big mistake. You won't know what to measure or how to define success. Don't skip this step in your community monitoring process, or you'll pay for it later when the political winds change amidst palace intrigues and someone second-guesses your work.

Once you have a genuinely supportive executive sponsor, you'll get the budget, resources, runway, and expectations to make the community flourish. In time, you'll repay your sponsor with hard evidence that the community effort is paying off—increased word-of-mouth sales, reduced call center traffic, lower hiring costs, and so on. This will give your sponsor the ammunition she needs to justify the gamble to less forward-thinking peers and marketing Luddites.

Ultimately, the executive sponsor's role is to help determine the business case for the community, to legitimize the effort, to overcome political roadblocks, and to decide which metrics and milestones will be used to determine success.

Administrators

Community administrators ("admins") link internal community members and external resources such as contractors or developers. They organize the structure of the community by creating topics, modifying areas, leading teams of moderators, and arbitrating disputes.

Administrators also adapt the community's terms of use when unexpected events happen, turning individual incidents like a flame war or the posting of inappropriate content into community-wide policies and providing a platform that allows the community to grow organically.

Finally, administrators are responsible for collecting community data, either from internal platforms or from whatever tools are available on external sites, and aggregating it into reports that the executive sponsor and the community team can understand.

Any community that you moderate or run yourself needs an authoritative administrator. The administrator will have the final word on contentious disputes that will inevitably arise. This person needs to be level-headed and professional, and must be able to anticipate flame wars and settle issues quickly, consistently, and decisively.

Moderators

Community moderators ("mods") are responsible for specific sections of the community. They read, maintain, and moderate platforms that are often related to their own interests or their roles in the organization. They answer basic questions that community members have, and make sure that new participants are welcomed.

Moderators' interactions happen on a public platform, so their work helping others is searchable. This can dramatically lower the support volume of your organization, because repeated questions can be answered with a search, which lets support scale more effectively.

Moderators also provide rough quantitative and qualitative measurements to administrators. They are your first line of defense against spam and trolls, and act as an early warning system for potential problems. Since community moderation is extremely subjective, mods are intimately familiar with your terms of service and are able to enforce them gently before escalating.

Don't Feed the Trolls

A troll is a community member who contributes controversial, inflammatory, or off-topic content to a community in the hopes of getting attention or provoking reactions. Trolls are one of the most difficult aspects of community management, since any response to their behavior is precisely the effect they're hoping to have.

In certain cases, legitimate and established community members can be seen trolling other users to get reactions out of them, but these incidents are often short-lived, done in jest, and superseded by the overall quality of content that these members provide. In other words, the more valuable a member is to the community, the more likely he will be able to troll without community backlash.

Wikipedia has dealt surprisingly effectively with trolls, considering the amount of subjective information on the site. Its primary rule, *do not feed the trolls*, simply states that trolling behavior should not be acknowledged. Some other techniques Wikipedia suggests include deleting offensive or illegal content quickly, enlisting the community's help to report trolling, and not undoing a troll's work immediately (provided that it's

---

not illegal) so that the troll loses interest. You can read more about Wikipedia's trolling policy at *http://meta.wikimedia.org/wiki/What_is_a_troll%3F#Dealing_with_trolls*.

Some social news aggregators hide responses that are rated below a certain level from everyone but the poster, so trolls don't know they're being ignored. Boing Boing has another approach: disemvoweling, in which offensive comments have their vowels removed, neutering them without deleting them entirely. Check out Cory Doctorow's web article, "How to Keep Hostile Jerks from Taking Over Your Online Community" at *www.informationweek.com/shared/printableArticle.jhtml?articleID=199600005*.

If you're running your own community platform, invite your best, most enthusiastic customers to help you run things. Strike a healthy balance between in-house moderators (employees who have access to moderation tools and administrative controls) and external moderators (third-party volunteers who care and are considered more "genuine").

Subject matter experts

Subject matter experts are occasional contributors who provide specialized knowledge, such as lists of frequently asked questions and answers, snippets of code, recipes, or technical configurations. They're effectively seeding this information into the community. Often, subject matter experts are product managers or certified professionals, such as doctors, who have in-depth experience dealing with the topics around which the community is organized, but have to be hounded for content or participation.

External Community Members

The community participants who really matter are those from outside. They're your content producers and consumers, your performers and your audience, your feedback, and your friends and your critics. They have different levels of involvement with your community and want different things from it.

There are many ways to classify the external members of your community. One approach is to look at their online behaviors. While many community members consume content, fewer share it, fewer still comment on it, and only a small portion produce or maintain it. Figure 13-1 shows one model of community participants and the roles they play.

There are many other ways to segment online contributors. Forrester Research, for example, splits them into creators, critics, collectors, joiners, spectators, and inactives, according to the actions they perform online. For the purpose of monitoring and responding to community members, we've separated them into six distinct groups according to their rates of contribution and engagement:

- The hyperinvolved, prolific few (what we call the *long tail of freaks*, or LTOF)
- The totally engaged, but less pathological, *fans and contributors*

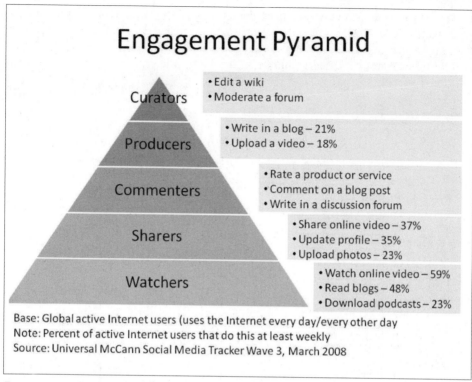

Figure 13-1. The Altimeter Group's model of community participants based on engagement; Altimeter's model counts community members by behavioral segments in the Pew Internet and American Life survey on Internet usage, where available

- *Bursty users*, who vary significantly in their contribution rates
- *Occasional participants*, who only interact in small ways
- *Lurkers*, who watch from the sidelines
- The *disengaged*, who you count as part of the community, but who no longer consume nor contribute content

To understand how we've separated these groups, let's look at a fundamental concept of information systems: power laws.

Power laws and community members

Most things in the information world follow a power law. A power law simply means that a system has a few very common things, and many, many rare ones. One power law you've probably heard of is the 80/20 rule (known more formally as the *Pareto distribution*), which says, for example, that 20 percent of people control 80 percent of the wealth.

Power laws show up in everything from bookstores (a few popular books sell millions, while many niche publications sell only a few copies) to last names (there are a few very common names and many rare names, as shown in Figure 13-2). On the Web, a small number of websites have nearly all of the Internet traffic, while there are millions of seldom-visited ones.

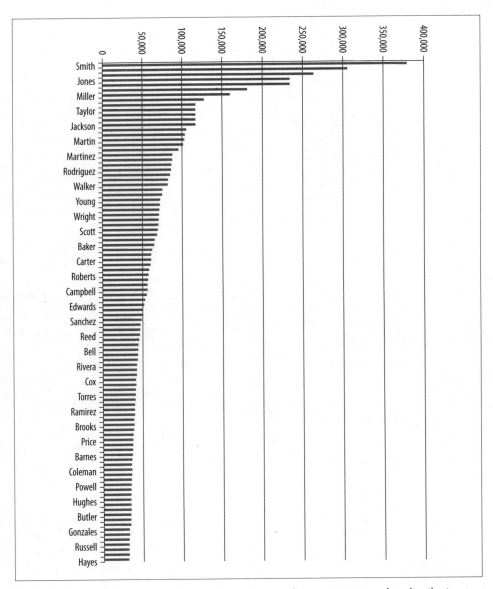

Figure 13-2. A graph of last names by popularity shows a characteristic power law distribution

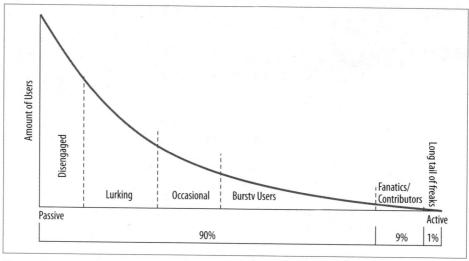

Figure 13-3. An analysis of community members by level of activity shows many disengaged members and a long tail of hyperactive participants

The long tail of power laws has been the subject of much discussion on the Internet, as information systems make it easier to find and analyze them (see "Power Laws, Weblogs and Inequality" at *www.shirky.com/writings/powerlaw_weblog.html* for an excellent discussion of web content and power laws in action). One place they turn up is within communities.

Many of your community members will be completely disengaged, having participated once and forgotten you (don't take it personally—we still love you). Some will be lurking, and still others will occasionally participate in discussions. You'll see one group of members become very active for brief periods of time, often on subjects they care a lot about. About 10 percent of your visitors will be heavy contributors, while a small subset will be so overactive they'll need to be politely but firmly restrained.

These numbers seem to be universal. Wikipedia has an extremely small number of active contributors—75,000, which is just over 0.01 percent of all its visitors—yet that tiny group has created over 10 million articles. Jakob Nielsen's research supports this distribution: "In most online communities, 90% of users are lurkers who never contribute, 9% of users contribute a little, and 1% of users account for almost all the action."

Figure 13-3 shows the relative distribution of these people across a community.

If you flip the axes (as we've done in Figure 13-4) and look at the volume of contributions from these users, nearly all content comes from the fanatics, freaks, and contributors. Think we're exaggerating? Amazon.com's top reviewer at the end of 2008, Harriett Klausner, had contributed over 17,900 book reviews.

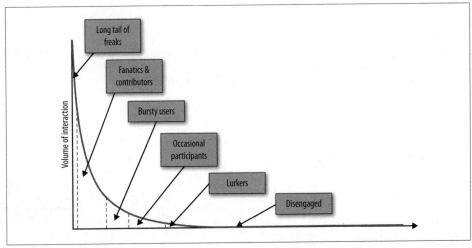

Figure 13-4. A histogram showing the number of posts by a contributor reveals that a very small number of participants are responsible for nearly all of the content in a community

 We're not implying that Ms. Klausner is a freak—her reviews have been rated "helpful" by over 70,000 customers. But to read 17,900 books, a 70-year old person would have to finish a book every 34.2 hours since birth, which seems a bit out of the ordinary. Amazon, perhaps reacting to this and other users who reviewed in volumes, changed their practices to rank based on the percentage of positive or "helpful" votes by Amazon.com users. For more information, see *http://en.wikipedia.org/wiki/Harriet_Klausner*.

The boundaries of the community change contribution rate. On less hierarchical platforms like Twitter, for example, there are millions of *emergent communities* that might coalesce around a hashtag or a group of friends. But there's still a power law at work: a few celebrities and highly followed personalities account for most of the content that's repeated and amplified.

For communities that you run, your goal is to try to get as many members of the community as possible into the "regular contributors" category by engaging the disengaged, coaxing the lurkers out of hiding, discussing topics that the occasional participants feel compelled to join in on, and dealing gently but firmly with the freaks and abusers.

 See Ross Mayfield's web article "Power Law of Participation" (*http://ross.typepad.com/blog/2006/04/power_law_of_pa.html*) for a discussion of contribution across population. Mayfield references the *Cornucopia of the Commons*, Dan Bricklin's term for open systems that grow because of their communities' contributions (*www.bricklin.com/cornucopia.htm*).

Long tail of freaks

The LTOF is made up of the few users who will do their best to be active within the community, *despite any barriers to entry, no matter how high*. These members may create, modify, and correct content, even to the point of undermining other contributors. They will report bugs and suggest feedback often and liberally to anyone who will listen.

The LTOF is a mixed blessing. They'll start to derive a sense of self-worth from your community, and they may be seen as authorities by their peers, but they can also make newcomers feel unwelcome and challenge the authority of legitimate administrators. On Wikipedia, for example, debates constantly rage between the core team of editors—who feel personally responsible for every entry on the site—and less fanatical writers, whose occasional entries are often removed by those editors, discouraging them from becoming more engaged.

Dealing tactfully with the LTOF is a big part of a moderator's job. These members want attention, praise, and sometimes rewards, but they can also be brittle and fickle, and will air their grievances openly. When they do, their familiarity with the community makes them hard to silence. They're one of the main reasons, outside of spam, that a well-understood and decisively enforced terms of service agreement is vital to the well-being of any community.

The first line of defense against the LTOF is to monitor for them. Set thresholds for rates of contribution, number of votes in a visit, volume of posts, and number of links in a message, for example, and you'll quickly identify the members who exhibit an overabundance of personality.

Fans and contributors

Fans and contributors are the ideal community members. They'll spend their time nurturing and pruning your community's content, and may even create new material that grows your audience and funnels members toward conversions. They might appoint themselves as moderators, spam catchers, or new user welcoming committees.

Their contributions often go beyond just posts. Contributors may use a website's APIs to develop tools that extend what your site can do, or they may find new ways of using your products and services that you hadn't intended to—or had time to—develop. They're your regular contributors and your grassroots support across the Web. They represent a small but valuable portion of your community. Love them and nurture them, and give them the authority they deserve.

From a monitoring standpoint, you need to analyze their behavior and use it as a "control group" against which to baseline other segments. For how long do they visit? How many comments a day do they make? Where do they link from? These measurements will give you clues about how to turn other segments of your community into the contributors you crave.

Lurking, occasional, and bursty users

The vast majority of your community's members fall into three groups:

- Bursty users, who only post when they feel passionate about something being discussed, but do so furiously for a brief period and disappear shortly thereafter
- Occasional contributors, who chime in from time to time, but don't feel a sense of ownership
- Lurkers, who occasionally watch the proceedings without contributing

Even relatively passive community members contribute to the community to a small degree by amplifying and selecting good content, or by flagging abuse. This may come in the form of reTweeting, digging, or upvoting, which reinforces the popularity of superior material and helps it rise to the top of search rankings or message lists.

Many community sites go to great lengths to make it as easy as possible to contribute. As a result, when a community feels passionate about something, lurkers quickly become active participants.

Consider Electronic Arts' release of the game *Spore*, which contained strong copyright protection. The game's Amazon.com page quickly yielded over 3,000 comments and ratings in just a few weeks, as shown in Figure 13-5. By comparison, every version of publisher Maxis' bestselling *SimCity* game, dating back over 10 years, has less than 1,000 reviews.

Disengaged users

Disengaged users are those who spend very little time within your community portal, entering and leaving almost immediately. They were either the wrong visitors in the first place—having arrived by accident—or have simply grown uninterested. Perhaps they signed up to get access to software or one-time content. Now, your email messages to them are bouncing.

If you're running a community, your goal is to optimize your points of entry so that users stay, subscribe, enroll, and bookmark the site, minimizing disengagement. Remember, too, that disengaged users can skew your metrics: you need to calculate your reach as the percentage of *active* community members who saw your message, not a percentage of all enrolled community members who saw it, to avoid counting disengaged users.

The Topics: What Are They Talking About?

Communities emerge to discuss something, whether within a Facebook group, a mailing list, an IRC channel, a blog's comment thread, or a Twitter hashtag. You want to be aware of any topics that concern you.

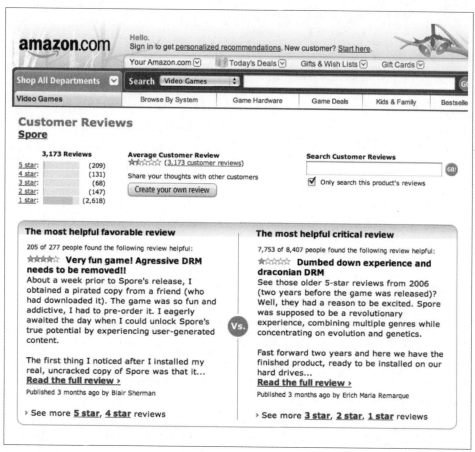

Figure 13-5. Objections to Maxis' use of strong copyright protection turned many lurkers into active community members for a brief period

You find these topics by searching for them in the aggregate (most popular Twitter hash tags or most active Facebook groups, for example) and in the individual (who's discussing a topic on Twitter or which Facebook users are members of a group). By tracking topics in the aggregate, you understand trends, community size, and growth. By tracking individual threads, you find evangelists and brand promoters who can help encourage the spread of your message and make others aware of the community to expand membership.

Platforms that you run yourself often have built-in tools to help identify popular subjects, and as an administrator you'll have access to user account information, so you can follow up directly with commenters. But if you're joining conversations elsewhere, you need to do some work to find out which topics are of interest to your business and your audience. Some obvious topics you probably want to track include company and product names—yours and your competitors'—and any topics that apply to you or

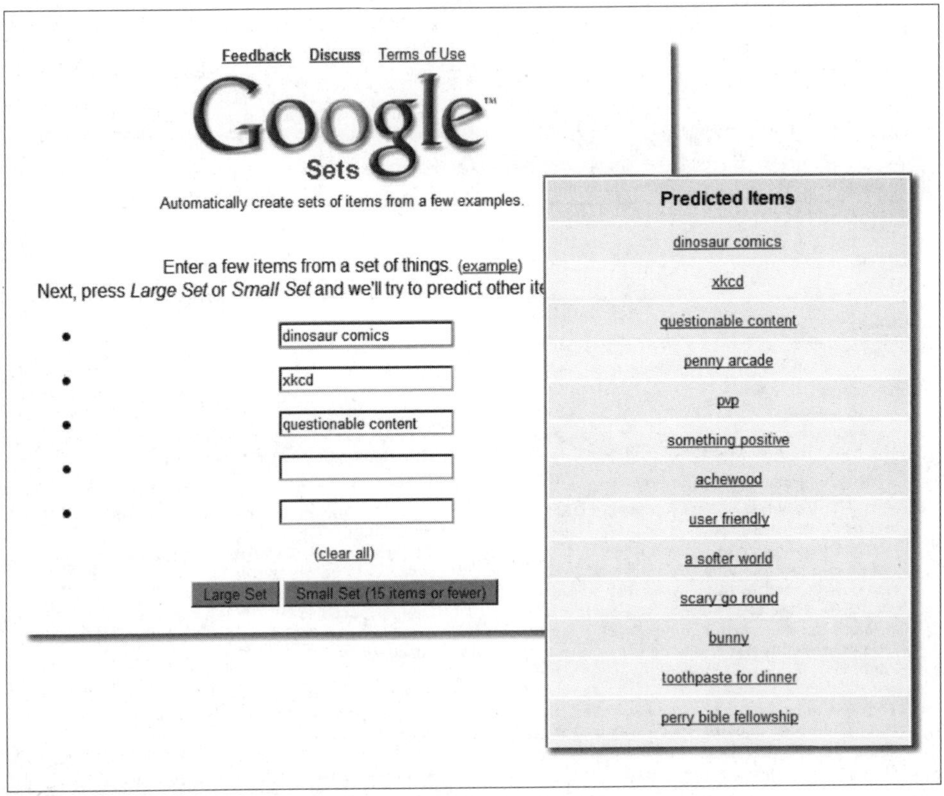

Figure 13-6. Goggle Sets suggest sets of terms that are related to those you supply

that you think are related, such as product categories, job descriptions, or geographic mentions.

If you're looking for inspiration about what to track, consider tools such as Google Labs' Google Sets tool (*http://labs.google.com/sets*), shown in Figure 13-6, which suggests related terms, companies, or products when you provide a few you know.

You should also check what terms others use to find you. A good list comes from organic search results within your analytics application, such as those shown in Figure 13-7.

It's not enough to look at the terms that are bringing *you* users—you should also look at companies that are competing with you for the mindshare of those terms. To do this, first find out which companies are also bidding on keywords that drive traffic to you. They may not be direct competitors, but they're competing for mindshare and attention.

Paid search uses an auction bidding model, so services like Google's AdWords offer a lot of detail about who's buying which terms. As a result, services like Spyfu, shown in Figure 13-8, can aggregate this kind of information.

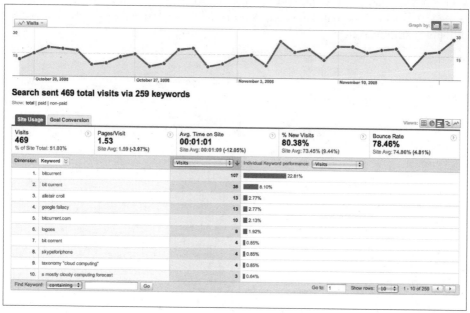

Figure 13-7. Organic search terms in Google Analytics

You can work the other way: starting with a known competitor (perhaps one that's stealing your mindshare), find out which keywords are sending traffic to that site. Figure 13-9 shows a list of organic and paid search keywords for analyst firm Gartner, generated by Spyfu.

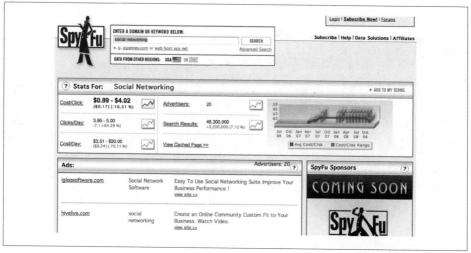

Figure 13-8. Tools like Spyfu show you who else is getting traffic for a particular keyword (in this case, "social networking")

You can then find groups and communities in which these keywords are being discussed. For example, you might enter a keyword into Twitter search to see what people think of it at the moment.

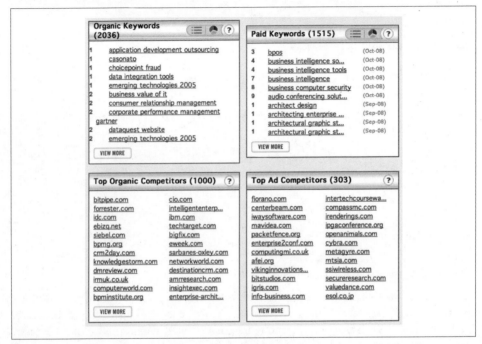

Figure 13-9. Using a competitor's name, you can find out which terms are driving traffic to its site, then include these terms in the topics you monitor across communities

You can also look at topics by geography. Google Insights, shown in Figure 13-10, will show you important news events, mentions of terms, and countries in which those topics are heavily discussed. You can use this information to check popular forums and portals in those areas.

Referring sites that are sending you traffic are often communities within which you're being discussed, as shown in Figure 13-11. Depending on your analytics tool and the amount of detail the referring site provides, you may be able to pinpoint the individual submission or comment that generated the traffic. You can then add that community to those you watch.

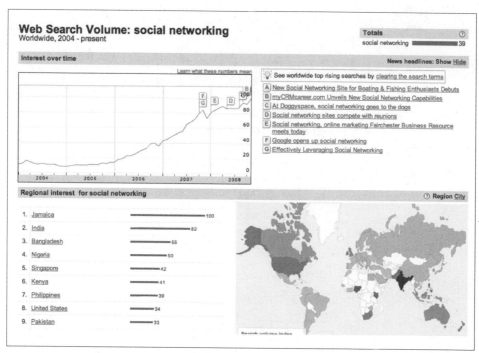

Figure 13-10. Geographical distribution of the term "social networking" in Google Insights

Unfortunately, analytics tools are often limited in what they can tell you about the sites that referred traffic to you. Many community sites won't identify themselves properly with referrers. This can happen for several reasons:

- The community platform may rely heavily on browser *bookmarklets* (Stumble-Upon and Twine are two examples), which interfere with referral mechanics.

- The community may want to make money from analytics and as a result may actively hide referrers, replacing them instead with a promotion of its advertising platform (StumbleUpon does this).

- The content may have come from a desktop client—most notably microblogging tools like Tweetdeck or Twhirl—that doesn't leave a referrer.

- The community may have third-party websites that post content on behalf of a user's account. For example, Twitpic, shown in Figure 13-12, lets community members upload an image and then posts it to the account on behalf of the member.

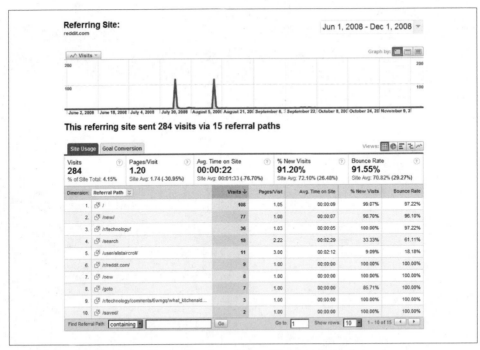

Figure 13-11. Traffic from reddit, broken down by source in Google Analytics

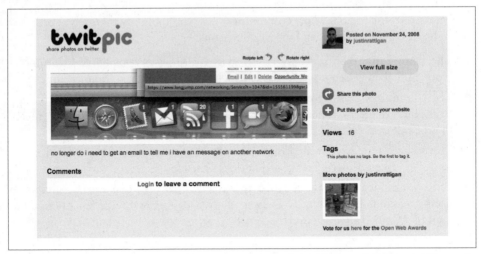

Figure 13-12. Referrals from third-party sites or desktop clients that use the Twitter API may be obfuscated, appearing to analytics tools as if they were typed in directly rather than linked from Twitter

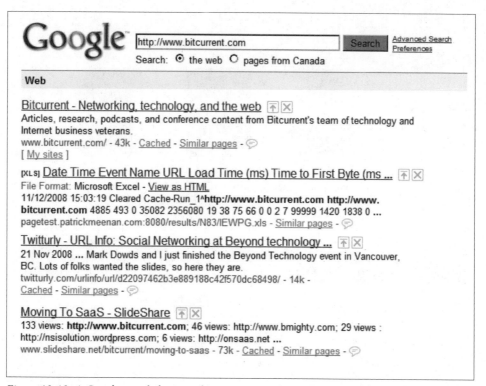

Figure 13-13. A Google search for www.bitcurrent.com shows mentions of the site on both Twitter and Slideshare

Because of how community sites often break referrers, if you want to find community mentions, you need to employ other techniques to track traffic back to the communities from which it came. You need to search for your URLs online and see where they're referenced. The search shown in Figure 13-13 shows not only the *www.bitcurrent.com* site, but also mentions of the site on both Twitturly and Slideshare.

URLs in their raw form are becoming less common. URL shorteners first emerged as ways to transmit long URLs, complete with parameters, through email without having them break across multiple lines of an email message, which would have rendered them useless. Shorteners have found new popularity in Twitter and other microblogging platforms where URL length is precious. Today, dozens of URL shortening services reduce long URLs into a simple, short string automatically using the process shown in Figure 13-14. This means that for many social networks and communities, your URL will be obscured when people mention you.

So how do you find places in which your URL has been mentioned online when it's been shortened? Here's one way.

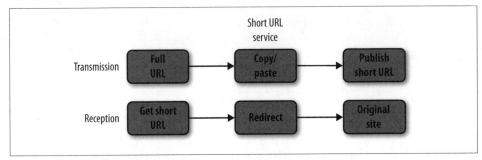

Figure 13-14. URL shortening services such as bit.ly and tinyurl abbreviate your URL, then redirect those who click on the shortened version to the original site content

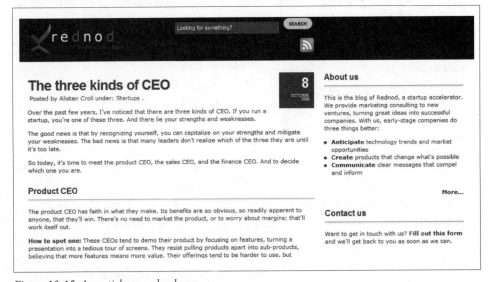

Figure 13-15. An article on rednod.com

Start with the online content whose incoming traffic you want to track back to a community, such as a blog post (Figure 13-15).

Enter the long URL of the content (in this case, *www.rednod.com/index.php/ 2008/10/08/the-three-kinds-of-ceo/*) into a URL shortening service (such as is.gd). The service will return a shortened URL, as shown in Figure 13-16. If someone has used the service to generate a short URL recently, you'll often get the same one that individual did (this depends somewhat on the service).

Enter the URL, in quotes, into a search engine (Figure 13-17).

The results will show you where your shortened URL has appeared on the Web. You can often find out more about the poster or the discussion thread in question,

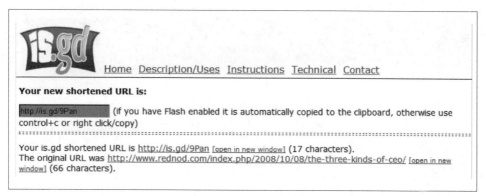

Figure 13-16. Generating a shortened URL using the is.gd shortening service

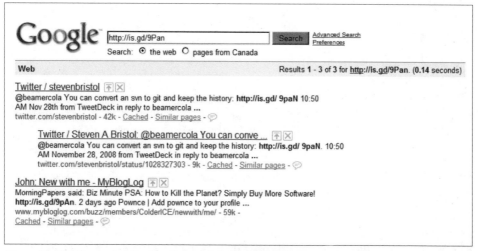

Figure 13-17. The results of a Google search for a shortened URL generated by a shortening service

particularly if that person is a member of a public community with a profile of some kind, as shown in Figure 13-18.[*]

Unfortunately, there are so many URL shortening services—each of which generates a unique shortened URL—that it's time-consuming to search for every mention you encounter online. Instead, use this approach judiciously to investigate particularly popular or contentious postings.

---

[*] We'd like to apologize to John for stalking him, featuring him in a book, and generally being creepy. As this example shows, you need to be careful when reaching out to your community. And we owe John some beers if he ever reads this.

Figure 13-18. Finding out more about the people who are amplifying your community message, what they talk about, and where they hang out

Shortened URLs do more than save space. They provide us with a method of embedding additional context into short messages. You can generate a unique URI for each recipient or embed parameters in the URL that help to track its spread with web analytics. We provide a complete example of how to do this in Chapter 14, in the section "The Mechanics of Tracking the Long Funnel" on page 520. For now, remember that short URLs will be an important part of microblogging analytics; the URL has become the new cookie. We recommend using cli.gs, bit.ly, awe.sm, or some other service that provides its own analytics capabilities.

The Places: Where Are They Talking?

We've looked at the people involved in a conversation, and at the topics you should be investigating. Where are those conversations happening?

There are hundreds of thriving communities, thousands of forums, and millions of comment threads to cover, but there are a small number of communities that will drive much of your traffic. Fortunately for us, popular communities also follow power laws.

Figure 13-19. The now-defunct usernamecheck.com allowed you to verify whether your name was taken on popular Web 2.0 sites

We also have tools such as search to help us, and many of the communities make their sites searchable through APIs and aggregators.

Your first instinct might be to register yourself on as many sites as possible, some of which are listed in Figure 13-19, and search for conversations involving your brand.

That's a good step, but while it might reserve your seat at the communal table, it won't let you sit down and interact. Before we delve into the ways we can track online activity, let's first cover the types of community on the Web today.

 Also check out *www.claimid.com* and similar offerings that manage all of your online profiles with OpenID.

Different Community Models

We can divide up models of online interaction according to two important dimensions: the complexity of their messages, and how openly those messages are shared.

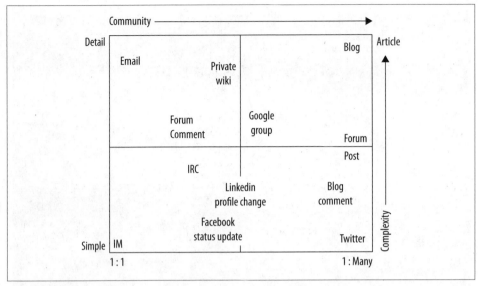

Figure 13-20. Classifying community platforms by message detail and degree of openness

Message complexity

> Some communications methods, such as IM and Twitter, are suited to simple thoughts and quick references. Others, such as email messages and blog posts, are better for detailed discussions and structured arguments.

Openness of transmission

> Some communications methods are one-to-one, such as IM, while others, like Twitter or blogs, are open to the whole world. Those that are more open encourage message spread, but make it harder to track individual users from a community message to a web analytics outcome the way we can with a one-to-one medium like email because we can't identify a unique conversation with them in order to start tracking their behavior.

Every community falls somewhere in this spectrum. A mailing list, for example, is a detailed message for a group interested in the same topic. A Facebook update is a relatively simple message for a group of friends. Twitter is a very simple message aimed at anyone who wants to hear from you. And so on.

Figure 13-20 shows how various community models fit within this classification.

All that's needed for a community is interaction and a group of people. The community can form in many places: groups and mailing lists, forums, real-time chat systems, social networks, blogs, wikis, micromessaging tools, and social news aggregators. Each has unique characteristics that are worth looking at in detail.

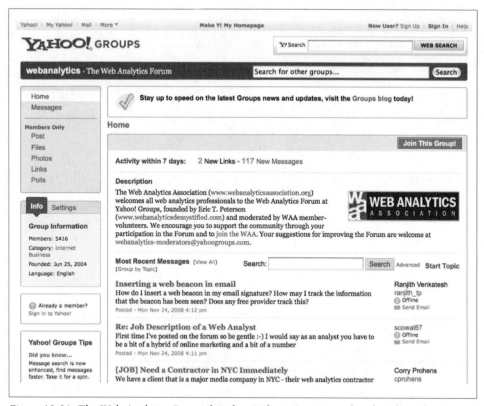

Figure 13-21. The Web Analytics Forum found on Yahoo! Groups reaches the inbox of many web analytics professionals

User Groups, Newsgroups, and Mailing Lists

At the start of the previous chapter, we saw several early forms of community, including mailing list servers (listservs). These systems are still alive and well, and have been reskinned for the Web by the major portals, such as Yahoo! Groups (shown in Figure 13-21). Many of these portals have comprehensive frontends for community organizers to manage members, change policies, and send out bulk messages.

Mailing lists are unsurpassed in their ability to distribute messages to a relatively large audience that's interested in the same topic.

Most groups have four main functions, as shown in Figure 13-22. They include:

- *Maintenance* of the group and its policies. The administrator creates the group and decides who can access it, how messages are distributed, and the degree of control he exerts over the community (for example, he needs to individually approve new participants).

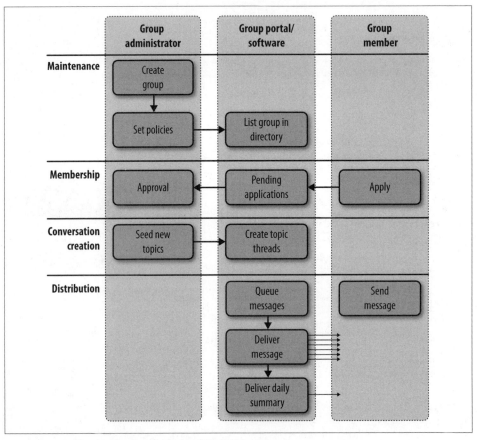

Figure 13-22. The core functions of a mailing list management system

- Managing *membership* to the group if applicable, including the approval of new applicants and blocking misbehaving members.
- *Creating conversation* by seeding new topics to spark conversation, and responding to messages from the community.
- *Distribution* of the day's conversations, either individually when they happen or as a daily summary, according to recipient preferences. This may happen via email or on the group's website.

Groups haven't changed much since the early days of the Internet. They're open and easy to join, and there are many of them out there. Some, such as the ones based on Usenet that are now available on Google Groups (Figure 13-23), are open and relatively unregulated. Others, such as mailing lists you run, are easier to oversee.

Figure 13-23. There are hundreds of thousands of newsgroups available through Google Groups

The most important changes to groups have been in their usability. At one time, the only way to search a group was by parsing its archives or digging through your inbox for a specific message. Now, however, these lists and discussions have been indexed by powerful search engines and are accessed through familiar web-based interfaces. As Figure 13-24 shows, search tools make it possible to find decades-old content quickly and easily across all groups.

Conversations on mailing lists and groups are comparatively slow-paced. Usually, email messages consist of longer, more well-thought-out messages. Many members elect to receive their lists only once a day when they subscribe to a list, as shown in Figure 13-25, meaning a conversation happens over several days.

The slower pace of mailing lists and groups makes them easier to administer, since you can review queued messages before they're sent out to remove content sent by spammers and trolls. If you hand-moderate posts, mailing lists often have higher-quality content than real-time models like comment threads. They're also easier to measure because you know how many subscribers you have; in some paid mailing list tools, you'll also have access to metrics such as open rate, which let you see how many subscribers actually read the content.

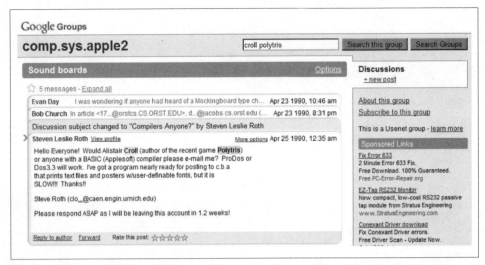

Figure 13-24. Searching Google's index of Usenet posts for a specific string over decades of content

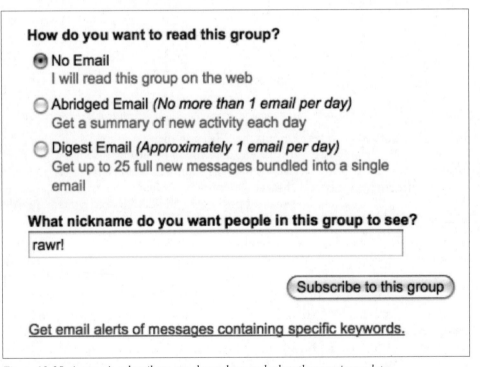

Figure 13-25. A group's subscribers can choose how and when they receive updates

These are the main roles you'll need to fill in a mailing list or group:

- *Administrators* maintain the mailing list or group. They may have created the group in the first place, and they define whether subscribers can join without moderator approval, whether the archives of the group are searchable, and how much editing and review happens before updates or comments are transmitted.

- *Moderators* deal with the content the group generates. They're responsible for enforcing the terms of service, weeding out abusive or illegal postings, and letting the administrator know about problems. They may be self-appointed or nominated subscribers who want to help, and have been given control by the administrator.

- *Subscribers* are the participants in the mailing list, and they may include the frequent participants, occasional contributors, lurkers, and disengaged members we saw earlier. If you're joining a mailing list or group that relates to your organization, you'll be doing so as a subscriber.

Forums

Close cousins of mailing lists, forums are online discussions that emerged from the BBS world. They consist of shorter responses that are viewed immediately within a web page, and are organized chronologically within topics. An example of a forum is shown in Figure 13-26.

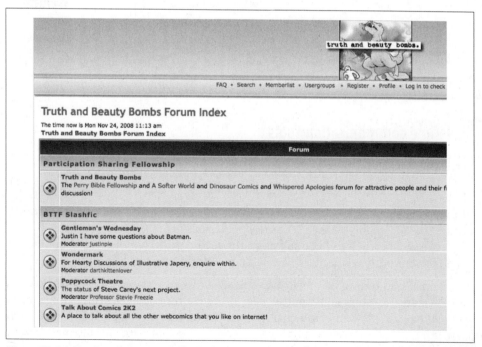

Figure 13-26. Truth and Beauty Bombs forum

Forums have low barriers to entry, often letting visitors read their content when they aren't enrolled, and asking for only basic information, such as a nickname and email address, in order to reply to an existing post or create a new one.

Some forums organize conversations into requests and responses in the form of threads, letting participants respond to responses and nesting the various threads hierarchically to make them easier to follow. Other forums simply organize all posts on a topic sequentially.

Forums are monitored by administrators and moderators:

- *Administrators* maintain the forum itself. They add, modify, and delete sections, patch the system, and collect and aggregate metrics for reporting. Their main involvement in content is when they must act as final arbiters when things get out of hand.

- *Moderators*, on the other hand, form the front line against spam and abusive contributors. Because forums are relatively open, the medium is untrusted. Moderators may also start new topic areas in a forum.

- *Members* respond to topics and to other people's posts. In a forum, members often rank their peers based on contributions. This ranking can be used when filtering content. In other words, a member might suppress comments from members with a low ranking. As a result, you need to add value and earn the respect of other forum members if you want what you say to have a wide audience.

Forums offer a variety of tools to keep you up-to-date with what's happening, so even if you're not administering the forum, you can still receive alerts when someone responds to your content or subscribe to an RSS feed on a particular topic. Most forums are also searchable. Though they provide a more interactive back-and-forth than mailing lists and groups, forums still aren't real-time. That's the domain of chat.

Real-Time Communication Tools

Instant messaging is a one-to-one medium, but real-time chat among groups is a common community model that can involve hundreds of participants across tens of thousands of topics. The biggest of these groups is *Internet Relay Chat*, or IRC.

IRC has existed since 1988, predating the Web. It has its own protocols that make it distributed and highly resilient. Entire books, such as *IRC Hacks* by Paul Mutton (O'Reilly), are devoted to IRC, and as you're reading this, hundreds of thousands of people are currently connected to an IRC network.

IRC is organized into topic "channels" hosted across servers that relay conversations to participants, as shown in Figure 13-27. IRC requires a greater degree of technical acumen than simple web-based chat, despite the efforts of sites like Mibbit to try to make using IRC easier. While this means IRC isn't a popular tool for interacting with

Figure 13-27. A FreeBSD support community on IRC (EFnet)

a mass-market audience, it can be an elegant and cost-effective community platform for technical groups.

Because IRC is its own protocol and platform, and can be completely anonymous, it's subject to exploitation:

- *Hackers* use IRC to control armies of infected PCs (botnets). The much-maligned IRC community has tried to address these problems itself, but its reputation has been tarnished and some carriers have tried to block IRC traffic in an effort to reduce the spread of botnets.

- *File sharing groups* can contact one another anonymously to share files that may violate copyrights.

- *Spammers* use scripts to join channels, send URLs, or advertising messages, then leave—a process known as *drive-by spam*.

- *Trolls* post inflammatory content in the hope of provoking a strong backlash, such as ruining the end of a movie ("Bruce Willis is dead") and then hoping for inflammatory responses that take the community discussion far off topic.

If used well, IRC can encourage customer self-support and provide easy interaction between your organization's technical team and knowledgeable end users, particularly in IT fields. Many employees in your organization can casually monitor a chat room, answering user questions as they come in throughout the day. IRC is also an excellent source of breaking news, given the real-time, fast-paced nature of the platform, so it can quickly communicate zero-day exploits or other technical concerns.

Recall, however, that IRC is an untrusted network. There are a high number of one-time visitors to an IRC community channel, and only a few recognized regulars. It takes time and patience to build a reputation, and channel operators are usually distrustful of newcomers. Problems happen far less often in topical, uncrowded channels, where it's harder for a botnet manager or file swapper to linger anonymously, and spam reaches a smaller audience.

Unlike forums and mailing lists, IRC doesn't have built-in monitoring tools or APIs. Administrators and operators rely on *channel bots*—scripts that eavesdrop on conversations and aggregate statistics on popular keywords, top talkers, and so on—to keep track of what's happening.

Existing communities, such as Facebook and phpBB, may also have simple real-time chat functions (as shown in Figure 13-28) to make them more interactive and current. This kind of chat may encourage members to stay engaged, but it's seldom the primary vehicle for communication within the community.

Simple chat platforms that are part of a larger social network inherit the trust model of their overall platforms. In other words, you've already got a profile on Facebook and a network of contacts within the broader community, so you're less likely to abuse Facebook's chat. Members will likely behave within the standards of the broader community (such as the phpBB forum) when using real-time tools such as its chat function. As a result, real-time interactions that are an extension of other community platforms require less policing and are less prone to abuse.

Real-time chatrooms have a three-tiered hierarchy of moderators and coordinators:

- *Administrators* focus on the technology. They often own the machine, manage traffic flows and routing, and maintain the latest versions of software. They understand peering and are often experts in routing protocols such as BGP (Border Gateway Protocol), because they must stitch together thousands of user sessions across hundreds of machines. They may also work with carriers to resolve security issues and IRC abuse.
- An *operator* ("oper") has the ability to disconnect users from the network (known as "killing" them), but in return for this power must arbitrate channel disputes,

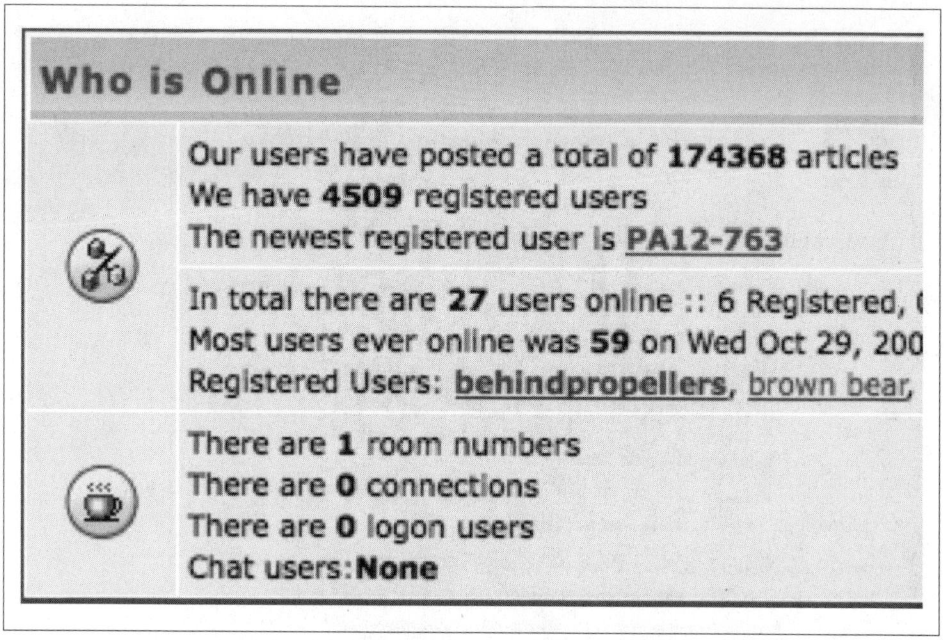

Figure 13-28. Extended chat functionality integrated into the phpbb forum software

work on spam prevention, and must try to kill botnets as quickly as possible by blocking bot traffic.

- Individual *channel operators* (known as "ops") control individual channels, changing permissions and banning users from their immediate channel as they see fit.

- For *participants*, joining IRC is easy; getting some reputation and the respect of your peers takes work.

Social Networks

Social networks are communities built around the interpersonal relationships of their members. By keeping track of who each participant knows—and how they know one another—members each create a private community through a whitelist of known friends.

While two members' friends lists might overlap considerably, very few people have identical social graphs. The community grows as members explore their friends' contacts, befriending them and growing their own social graphs. Requests from strangers are discouraged, which improves the quality of interactions and reduces traditional spam and trolling.

Facebook, in particular, differs from other social networks because it's a place for connecting with real-world friends. Facebook social graphs look more like their real-world counterparts.

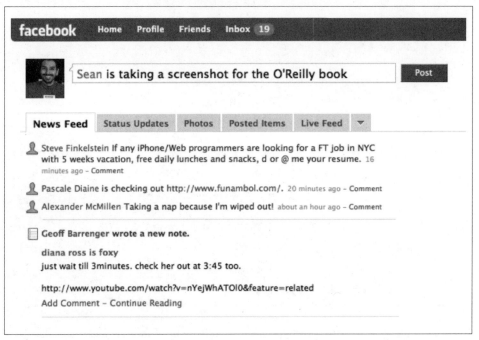

Figure 13-29. A Facebook lifestream

Social networks like Facebook and Friendster began as hyperlinked online contact lists, but have increasingly shifted toward sharing small updates among friends, as shown in Figure 13-29. This interaction, commonly referred to as a *lifestream*, contributes to a feeling of ambient awareness within the community, much as microblogging does.

Social networks often include many other community features, such as image sharing, group games, discussion groups, and real-time chat, all built around the central idea of the friend list and the social graph. Messages may be open to the community (as is the case with Facebook's "write on my wall" and "My News" features) or limited to only the sender and receiver.

Behind the few very large social networks are many smaller, more focused ones. *Compete.com* estimates that only five social networks have over five million visitors a month, but there are many smaller ones, as shown in Figure 13-30.

Outside these top five, niche communities thrive. Community Connect operates several smaller sites (BlackPlanet, Glee.com, MiGente, and AsianAvenue) that target specific groups based on ethnicity and sexual orientation. And Disaboom provides a social network for people dealing with or caring for those with disabilities.

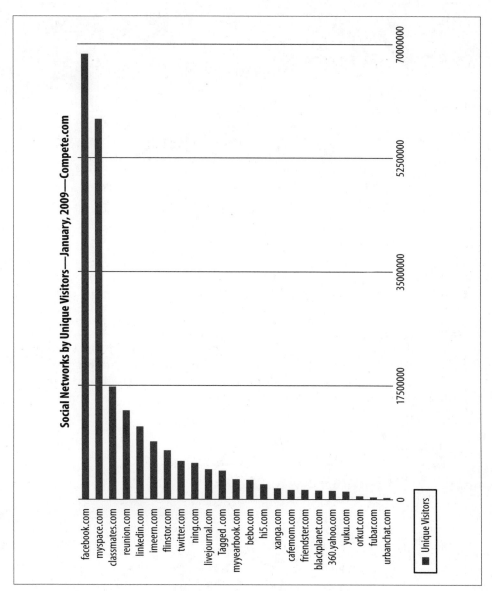

Figure 13-30. The top 20 social networks according to Compete.com

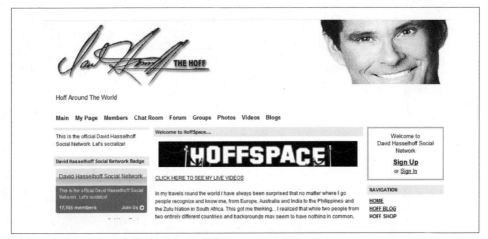

Figure 13-31. *A social network built around an individual*

Many of these social networks contain groups in which a member may occasionally be the topic of conversation. In some cases, the platform may arise around a particular person, brand, or company, as shown in Figure 13-31.

At the other end of the social network spectrum, companies like Ning, shown in Figure 13-32, allow users to create small social networks with features similar to those found on Facebook. This can be an easy alternative to building your own platform—you won't have to build and deploy your own software, but you'll be limited in the data you can collect about your members and their discussions, as well as your ability to moderate and control what's said.

Figure 13-32. *Dell-related social networks created within Ning*

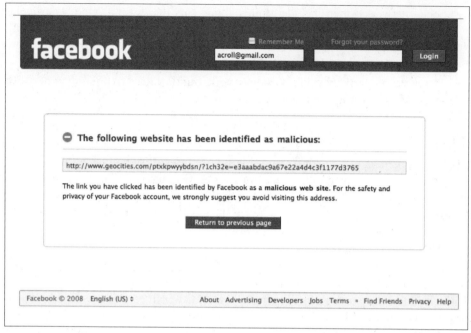

Figure 13-33. A Facebook warning about a malicious site invitation sent from an infected friend's computer

Platforms built around a social graph strike a balance between openness and trust, which helps to prevent abuse. But spammers still work on social networks, either by sending friend requests that encourage users to check the spammer's profile or by infecting users' PCs and then using their accounts to send spam and malware links to their social groups, as shown in Figure 13-33.

There's a conflict here for community managers. The more open your social platform, the easier it is for casual users to contribute and join. However, the more closed it is—and the more of it you control—the better the content, the greater your analytical visibility, and the less likely the network is to be abused by spammers and trolls. If there's a closed network whose subject matter you care about, you have to join the conversation in earnest.

Here are the roles in a social network:

- The *platform operator* (the person running the servers) has complete control over the tools, functions, and messages within the system. Unfortunately, if you run your own platform, you'll need to convince others to join you there—and getting a community's attention is often the hardest part.

- On a site like Ning, the *social network administrator* creates a new social network and defines which features are available to whom.

- Social networks often include groups and events that form around a particular topic. *Group administrators* have a certain degree of control around membership and policies, as well as the ability to communicate with the group and to create events associated with the topic.

- *Members* can participate in a social network, but they need to build their network of friends by inviting others and connecting with people already using the network.

Blogs

Blogs are content management systems that simplify the task of publishing on the Web. They often focus on a particular topic or community of interest. Blogs handle the distribution of content via RSS feeds, automatically update aggregation tools like Technorati, and structure posts chronologically in an archive. They also have comment threads and several other features to promote interaction between the blog's authors and its audience.

A blog is often the public face of a company to its community. Unlike other social networks seen here, in which all participants have a roughly equal voice, in a blog, you have clear authority to create content and to decide which comments to post. It's up to you to pen insightful posts that will encourage your audience to comment, tell others, and return. For an excellent list of tips on writing and maintaining blog content, see Chris Brogan's article at *www.chrisbrogan.com/40-ways-to-deliver-killer-blog-content/*.

If you're writing your own blog, know that it's not a magazine. The goal is to get in, make a point—often provocatively—and get out fast. Put another way, on a blog, a mystery story would begin with the words, "The butler did it. And he was right to. Here's why." An effective blog posting is the starting point for a conversation, offering a viewpoint with some backup material and inviting the community to respond within the comment threads.

Blogs have their own challenges, primarily spam and trolling, but there are many automated tools that can reduce spam and bulk moderate unwanted comments.

Here are the main roles you'll need to fill when running a blog:

- The *operator/administrator* implements analytics and measures uptime for the system, handles upgrades, adds and removes others from the system, and maintains the stylesheets and blog layout.

- The *editor/publisher* gets the final say over what goes live, and it's her responsibility to validate what you're writing as well as to ensure it's not libelous. Most blogging tools support an editorial cycle in which only editors can publish content.

- *Contributors* generate content, which is saved as drafts for review by the editor. Contributors probably have several posts in the queue pending publication. They should also read and respond to comment threads for posts they've written and, in conjunction with the editor, correct or retract information.

- Anyone can be a *commenter* on a blog. If the blog uses a consolidated comment system like Disqus or Mybloglog, the commenter may have a "persona" that ties together comments across multiple sites or over time.

A note on megablogs

Power laws tell us that while there are millions of blogs on the Internet, the majority of traffic goes to only a small number of them. These are the megablogs.

In recent years, the Huffington Post and Engadget have dominated Top 100 blog lists. In every sector, megabloggers have risen to prominence. For technology, it's Mike Arrington, Om Malik, Tim O'Reilly, and Robert Scoble. They have the ability to create a community within the comment threads of a single post, and have tens of thousands of micromessaging followers. When these sites and bloggers talk, others listen.

Does that mean you should focus community efforts only on the "big talkers"? Should your community strategy focus on trying to convince Kara Swisher or Walt Mossberg to blog about you?

The short answer is no, for three reasons:

- The megablogs are increasingly like mainstream media. This is a simple fact of being big. They're businesses, with bills to pay and an editorial beat to cover. As their audiences grow, it's harder for them to find a niche (though notably, GigaOm has worked hard at creating a family of smaller, topical blogs within its network). They're inundated with messages from traditional PR firms, who see them as just another outlet. It's impossible for them to read as much as they write.

- The big bloggers are wise to the games PR and promoters play to get their attention. They'll use their own networks of informants and experts—the less vocal "Interesting Middle" of the long tail—to find topics and validate information. So if you don't focus your attention on this secondary group, what you're saying to the megabloggers won't ring true when they check their sources.

- On megablogs, comment threads are where much of the community interaction happens. Every writer reads her own comment threads (those who tell you they don't are lying), and contributing useful information such as statistics or links to useful additional content is a great way to join the conversation without "pitching."

What you're after are the newcomers who are quickly gaining influence. For that, you need to engage with them on topics they care about—giving content to gain attention.

Wikis

A wiki is a collection of pages written by users that can be edited by other users. While other social platforms have a hierarchy (topics for groups, threads for forums, time for blogs, and a social graph for social networks), wikis have no rigid hierarchy; rather, they are a collection of documents linked to one another by topics.

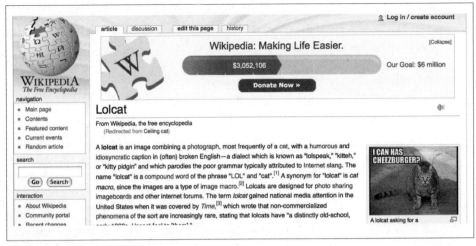

Figure 13-34. User-generated content on Wikipedia

The main focus of a wiki is collaboration, and a single page may have hundreds of edits, all preserved chronologically. Wikis make it easy to link to a new topic from within an entry and then return to complete the new topic later. The most popular example, shown in Figure 13-34, Wikipedia, is the world's largest and most comprehensive encyclopedia, but a wiki can be used for many kinds of content.

Wikis often contain reference material, but their openness makes them inherently less trusted than more formal reference sources. Wikipedia deals with this problem constantly; in some cases, the Wikipedia team has had to lock certain entries due to warring factions of editors or social pranks. For example, on July 31, 2006, Stephen Colbert urged his viewers to modify the Wikipedia entry about elephants, declaring them no longer endangered, and Wikipedia had to halt edits to the page when thousands of edits ensued, as shown in Figure 13-35. Only weeks later could the page be safely unlocked.

Despite this, and thanks in no small part to the efforts of hundreds of dedicated content auditors who check references and flag questionable content, Wikipedia has become one of the world's largest online information resources. But it's allergic to subjective or promotional content, so any content you contribute to it must be independently verifiable, contain reliable references, and be considered nonpartisan.

Wikipedia, which runs atop the MediaWiki platform, is a good model for any wiki you create or moderate. There are many other wiki platforms out there that you can use or run yourself, including TikiWiki, Wetpaint, and Socialtext.

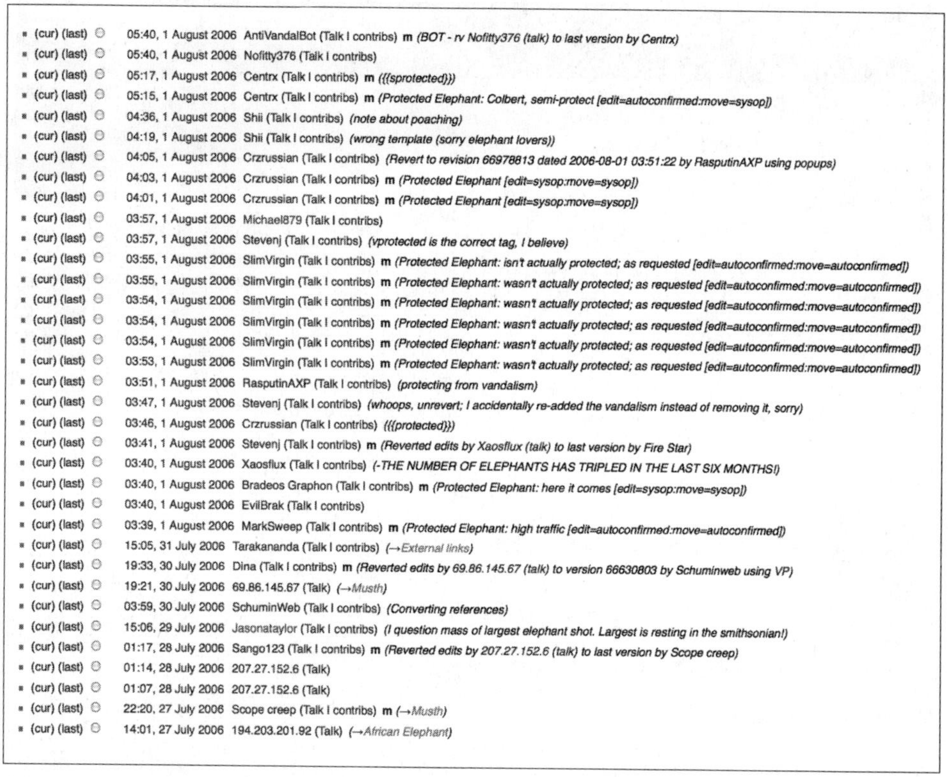

```
■ (cur) (last) ○   05:40, 1 August 2006  AntiVandalBot (Talk | contribs)  m  (BOT - rv Nofitty376 (talk) to last version by Centrx)
■ (cur) (last) ○   05:40, 1 August 2006  Nofitty376 (Talk | contribs)
■ (cur) (last) ○   05:17, 1 August 2006  Centrx (Talk | contribs)  m  ({{sprotected}})
■ (cur) (last) ○   05:15, 1 August 2006  Centrx (Talk | contribs)  m  (Protected Elephant: Colbert, semi-protect [edit=autoconfirmed:move=sysop])
■ (cur) (last) ○   04:36, 1 August 2006  Shii (Talk | contribs)  (note about poaching)
■ (cur) (last) ○   04:19, 1 August 2006  Shii (Talk | contribs)  (wrong template (sorry elephant lovers))
■ (cur) (last) ○   04:05, 1 August 2006  Crzrussian (Talk | contribs)  (Revert to revision 66978813 dated 2006-08-01 03:51:22 by RasputinAXP using popups)
■ (cur) (last) ○   04:03, 1 August 2006  Crzrussian (Talk | contribs)  m  (Protected Elephant [edit=sysop:move=sysop])
■ (cur) (last) ○   04:01, 1 August 2006  Crzrussian (Talk | contribs)  m  (Protected Elephant [edit=sysop:move=sysop])
■ (cur) (last) ○   03:57, 1 August 2006  Michael879 (Talk | contribs)
■ (cur) (last) ○   03:57, 1 August 2006  Stevenj (Talk | contribs)  (vprotected is the correct tag, I believe)
■ (cur) (last) ○   03:55, 1 August 2006  SlimVirgin (Talk | contribs)  m  (Protected Elephant: isn't actually protected; as requested [edit=autoconfirmed:move=autoconfirmed])
■ (cur) (last) ○   03:55, 1 August 2006  SlimVirgin (Talk | contribs)  m  (Protected Elephant: wasn't actually protected; as requested [edit=autoconfirmed:move=autoconfirmed])
■ (cur) (last) ○   03:54, 1 August 2006  SlimVirgin (Talk | contribs)  m  (Protected Elephant: wasn't actually protected; as requested [edit=autoconfirmed:move=autoconfirmed])
■ (cur) (last) ○   03:54, 1 August 2006  SlimVirgin (Talk | contribs)  m  (Protected Elephant: wasn't actually protected; as requested [edit=autoconfirmed:move=autoconfirmed])
■ (cur) (last) ○   03:54, 1 August 2006  SlimVirgin (Talk | contribs)  m  (Protected Elephant: wasn't actually protected; as requested [edit=autoconfirmed:move=autoconfirmed])
■ (cur) (last) ○   03:53, 1 August 2006  SlimVirgin (Talk | contribs)  m  (Protected Elephant: wasn't actually protected; as requested [edit=autoconfirmed:move=autoconfirmed])
■ (cur) (last) ○   03:51, 1 August 2006  RasputinAXP (Talk | contribs)  (protecting from vandalism)
■ (cur) (last) ○   03:47, 1 August 2006  Stevenj (Talk | contribs)  (whoops, unrevert; I accidentally re-added the vandalism instead of removing it, sorry)
■ (cur) (last) ○   03:46, 1 August 2006  Crzrussian (Talk | contribs)  ({{protected}})
■ (cur) (last) ○   03:41, 1 August 2006  Stevenj (Talk | contribs)  m  (Reverted edits by Xaosflux (talk) to last version by Fire Star)
■ (cur) (last) ○   03:40, 1 August 2006  Xaosflux (Talk | contribs)  (-THE NUMBER OF ELEPHANTS HAS TRIPLED IN THE LAST SIX MONTHS!)
■ (cur) (last) ○   03:40, 1 August 2006  Bradeos Graphon (Talk | contribs)  m  (Protected Elephant: here it comes [edit=sysop:move=sysop])
■ (cur) (last) ○   03:40, 1 August 2006  EvilBrak (Talk | contribs)
■ (cur) (last) ○   03:39, 1 August 2006  MarkSweep (Talk | contribs)  m  (Protected Elephant: high traffic [edit=autoconfirmed:move=autoconfirmed])
■ (cur) (last) ○   15:05, 31 July 2006  Tarakananda (Talk | contribs)  (→External links)
■ (cur) (last) ○   19:33, 30 July 2006  Dina (Talk | contribs)  m  (Reverted edits by 69.86.145.67 (talk) to version 66630803 by Schuminweb using VP)
■ (cur) (last) ○   19:21, 30 July 2006  69.86.145.67 (Talk)  (→Musth)
■ (cur) (last) ○   03:59, 30 July 2006  SchuminWeb (Talk | contribs)  (Converting references)
■ (cur) (last) ○   15:06, 29 July 2006  Jasonataylor (Talk | contribs)  (I question mass of largest elephant shot. Largest is resting in the smithsonian!)
■ (cur) (last) ○   01:17, 28 July 2006  Sango123 (Talk | contribs)  m  (Reverted edits by 207.27.152.6 (talk) to last version by Scope creep)
■ (cur) (last) ○   01:14, 28 July 2006  207.27.152.6 (Talk)
■ (cur) (last) ○   01:07, 28 July 2006  207.27.152.6 (Talk)
■ (cur) (last) ○   22:20, 27 July 2006  Scope creep (Talk | contribs)  m  (→Musth)
■ (cur) (last) ○   14:01, 27 July 2006  194.203.201.92 (Talk)  (→African Elephant)
```

Figure 13-35. "The number of elephants has tripled in the last six months"

Here are the roles in a wiki:

- The *operator* runs and maintains the wiki platform and servers, and manages the "blunt instruments" of content management, such as locking down topics to prevent change. The operator also decides on layout and functionality. In Wikipedia's case, the operator is Jimmy Wales' Wikimedia foundation.

- *Authors* can be anyone with a claim to expertise; their authority should be supported by third-party material and references, or content they generate they will be quickly questioned. If you're running a wiki as a reference source for your customers, you may limit authors to employees or accredited subject matter experts. For example, genetic analysis site 23andMe has a reference wiki with articles authored by doctors built into the service.

- *Editors* are the content police of a wiki. They correct and update content, and flag problems such as poor syntax or missing references so that authors can fix them. They also have the ability to revert contributions to a previous version and try to identify bad behavior that an operator needs to address.

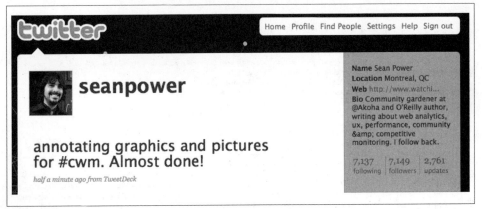

Figure 13-36. A Twitter feed, showing the number of followers and updates

Figure 13-37. Facebook's "status" is a micromessaging tool

Micromessaging

Micromessaging, or microblogging, applications are streams of small messages shared with a group. There are dozens of platforms, but Twitter, shown in Figure 13-36, is by far the largest independent microblogging tool in terms of adoption and message volume.

Micromessaging does exist within larger social networks such as Facebook in the form of status updates like the one shown in Figure 13-37. In these models, however, interactions generally occur among a preexisting social graph rather than between strangers.

Micromessaging is a relatively new communications model, with unique characteristics that affect how we use it and what's appropriate. It's an RSS feed for people, a way to direct the attention of audiences, and a means of reaching the famous without burdening them with an obligation to respond. In short, *Twitter is an API for human attention.* While the rules of micromessaging are still being written, it has some unique features worth noting.

Asymmetric following

On Twitter, following is politics. Well-known Twitter members have huge followings, and often use this network to promote blog postings or amplify messages they find interesting. We often judge members according to their ratio of followers to people they're following (see *http://twitter.grader.com/* for one ranking system created by marketing consultancy Hubspot). To build a following, you need to strike a balance between creating regular, interesting updates and replying to those you follow in the hope that they will return the favor.

Twitter is neither one-to-one (unicast) nor one-to-many (broadcast). Call it "sometimescast". This produces strange behaviors. People who are highly followed selectively amplify messages, and people who are less followed send nuggets of wisdom to the more followed in the hopes of getting their attention.

When Twitter Followers Become Botnets

British author, actor, and technology aficionado Stephen Fry (@stephenfry) has hundreds of thousands of followers. We asked him how being such a Twitter heavyweight affects his online behavior.

CWM: Is there a burden of responsibility that comes with being a widely followed community member?

Stephen Fry: I have learned to be very careful about revealing frustrations with any individuals, companies, or institutions. I once—almost casually—mentioned an individual who had expressed bizarre antiscience opinions that I found absurd and some followers hounded this poor person with really unkind and abusive emails and posts. That's a worry. It's a tiny minority, but a small percentage of a very large number is a force to be reckoned with.

The same can happen with people who tweet me rudely: if I, as it were, raise my eyebrows at their posts in a public reply their Twitter stream can be looked at by my followers and they can be harried for their eccentricities, politics, or general manner. If, therefore, someone is rude to me, they have to be very rude or unreasonable for me to draw attention to it.

CWM: What's the most havoc you've (unintentionally) wreaked on a site, now that you're a human denial-of-service attack and Twitter is your bot network?

Stephen: Perhaps justly the most havoc I've wrought has been on a site of my own. In a vain effort to get a revenue stream from the increasing costs of hosting my website I tweeted the arrival of a t-shirt store which carried some "I tweeted @stephenfry and all I got was this luxurious t-shirt" shirtings—within seconds the site was down. I didn't have the nerve to reTweet when it was finally up and running—looked too much like huckstering. "What a waste of time!"—as Bill Murray expostulates in *Groundhog Day* on the subject of French poetry.

CWM: Counting followers seems awfully "high school yearbook" as a way of measuring online success. What metrics do you care about beyond followers?

Stephen: Yes, the pecker test really won't do. In the early days it was all one had and that was before institutions like TV networks, musicians, and PR machines clambered onto the tweetwagon. Nowadays, I suppose one measures by the levels of genuine interactivity. I think my followers know that it is always me tweeting, that I don't use Twitter to sell anything (with the exception of aborted t-shirts from time to time!) and

---

that such views, observations, banalities, and quotidian mundanities as issue forth are always my own.

CWM: What tools can you possibly use to monitor a conversation with hundreds of thousands of followers? Are you forced to resort to hashtags? How do you collect the zeitgeist of a following?

Stephen: I do use hashtags for what one might consider important purposes. I have a hashtag bin (I'm using bin in the film editor's way, not in the sense of trashcan!) for those wanting me to retweet a message or highlight a charitable website. I have another hashtag bin for those who are anxious for me to follow them, and yet another which explains the general ethos of Twitter as I see it.

As to how one can monitor, well, with tens of thousands of new followers arriving weekly I do hope they are able to understand that the chances of me seeing and reading a tweet are very low. It's a simple question of probability.

Imagine a gusty autumn day: I'm standing in a forest as thousands and thousands of leaves stream past me borne on the wind. I snatch at one or two to look at them as they pass. That's my position. There's the option of direct messaging, which increases the chance of me getting a glimpse (which is why people want me to follow them, I guess, as that means they can DM me), but they're coming in pretty thick and fast, too, now.

So long as people understand that it's a kind of lottery as to whether I respond, that's fine. But Twitter is not email, it is not about access and chat. It is, as it tells anyone, about what the tweeter is tweeting. It is "microblogging." I like reading others' tweets and get a little frustrated if all I get are questions (99% of which don't need to be asked, but are clearly a pretext for initiating a conversation—which is cool, but hard to act upon)—but it's not a frustration that matters, and can be set down as the grumpiness of a busy old man.

Fluid relationships

Twitter's explosive growth is due in part to the fact that you can follow anyone and they can follow you. *It's just as easy to "unfollow" someone*, and we'll soon see more sites that track who's being followed or unfollowed the most. This keeps people honest; there's always a lingering fear of being abandoned by your flock.

The notion of whether you care if you're followed is an elephant in the Twitter room. People want to seem smart. They want the affirmation of reTweeting. They want to be noticed, to be perceived as having authority. Like it or not, Twitter's fluid social graph unearths a yearbook psychology buried deep in our high school psyches, and has more of an impact on our behavior than we think.

Follower count

Users who follow many people but have few followers themselves are often seen as sources of spam. And those who have many followers but aren't following others are assumed to be simply broadcasting messages rather than interacting. During the 2008

Figure 13-38. A following: followers' comparison of Hillary Clinton and Barack Obama during the 2008 democratic elections

Democratic party campaigns, Hillary Clinton's team was criticized for not reciprocating Twitter follows, as shown in a comparison of Clinton and Obama's Twitter profiles in Figure 13-38.

It isn't really possible to read what 146,000 people say (Obama's numbers in Figure 13-38) unless you're using tools for automation. Many Twitter users rely on clients like Tweetdeck to filter traffic from the people they're following, sometimes going so far as to create their own groups of "real" friends they actually follow. This is disingenuous, because it makes people appear to be listening to more people than they really are, but in a world of reciprocal following, it's standard practice. Consequently, *seeing who's mentioning others is a far better indicator of actual social engagement than follower count is.*

Limited context

Twitter doesn't thread messages, and the messages themselves are brief. As a result, there's less context for a conversation, since third-party observers don't easily see who's talking to whom. While back-and-forth chats happen, they're usually brief.

Twitter messages tend to remain a mix of instant messaging, URL referrals, bumper-sticker humor, and status updates. *The lack of conversational context stops Twitter from devolving into a party line.*

Extensible syntax

At the outset, Twitter didn't have much functionality built in (prefacing a member's name with the letter "d" for direct messaging was its only "feature"). This meant that *the community created and selected its own conventions:* reTweeting (prefixing a message with RT), @naming, and #hashtags started out independent of Twitter, and only later became part of the system.

One convention—flagging topics with a hashtag ("#")—is similar to the IRC channel naming convention. When an IRC user talks about a channel called "#channel," he's referring to a chatroom with that name. Similarly, if a Twitter user includes the tag "#memes," she's tagging the message with a particular topic. Users can then converge around that tag for something approximating real-time chat on sites such as Twemes, or form a spontaneous community on thread.io. While hashtags aren't formally part of Twitter, some clients, such as Tweetdeck, will persist hashtags across replies to create a sort of message threading.

Hashtags can also serve as backchannels or a means of circumventing Twitter problems, since they're another way for groups to organize and communicate without being one anothers' followers, as shown in Figure 13-39.

The constraints of brevity

Twitter's founders attribute much of their success to self-imposed constraints like message size (*www.inc.com/magazine/20080301/anything-could-happen_pagen_2.html*). Developing for a constraint like 140 characters has an important side effect: short messages make traditional brand marketing hard. Spamming a slogan on Twitter is useless, and bad behavior (such as letting companies like Magpie send messages on your behalf) gets you quickly unfollowed.

This has kept the medium relatively spam-free despite the fact that it's so open. Other, richer social networks have to contend with far more spam; *it's hard to hide a virus in only 140 characters* (or is it 255 characters? Or 920? Check out *www.radian6.com/blog/ 101/learnings-from-twittersecret/*). This limitation also makes users prune and clarify their thoughts, which means your audience can process more ideas with less effort.

An open API

Twitter has made it very easy to extend its functionality, in part because it hasn't focused on monetizing the system (yet). This means there are quite literally new Twitter applications going live every week. There are also several popular desktop clients and web plug-ins.

Figure 13-39. Using hashtags to circumvent Twitter's follow mechanism and identify members of a group

Some of these applications are hashtag trackers, some are tag clouds, and some are ranking tools. There's a vibrant development ecosystem around micromessaging, and it's creating new ways to extract meaning from the thronged masses. These days, even if individual conversations won't get your attention, a groundswell of objections will force you to get involved.

Micromessaging nay-sayers should remember that HTTP was a nascent protocol once, too. It had a few very simple verbs (GET and POST, mostly) and a lot of possibility. Developers and innovators came up with cookies, embedded content, URI parameters, and so on. They were optional, which allowed browsers and web servers to evolve independently rather than having to be in lockstep, resulting in much faster adoption and innovation.

As a communication model, we're still defining micromessaging. If early adoption is any indication, we'll build rich functionality atop it just as we have with HTTP.

The following roles all exist in micromessaging:

- While Twitter is a privately operated service, micromessaging platforms like La-coni.ca are available that can be run independently. The *operators* of these plat-forms have considerable control (and in Laconi.ca's case, the ability to modify the platform itself, since it's open source). On the other hand, operators have to convince a community of microbloggers to join them.

- While micromessaging platforms have very little hierarchy, one of the ways that people can rally others around a topic is to create a hashtag that others can use to flag their messages with a particular topic. The person who creates this group is a *group/hashtag creator*. People don't get any special privileges as the creator of the hashtag, but they can track other posts that include it. Some sites also invite hashtag creators to claim a particular hashtag and define it, so there's an advantage to being the first to claim a hashtag for your community.

- For *participants*, the bulk of micromessaging interaction is simply participation— following others and interacting with them. Participants may also follow certain automated services so that those services can track and analyze their content over time.

Social News Aggregators

With hundreds of millions of people posting their thoughts to the Web, we rely on the wisdom of crowds to decide what's valuable. Social news aggregators let members nominate content they find interesting, allowing others to upvote (promote) it if it's worthy. These sites also encourage commenting on the submitted content, and the comment threads themselves are subject to voting so that the most popular topics rise to the top.

Digg, reddit, and Slashdot are three of the biggest social news aggregation platforms. Each has a community of fanatics responsible for many of the submissions. Other news aggregators, such as StumbleUpon, Twine, and ShareThis, rely more on browser plugins that make it easy to bookmark and nominate sites, but are less about member comments on submissions. As URL redirection becomes more commonplace, still more link aggregators will emerge to show which sites web users are sharing the most within social networks and micromessaging platforms.

If your website gets noticed by a social news aggregator, you'll get a huge number of visitors in a very short period of time. This means a heavy burden on your servers. To make matters worse, few of the new visitors attracted by your newfound popularity will have been to your the site before, so they will all be retrieving every object from your servers (since none of it was cached on their browsers). The effect will often break a website entirely, as shown in Figure 13-40.

News aggregators have different reputation systems. On some systems, such as reddit, submitters receive "karma" for popular submissions, but that karma has no effect on a person's votes; it's simply a mark of community respect. On other sites, past performance gives more power: a well-regarded Digg user's votes count for more.

This is a controversial matter, because reddit's one-voice-one-vote means the site is less prone to having a few very influential users—something that has plagued Digg—but

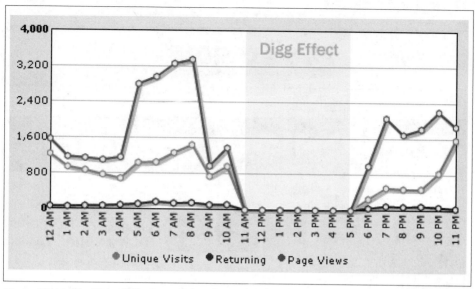

Figure 13-40. The "Digg Effect" on Nick La's N.Design Studios portfolio site—after a sudden burst of interest, this site became inoperable

spammy content is harder to filter out automatically on a site like reddit where every submission, whether from a one-time spammer or a veteran, has the same initial merit.

These are the roles in social news aggregation:

- *Operators* may run platforms of their own on software such as Pligg. Because the power of aggregators comes from the collective wisdom of a large community, however, these will probably be part of an intranet.

- A *group creator* creates a community of interest within an aggregator (on reddit, this is called a subreddit). Group creators don't get additional visibility into the communities they create, but they will have a place to unite like-minded contributors.

- When *submitters* submit links, those submissions become part of their personal histories, and upvotes for something they've submitted contribute to their reputations. Submitters are expected to explain why the submission is interesting and summarize it for others. If submitters want the subject to get attention, they need to make the summary compelling, controversial, and brief.

- *Commenters* comment on submissions, and the comments themselves are subject to voting by the community. Commenters may receive some kind of ranking for content that rises to the top.

- *Voters* alter the ranking of content and comments.

Figure 13-41. A community site built entirely in Drupal that includes many social network features

Combined Platforms

Major social networks have incorporated many of the communications models we've seen above. Facebook, for example, has real-time chat, open message threads (walls), one-to-one messaging, groups, mailing lists (messages to groups), and so on.

On the private side, servers like Drupal combine blogging, wikis, and other components in a single platform that many developers have extended with plug-ins and new functionality, as shown in Figure 13-41.

Why Be Everywhere?

Clearly, there's a daunting variety of community platforms on the Internet. Can you just pick a few?

It depends.

If you have a specific audience with a strong preference for one type of community, one platform may be enough. An extremely technical, always-connected set of server administrators, for example, may be content with just an IRC channel. But if you're trying to engage a broader market, particularly for consumer products or public causes, you need to be on all of the dominant platforms. Here's why:

Reach

> By being active on many platforms, you increase the chances that your messages will be picked up by the community and amplified.

Early days

> It's still hard to know which social platforms will dominate the Internet, and you need to hedge your bets. Will Twitter stand on its own? Will it fracture into several federated micromessaging platforms? Or will micromessaging simply become a feature of social network sites? Because it's too soon to tell, you need to engage on all of them.

Different platforms for different messages

> A last-minute announcement about a party goes best on Twitter, while a detailed list goes in a blog posting, and a question looking for responses goes to a mailing list.

Awareness

> You never know when you're going to become the subject of discussion, so you should be monitoring as broad a range of sites as possible in case conversations about you or your competitors emerge.

Multiple audiences

> You engage with multiple people in an organization. The accounting department may not use the same social networks as the executive team, who may work with different tools from the folks in support. Different audiences gravitate toward different platforms.

The media you choose strongly influences the messages that are discussed, the people who engage, and the detail of their responses. Real-time platforms require quick thinking, wit, and a thick skin. Micromessaging tools demand bumper-sticker cleverness, pithy quips, and an understanding of shorthand conventions and abbreviations. Blogs have to engage visitors with a single thought early on to capture attention. Wikis need deeply linked, comprehensive, defensible content.

Monitoring Communities

Regardless of the community platform, several fundamental building blocks keep emerging. These include the following:

User reputation

> This is a score of how well respected you are within the community.

Threading

> Threading is a way of organizing replies and responses between community members, often with comments.

Voting by the community

> This helps point out what's good, in the form of reTweeting, upvoting a submission, or flagging an inappropriate addition.

Notifications
> These let members know about what's happening within the community and keep community members engaged while fighting attrition.

Lifestream
> A lifestream provides status updates on community members for those who choose to follow them.

Social graph
> A social graph describes the relationship of members to one another, either in a whitelist (Facebook-like) or blacklist (Twitter-like) model.

Searching
> You can search across historical interactions, often using tags and other filters.

Topics, subgroups, and events
> These are groupings on which a community can converge.

Befriending and unfriending
> This can change the structure of the social graph.

These building blocks can all be measured and analyzed. They represent the KPIs and metrics of community monitoring. In the next chapter, we'll turn our attention to how to monitor them.

CHAPTER 14

Tracking and Responding

Engaging a community is one thing; understanding the results of those engagements is another entirely. If we're going to link community activity to business outcomes, we need to track conversations and respond when appropriate.

Our ability to track what's said online varies widely depending on where that conversation is taking place. Some community platforms, like Twitter, are wide open, myriad tools for analysis and search. Others, such as Facebook, are firmly closed—third parties are forbidden from extracting information and publishing it elsewhere, and while the company is putting some carefully designed gates in its walled garden, Facebook can't easily be crawled, curtailing the number of community management tools that can work with it.

It's not just Facebook. StumbleUpon, shown in Figure 14-1, shows a promotion for marketers when you follow a referring link from its site. There are ways around this, primarily by installing and using the StumbleUpon toolbar, as described by Tony Adam at *http://tonyadam.com/blog/checking-stumbleupon-referrals-reviews-urls*. For community operators, there's a constant tension between the desire to be open (with the community development, rapid growth, and innovation this entails) and the desire to monetize the community somehow by controlling who has access to information on its activity.

The way in which you engage a community will depend on how much visibility you want into that community's activity. In the coming pages, we'll look at four distinct levels of engagement you can have with a community.

- You can *search* the Web using tools like Google, or search within a specific community using the platform's search functions. In this case, you're simply a passive observer, and this is a hands-off approach. It's also the one that yields the least information, because you can only search what you're allowed to see as an outsider.

- You can *join* communities, seeing what's happening from within. To do this, you must get an account and reciprocate interactions (follow users who follow you, comment on others' posts, or join mailing lists they create).

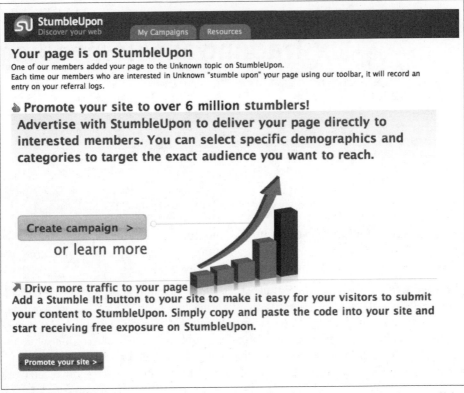

Figure 14-1. StumbleUpon's referral landing page doesn't show you which content drove traffic to your site

- You can set up and *moderate* a community within an existing platform, inviting users to join you while remaining responsible for the operation of the group. This may be a Facebook group, a Yahoo! mailing list, a subreddit, or even a hashtag you define on sites like thread.io.

- You can *run* your own community platform, giving you complete control, but requiring that you convince others to join you. This works best when others have a reason to join you, for example, when they want product support or access to people within your organization.

Figure 14-2 shows these four levels of engagement, along with the degree of commitment required, the level of analytical detail you can hope to extract, and the number of existing users you'll find with the community when it launches.

In the preceding chapter, we looked at eight distinct community models: groups and mailing lists, forums, real-time chat systems, social networks, blogs, wikis, micromessaging tools, and social news aggregators. By combining these eight models with the four levels of engagement, we can put together a community monitoring strategy.

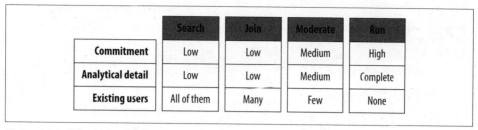

	Search	Join	Moderate	Run
Commitment	Low	Low	Medium	High
Analytical detail	Low	Low	Medium	Complete
Existing users	All of them	Many	Few	None

Figure 14-2. What you put in and get out of various levels of community engagement

	Search	Join	Moderate	Run
Group/mailing list	Mailing list search tools	Subscribe to Yahoo Group	Set up Yahoo Group	Operate a Listserv
Forum	Forum search tools	Post in online forums	Start a thread in a forum	PHPBB
Real-time chat	N/A	Post on an IRC channel	Create an IRC channel	Operate an IRC node
Social network	Search Facebook	Subscribe to Facebook	Create a Facebook group	Drupal
Blog	Blog search tools	Comment on a blog	Contribute to a blog	Wordpress.com installation
Wiki	Wiki search tools	Contribute/edit Wikipedia	Create a Wetpaint wiki	Mediawiki
Micromessaging	Twitter search sites	Get a Twitter account	Create a group or hashtag	Indenti.ca installation
Social news aggregators	Reddit, Digg search	Vote & post content	N/A	Pligg

Figure 14-3. Examples of engagement across each of the eight community models

Figure 14-3 provides some examples of how we might search, join, moderate, or run each of the eight types of community.

Searching a Community

Web crawlers are constantly crawling the public Web, indexing what they find into dozens of online services, such as search engines, plagiarism detectors, alerting tools, or social sentiment collectors. Search is your main way to find conversations that concern you, especially when you don't actually want to join or run the communities yourself. There are two kinds of search: traditional manual searches, known as

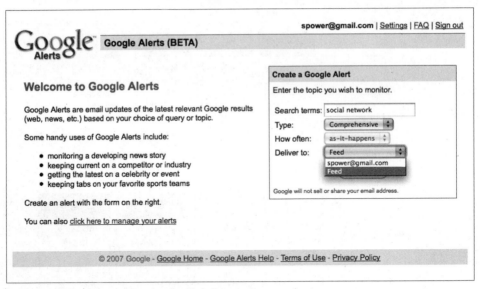

Figure 14-4. Setting up a Google Alert for a prospective search

retrospective searches because they look for data that's already been published, and alerts, known as *prospective searches* because they let you know when new data appears.

You need to search across all sites that matter to you—industry publications, analysts, and readers. Once you find a search that's useful, store it so you can run it again. You can store all of the searches as bookmarks, so if you want to share them with your team, consider a social bookmarking site like Delicious.

Stalking yourself daily with search engines might be fun and flattering, but it gets old fast. Fortunately, you can get alerts to notify you when content that you care about appears, using prospective search. To do this, set up a search term in advance using tools like Google Alerts (shown in Figure 14-4). In this case, Google will let you know whenever that term is seen.

You can also receive updates via RSS feeds. While these will only pull in content from postings and feeds that are published, anyone tasked with monitoring a community needs to run an RSS feed reader to keep track of influential posters or relevant posts.

Finally, you can also use alert services that check web pages for you and let you know when their content changes, such as WatchThatPage.com (shown in Figure 14-5) or ChangeDetection.com. Then you can dig deeper to see if the changes matter.

Searching Groups and Mailing Lists

Mailing lists that are archived online may be searchable without your participation if they're public. Yahoo! Groups, shown in Figure 14-6, lets you view archives and search

Figure 14-5. A report from WatchThatPage.com showing changes to a monitored website

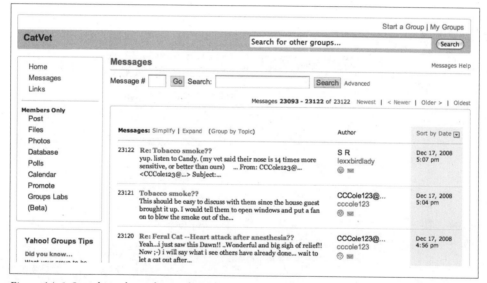

Figure 14-6. Searching the archives of a Yahoo! group

for strings in many mailing lists for which the administrator has made the archives available.

You can also use a prospective search portal like Rollyo to search for specific topics within the constraints of specific sites that contain mailing list archives, as shown in Figure 14-7.

Some mail tracking tools index mailing lists. Markmail (Figure 14-8), for example, indexes roughly 500 Google Groups lists (about 3.8 million email messages) and thousands of other mailing lists, and makes them searchable through a web interface that also provides some keyword analytics showing which mailing lists and authors mention a topic most often.

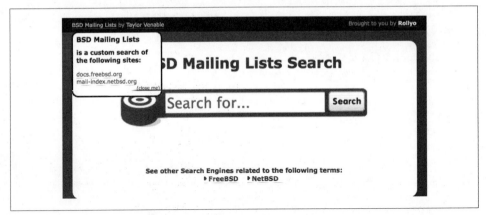

Figure 14-7. A prospective search query on Rollyo for the NetBSD and FreeBSD mailing lists

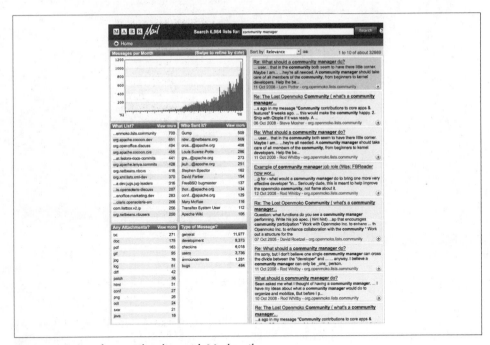

Figure 14-8. Searching mailing lists with Markmail

In addition to finding topics and content, Markmail reports on how popular specific topics or keywords are becoming by graphing the percentage of messages that contain a particular term (Figure 14-9).

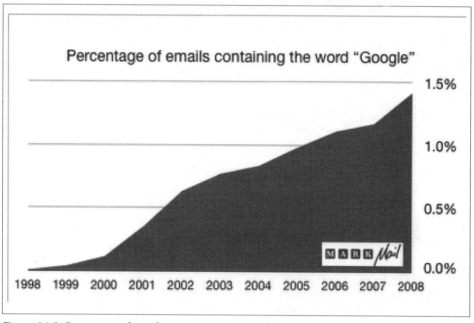

Figure 14-9. Percentage of email messages containing the word "Google" across 3.6 million email messages in 10 years

Searching Forums

Public forums are often indexed by web crawlers, so you may be able to search them by narrowing the scope of a Google search to just the forum in question. Using Google's advanced search, shown in Figure 14-10, you can constrain a search to a single site (in this case forums.worldofwarcraft.com) and find specific keywords.

Forum ranking sites, like Big Boards and Boardtracker, index many forums on the Web and may also help you to find the forums you should be tracking (Figure 14-11).

Once you're on a forum, you can use its internal search tools to look for specific content. But many forums require you to be a member, so you may be forced to join the group, forum, or mailing list in order to access its search features or to view archives.

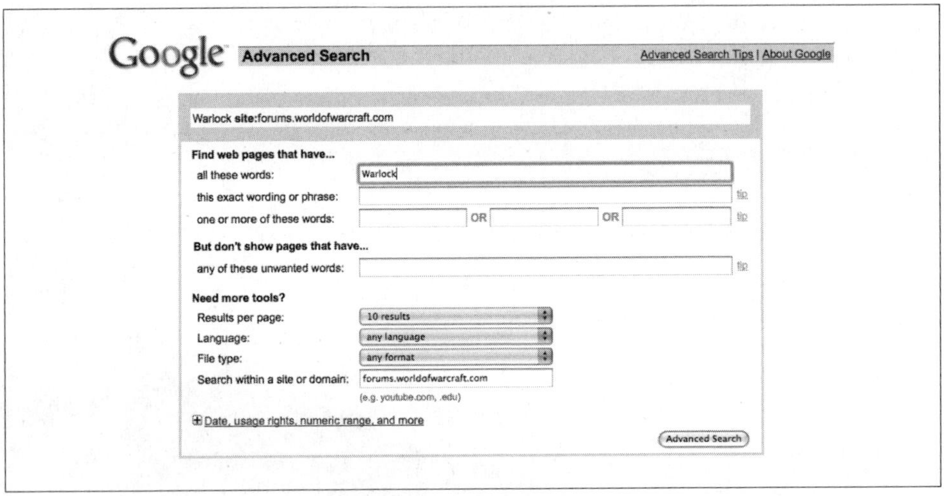

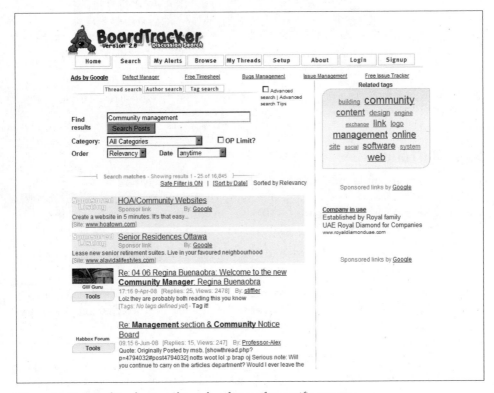

Figure 14-10. Using Google's advanced search to find content on a specific forum

Figure 14-11. BoardTracker searches online forums for specific content

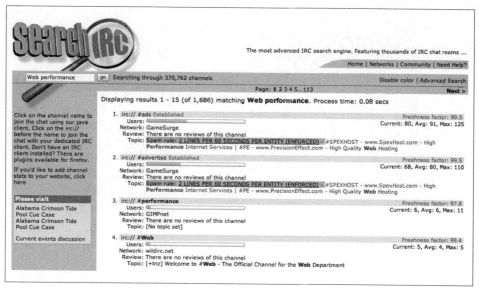

Figure 14-12. SearchIRC indexes and searches content on selected IRC channels

Searching Real-Time Chat Systems

Real-time chat isn't archived in the same way that groups and forums are. To "crawl" an IRC channel, an indexing service requires a bot that listens to traffic on the channel and indexes it. If you don't feel like joining a channel yourself and indexing content, you can use sites like SearchIRC, shown in Figure 14-12, to look for specific terms.

IRC isn't the only chat platform on the Internet, of course. Many instant messenger platforms today (such as Skype and MSN Messenger) give users the ability to create ad hoc group chats, but if you're not party to the conversation, you can't track it.

There are some public conversations that resemble chats. Twemes, for example, strings together micromessages around a specific topic using a hashtag (Figure 14-13 shows the Twemes analysis of the #food hashtag). You can stream the content of a topical discussion to an RSS feed and examine it later, or use Twemes itself to search a topic's conversation.

Searching Social Networks

Social networks come in two flavors: open models and walled gardens. Simply put, you can't browse the walled ones from outside their walls.

Figure 14-13. Twemes looks like IRC or near-real-time chat, but is based on Twitter

A social network is only as valuable as its social graph. So some sites, such as Facebook, LinkedIn, and MySpace, have guarded them jealously (though they've opened up considerably in the face of rapid growth from competing communities like Twitter). These sites don't let you view profiles or navigate relationships unless you're logged in. And even if you're logged in, it's a violation of their terms of service to try to map the social graph with outside mechanisms. In other words, you need to be a Facebook member to access your social network, and even then you can only do so *within* Facebook's website.

Facebook does provide some data on popular topics, using Lexicon analytics, shown in Figure 14-14. Lexicon measures the popularity of terms within the public portions of Facebook, such as wall posts and status updates.

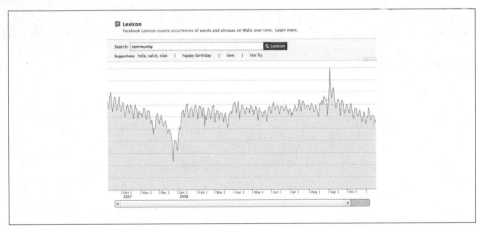

Figure 14-14. Facebook Lexicon shows occurrences of particular words over time within the Facebook Walls application

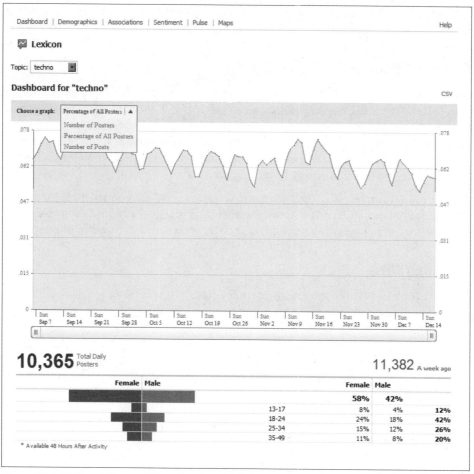

Figure 14-15. The public view of Facebook's Lexicon only indexes a few terms and doesn't offer the capability to analyze terms you specify

However, Facebook doesn't provide much detail within this model. You can't see who's talking or what they're saying. And they provide deeper analysis only for specific terms they dictate, as shown in Figure 14-15. Facebook's advertising platform offers additional visibility for paying customers, as the company is sitting on a motherlode of public sentiment data.

By contrast, more open social networks (most notably Google's OpenSocial model and those based on OpenID) make it easier for anyone to browse a social graph through a set of well-defined APIs.

There's one exception to this walled garden limitation, however. Business-oriented social networks are only partly walled. Sites like LinkedIn and Spoke have to be somewhat open, since one of their main functions is to help their users promote their profiles and extend their business networks. This means you can usually search and view profile information on a site like LinkedIn without being logged in.

Searching Blogs

Blog content is fairly persistent (unlike chat), seldom password protected (unlike groups), and open to everyone (unlike social networks). Consequently, most blogs are indexed by search engines and you can track them using the prospective and retrospective search methods outlined earlier. You can also collect and archive blog content via RSS feeds or mailing list subscriptions, and many blogs have their own search functions that let you find content by keyword or author.

Google's Blog Search service (*http://blogsearch.google.com*) is a good place to start. Sites such as Technorati and IceRocket also rank blogs and categorize them by topics, providing a good way to identify websites that cover topics in which you're interested.

Searching Wikis

Wiki platforms are searchable, as shown in Figure 14-16, and provide tools for tracking content you care about, letting you know by email or RSS feed when something has been edited.

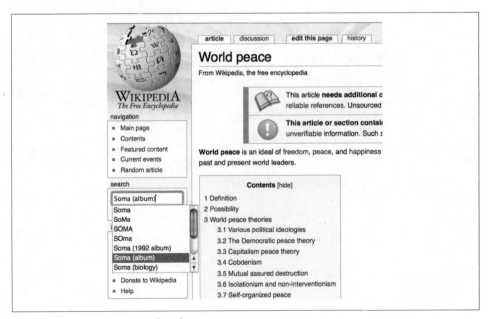

Figure 14-16. Searching on Wikipedia

Figure 14-17. Tracking several live search results with Tweetdeck

Wikis are also indexed by search engines, but if a particular topic interests you, it's a good idea to subscribe to changes to that topic, because a heated debate on a particular subject may mean that the topic changes more frequently than the search engines can index it.

Searching Micromessaging Tools

Twitter is notable for its open APIs, which make searching and monitoring specific topics relatively easy. Many desktop Twitter clients let you search Twitter for content. Tweetdeck, shown in Figure 14-17, can maintain multiple search windows that are refreshed periodically.

In addition to desktop clients, Twitter's open APIs have also given rise to a huge number of other sites for searching and analyzing micromessaging content (for a list, see *http://mashable.com/2008/05/24/14-more-twitter-tools/*), so you can use these to see what's going on, even without an account. Sites like Twitscoop, shown in Figure 14-18, provide visual information on popular Twitter topics.

Twitscoop can also show the popularity of specific topics in recent days, as shown in Figure 14-19.

Figure 14-18. Twitscoop provides tag clouds of specific topics within Twitter, segmented by username or keyword

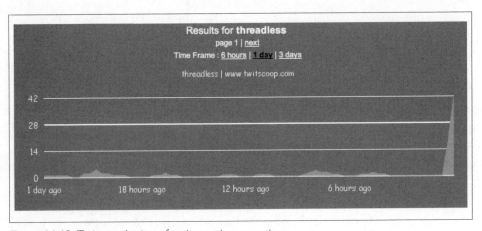

Figure 14-19. Twitscoop's view of topic mentions over time

Tweetstats, shown in Figure 14-20, can show how prolific a user is, who he talks with, and what systems he uses to send messages.

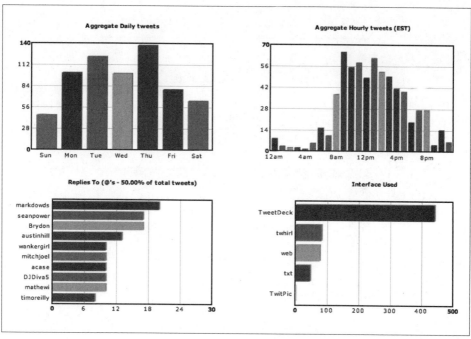

Figure 14-20. Tweetstats report on a Twitter user's volume of messages

Tweetstats also shows a profile of when a user is most active, shown in Figure 14-21. If you're trying to get noticed, this kind of analysis can give you an idea of when your target audience is most likely to be on Twitter and therefore to see your message.

If you want to find topics and posters, you can search for them within Twitter or on directories such as Twellow, shown in Figure 14-22.

Unlike social networking sites, micromessaging platforms are so open that there's little you can't analyze, even when you don't have an account. If you join a social network, you get much more visibility, but if you join a micromessaging platform that has open APIs, the only additional visibility you'll gain is detail about your personal profile.

Searching Social News Aggregators

Social news sites that rely on upvoting are important sources of sudden attention, and their content isn't easily crawled by search engines, because it changes rapidly. You'll need to use built-in search interfaces (shown in Figure 14-23) or follow referrals to find content that relates to you. The good thing about this is that it will show you not only stories, but also comments and other content you should be tracking.

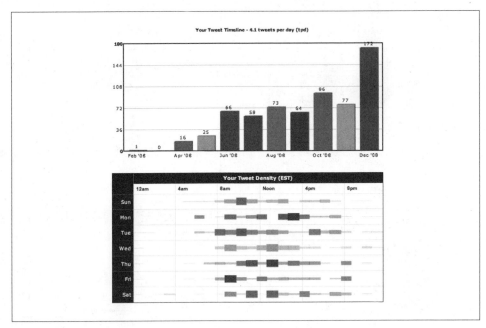

Figure 14-21. Tweetstats report on patterns and timing of Twitter use

Figure 14-22. Twellow lets you browse a directory of topics, then see who's discussing the topic most actively

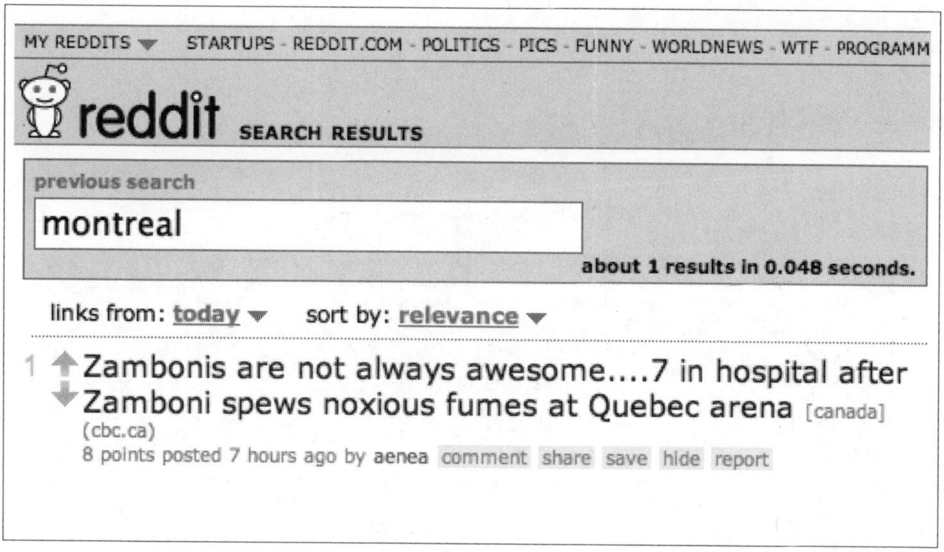

Figure 14-23. Searching social news aggregator reddit using built-in search tools

Cross-Platform Searching

With so many communities generating content, sites like Buzzfeed assemble a "river of news" that shows messages across many communities. You can search this feed. Unlike Google, which focuses on slower, more comprehensive search, sites like Buzzfeed and Serph update themselves more frequently, but don't try to index the entire Internet.

You can also use keyword popularity tools like SiteVolume, shown in Figure 14-24, to find and compare topics against one another across multiple platforms. This will show you how well you rank against competing terms, companies, or individuals.

By keeping a close eye on sites that refer to you, you can add new places to search. Maintaining a comprehensive list of social sites that should be searched on a regular basis and keeping prospective search alerts up-to-date is a key job for community managers.

Despite all of these tools, you need to join a community. Not just because it will give you greater visibility—though that's reason enough—but because the whole purpose of communities is to build interaction with your audience.

Joining a Community

To really analyze and track communities, you need to marinate in them. This means getting an account, connecting with others, and analyzing the results. Often, you'll be rewarded with greater visibility as a result of signing up.

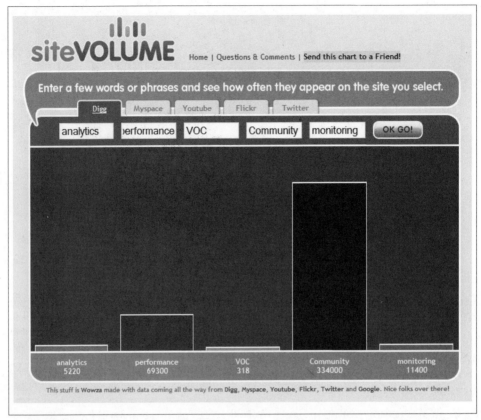

Figure 14-24. SiteVolume compares the popularity of mentions across several social platforms

When you join a community, you'll have little control over the conversations that take place and you'll be at the mercy of the tools that the community makes available. Once you're a part of the network, however, you can mine your own social graph and use internal tools such as groups to better understand your audience. Many of these tools are still rudimentary and are often constrained by how much data the platform owners are willing to share with users.

Unlike searching, which leaves no trace, becoming a part of the community has consequences. When you sign up, others will know you're there. You need to decide whether you want to identify yourself as a member of your organization (we strongly recommend that you do—these things have a way of leaking out) and understand what legal ramifications your participation will have. For example, if you work for a car manufacturer, you may be subject to lemon laws that require you to respond to complaints—so tread carefully.

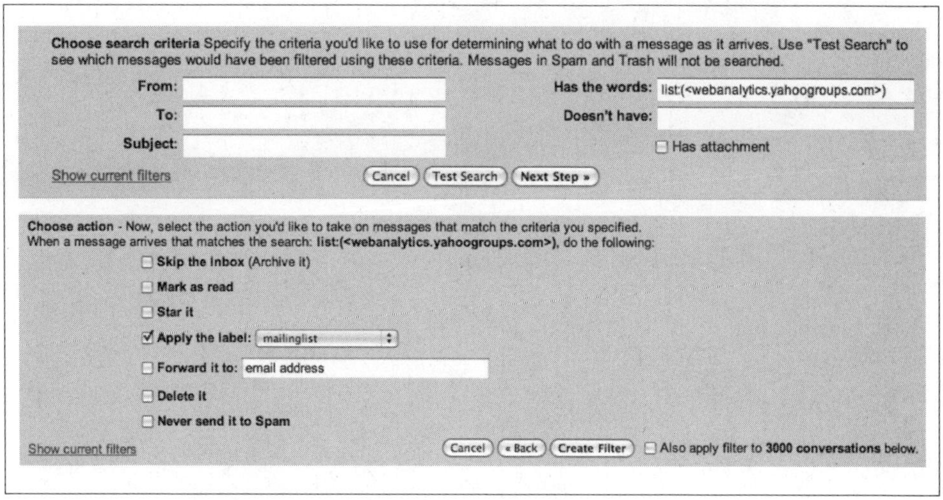

Figure 14-25. Defining custom filters in Gmail to label inbound messages

Joining Groups and Mailing Lists

For groups and mailing lists that are members-only, you can join through a mailing list management portal, after which you'll be able to read messages. Depending on how the moderator has configured the mailing list, you may have access to historical archives as well.

If you join many mailing lists and groups that are relevant to your community, you can use webmail as a form of prospective search. Using Gmail, you can append topic labels to a username (in the form *username+topic@gmail.com*) and have messages automatically tagged when they arrive, making it easier to search them later. To do this, first send all of the mailing list content to a dedicated Gmail inbox and create custom filters that label messages according to their groups (Figure 14-25).

Next, create a second set of filters that watches for specific topics and either flags them for review or forwards them to your primary mail account, as Figure 14-26 illustrates.

With a bit of work, you can create your own searchable archive of relevant mailing lists and define your own alert system to handle thousands of messages a day, letting you search many mailing lists with a single query and archiving list content indefinitely.

Joining Forums

While some forums are indexed by web search engines and let anonymous visitors search for content, some require that you be a member of the forum in order to see content.

Once you're logged in, you can interact with other participants and engage the community that's already in place. Web forums are relatively mature and have good

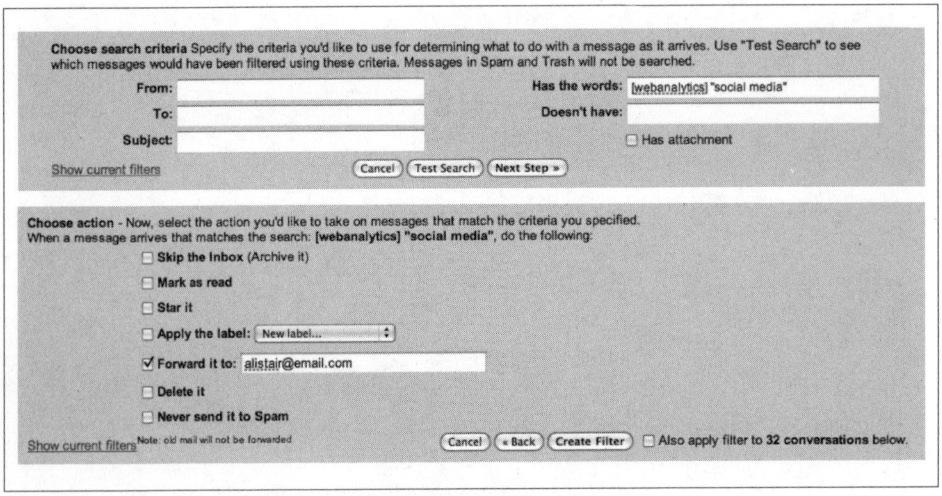

Figure 14-26. Defining actions for messages in Gmail for archiving and management

notification features, letting you know when specific topics are discussed, as well as allowing you to subscribe to content via email updates or RSS feeds.

A more mainstream kind of forum is a review site. These sites are a cross between forums and blogs, and often revolve around ratings systems. Yelp and TripAdvisor are examples of these. When you join review sites, you face a dilemma: if you identify yourself, others may be able to see your actions and accuse you of falsely inflating scores. On the other hand, if you choose to use a pseudonym and are discovered, you may face accusations of dishonesty. Most of these sites allow you to view their content anonymously, so you don't need to join them to see content and ratings.

Joining Real-Time Chat Systems

With IRC, you simply subscribe to a channel in order to analyze it. Some channels may be available only to members, and if you want to join them, you'll need permission from the moderators. Once you're in, you use the same bot tools to eavesdrop on what's said and analyze the conversation.

Joining Social Networks

As we've seen, social networks are hard to measure from the outside because of their walled garden design and their restrictive terms of service. Even once you're inside the walls, the tools aren't very flexible and are usually limited to your immediate social network. Some Facebook applications, such as Socialistics (shown in Figure 14-27), provide you with rudimentary information about your social graph.

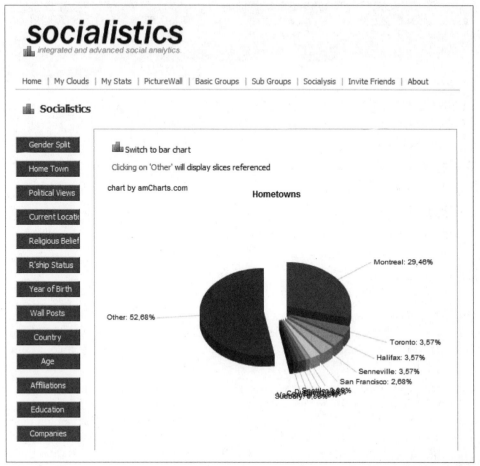

Figure 14-27. Socialistics is a Facebook application that provides basic segmentation of your social graph

The only way to find out what's going on within a social network is to befriend people, and with growing concerns about spam and malware on these sites, users may be reluctant to accept friend requests from strangers. Facebook is starting to offer marketers data on members as a part of its advertising system, but these are paid services.

Joining Blogs

Blogs are public, so you don't "join" them the way you do a forum. By posting to a comment thread as part of a blog, you're joining a transient conversation on a particular topic.

Comment communities such as MyBlogLog, BackType, and Disqus consolidate an individual's comments and make them searchable, and if you use them you'll have a

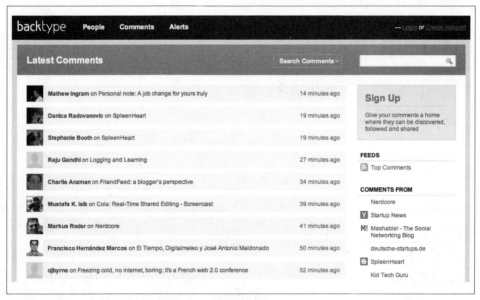

Figure 14-28. BackType monitors your comments and those of others

profile others can see. These services offer limited aggregate analysis, but they can be useful for looking into individual commenters and seeing where else they're posting. In other words, if you run a comment community on your blog, when someone comments on it, you'll be able to see what he has said elsewhere.

There may be situations in which you simply want to chime in on a comment thread, or even become a contributor in a blog network. The comment threads for an active post are the best places to join the conversation. When you comment on a blog post, you may get the option to be notified via email if someone responds to something you've posted. BackType, shown in Figure 14-28, will let you track your comments and those of other community members.

Joining Wikis

When you contribute content to a wiki, you're effectively joining it. You'll have an account—or at least an IP address—that can be tracked back to you. You're also implicitly agreeing that your content is subject to the modifications of others.

You don't get additional visibility into wiki activity when you start creating content, since transparency is one of the cornerstones of a collaborative reference. However, you do have the opportunity to correct issues or add information, provided it's nonpartisan. If you believe content is false, you may want to enlist the assistance of editors or flag it as problematic. If you react directly within the entry by posting a rebuttal or deleting someone's work, you risk an escalating battle.

If you become a contributor to a wiki, you should make sure your user page is complete and discloses any interests or affiliations you have.

Joining Micromessaging Tools

Twitter's APIs allow you to get detailed information about activity without joining. The main reason for being a part of a micromessaging platform is to interact with others and to build a social community.

Once you've got an account, you can track your own statistics: how many posts you've sent, how many friends you have, and how quickly that group of friends is growing. You can turn the many micromessaging analytics tools on yourself and analyze data, such as direct messages, that aren't publicly available.

Joining Social News Aggregators

Participating in a news aggregator means submitting stories, commenting, and upvoting. As with a wiki, you're subjecting yourself to the whims of others and need to be transparent about who you are and what you represent. You can use the comment threads associated with an entry to provide additional details or differing viewpoints.

The people who run social news aggregators are constantly adjusting the algorithms by which votes are tallied in order to fight paid upvoting and spammers.

Social news aggregators have many clues they can use to figure out whether a submission's popularity is legitimate, and simply submitting good content to the right group with an appealing title is your best policy. By all means, join in and vote on stories you think are worthwhile, and leave comments on submissions when appropriate. Just don't be disingenuous, or the community will find out.

Moderating a Community

Search is hands-off community monitoring, and joining communities only grants you more visibility if you're entering a walled garden you couldn't otherwise access. If you *really* want to see what's going on within a community, you need to help it form. As moderator, you'll have access to more details about the community than you would have as an outsider or even a member.

Many community platforms let you set up subgroups within the community and manage membership. The only problem with running or moderating a community, rather than joining an existing one, is convincing others to join it.

People will join your community if it provides value or gives them information they can't get elsewhere. For example, Dell runs a mailing list to which they send special offers. Subscribers get discounts, and Dell can drive traffic to its site and dispose of overstock.

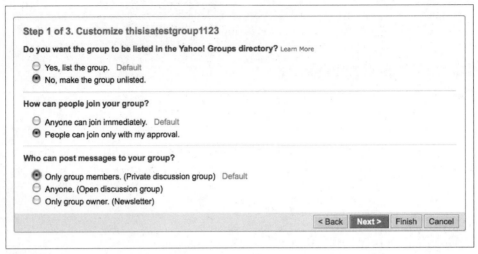

Figure 14-29. Configuring a Yahoo! group's privacy settings

Similarly, people will visit a forum, Facebook group, or IRC chatroom if they know they'll get access to subject matter experts or celebrities.

Moderating Groups and Mailing Lists

Google Groups, MSN Groups, and Yahoo! Groups all offer hosted tools to create and moderate a mailing list (Figure 14-29). Unless you have specific privacy or security needs that require you to manage your own mailing systems, consider these services. Not only will they let you control access to the mailing lists, they'll also remove the burden of things like spam blocking and unsubscribe requests. Additionally, the group directories that each site maintains will automatically index them.

As a moderator, you'll have considerable control over which features and functions members can use, as well as how easily they can search through content (Figure 14-30).

These services are lacking in analytics, however. If you want to tie mailing list activity to website outcomes, such as purchases or enrollments, you may need to run the group yourself. At the very least, you'll need to use custom URLs built with a tool like Google Analytics' URL Builder for the links you share with the mailing list so that you can track them through to conversion on your website.

Moderating Forums

Forums are made up of topic areas, and you can create and moderate a topic without running a forum yourself. If you're a vendor, you may want to start a topic thread within a larger forum. In this way, you'll already be close to your audience. As topic moderator, you'll have additional control over what people can say and whether others can search the forum.

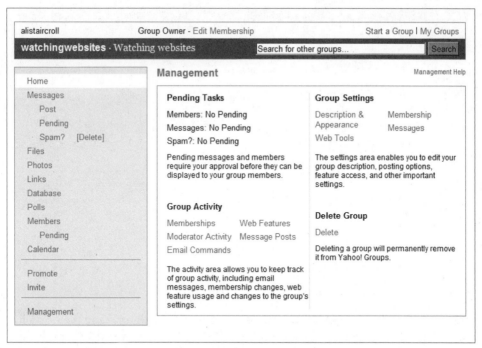

Figure 14-30. Administrative controls in a Yahoo! group

Moderating Real-Time Chat Systems

By creating an IRC channel, you launch a discussion around which others can rally. If IRC users know that they'll have access to expertise on that channel, they're more likely to seek it out.

Creating channels on IRC is easy; you simply need to join a channel and hope that no one is already there. If you are able to join an unused channel, you will usually be automatically promoted to channel operator and you can begin to set up the channel topics and bots before you start funneling traffic to it. IRC channels are examples of an *emergent community* within a social platform; hashtags on Twitter have a similar function.

As a moderator ("@" or channel operator), you'll have the ability to enforce any rules that you see fit. You can do this by either moderating the channel when conversations veer away from what you deem acceptable or by kicking and banning a user from the channel for a designated period of time. These actions are generally reserved for times when things truly get out of hand, or when spammers begin to join the channel.

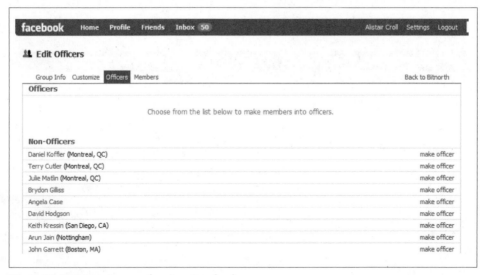

Figure 14-31. Managing members in a Facebook group

Moderating Social Networks

You can create groups within social networks like Facebook (an example of such a group is shown in Figure 14-31). Groups can have events associated with them, and you can use the system to send messages to members, manage membership, and so on.

However, in keeping with the walled garden model of social networks, you have limited access to analytical details, such as whether people are reading your messages or accessing the group.

Moderating Blogs

If you want to do more than comment on a blog, but don't want to run one outright, you may be able to contribute to an existing blog on occasion. For example, members of both U.S. political parties were given access to post to the site BlogHer.com during the elections, in order to foster an ad hoc community. Michelle Obama posted, as did a Republican chairwoman.

The blog operator may not give you access to information on your posts, such as how many people read the message or subscribed to an RSS feed, but you'll create a transient community in the comment thread that follows the post itself.

You can also encourage your community to let you know when they comment on you. If you have a blog, you'll get linkbacks whenever people reference you in their own blog postings. If someone mentions you on a community platform, acknowledge it and link to it when possible—this will encourage that person to engage more with you.

Figure 14-32. Popularity of specific terms on Hashtags.org

Moderating Wikis

While Wikipedia is an open system, there are other topically focused wikis that may welcome your assistance in moderating and editing content. This is particularly true if you have subject matter expertise. As a moderator you'll have additional privileges, such as the ability to block contributors and undo changes; whether you'll get additional visibility into the activity of the community (e.g., how many people have read an entry) depends on the platform itself.

Moderating Micromessaging Tools

You don't have to run your own microblogging site to create a community. Something as simple as a hashtag can be a rallying point for a community. For example, when the organizers of the Mesh conference wanted to launch a holiday party in Toronto, they used the hashtag #hohoto. Hashtags.org, a site that tracks hashtag use, immediately started showing the popularity of the term as shown in Figure 14-32.

Just remember that micromessaging is so open, you don't have any control over who joins or leaves the group (though you can claim a particular hashtag on sites like Thread.io).

Moderating Social News Aggregators

There's not a lot you can do to moderate news aggregators. While the systems for submitting, commenting, and voting on content are relatively open, most of these sites handle the backend moderation themselves.

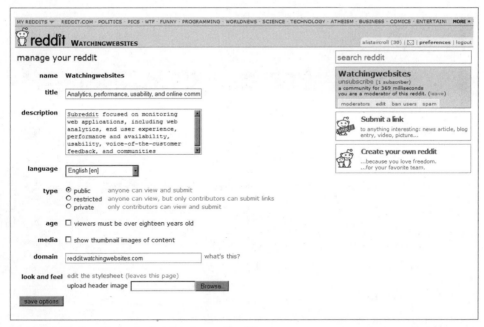

Figure 14-33. Creating and administering a subreddit

You can, however, launch a subgroup. In reddit, these are known as subreddits, as shown in Figure 14-33.

These subgroups act as communities of interest in which members submit and vote on content. You can add moderators, block users, and see the size of the community, but you don't have access to analytics such as how often links were clicked.

Running a Community

There are times when you'll need to operate the community yourself. Much of what we've already seen regarding watching your own websites applies to communities, and as a result, you have access to everything—server logs, analytical details, and more. Because you'll control design and layout, you can often embed JavaScript for analytics. You can also more tightly control the look and feel of the community and enforce terms of use that suit your business. In some cases, you may be forced to run a community yourself for legal or privacy reasons.

Note that we're talking here about *communities that are still intended for the public*. You may have internally facing communities intended for collaboration and project work, which we'll look at in the next chapter.

In addition to the headaches of setting up and running a community platform, you'll also need to convince others to join you. If you can manage to get your customers and fans to use a community that you run, you'll have an important asset for your organization. You'll also have unbeatable insight into the community's behavior. Best of all, you'll be able to tie community activity back to business outcomes, such as viral spread, contribution, purchases, and so on.

Running Groups and Mailing Lists

There are a number of open source and commercial mailing list servers. *majordomo* and GNU *mailman* are both mature listserv implementations, and while you can configure Microsoft's Exchange Server to support and run mailing lists, most administrators prefer to manage them through Microsoft's Live Communication Server.

If you don't want to run the servers, but still need control over mailing lists and the resulting analytics, consider a hosted mailing list service company such as Constant Contact or a private-label service like CakeMail, which can provide greater customization and analytical detail.

Running Forums

There are two major kinds of forum software: general-purpose and commercial platforms.

The first kind of forum grew out of the open source world. These are general-purpose forums that aren't targeted at a particular business problem or industry. They're relatively mature and have ecosystems of developers who have released various plug-ins and extensions to add functionality such as analytics, topic voting, and spam filtering. Some of the most popular include phpBB and vBulletin, although there are many alternatives (see *http://en.wikipedia.org/wiki/Comparison_of_Internet_forum_software*).

The second kind of forum platform consists of commercial software that you can use for internal or external communities. Vendors such as Jive Software, Passenger (shown in Figure 14-34), Telligent (Figure 14-35), LiveWorld, and Mzinga fall in this category.

Because you're running the forum, you get far more visibility into what users are up to, including keyword popularity, sources of visitors, and more.

Running Real-Time Chat Systems

If you want to run a private chat system, you have several options. You can use the IRC stack, which is a mature and reliable platform for real-time messaging but requires considerable know-how and demands that your community use client software to connect. These systems scale to hundreds of thousands of concurrent users, but they require dedicated clients such as *irssi*, *epic*, and *mIRC*, and aren't particularly user-friendly.

Figure 14-34. Running a forum yourself gives you far more insight into what's happening and who's participating

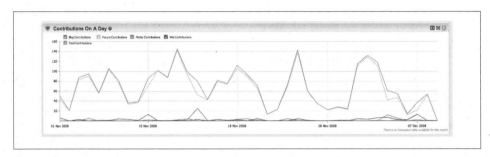

Figure 14-35. Contributions per day in Telligent's Harvest Reporting Server

```
z blockheap auth_heap elements used: 0 elements free: 256 memory in use: 0 total memory: 11264
z blockheap monitor_heap elements used: 13 elements free: 1011 memory in use: 416 total memory: 32768
z blockheap member_heap elements used: 226217 elements free: 35927 memory in use: 12668152 total memory: 14680064
z blockheap topic_heap elements used: 11716 elements free: 572 memory in use: 1171600 total memory: 1228800
z blockheap ban_heap elements used: 58329 elements free: 3111 memory in use: 1633212 total memory: 1720320
z blockheap channel_heap elements used: 25419 elements free: 7349 memory in use: 3863688 total memory: 4980736
z blockheap away_heap elements used: 5895 elements free: 761 memory in use: 565920 total memory: 638976
z blockheap user_heap elements used: 53043 elements free: 12493 memory in use: 1697376 total memory: 2097152
z blockheap lclient_heap elements used: 536 elements free: 488 memory in use: 237984 total memory: 454656
z blockheap client_heap elements used: 53100 elements free: 12436 memory in use: 14868000 total memory: 18350080
z blockheap nd_heap elements used: 0 elements free: 512 memory in use: 0 total memory: 24576
z blockheap confitem_heap elements used: 48298 elements free: 9046 memory in use: 2511496 total memory: 2981888
z blockheap librb_linebuf_heap elements used: 12 elements free: 2036 memory in use: 6336 total memory: 1081344
z blockheap librb_dnode_heap elements used: 202112 elements free: 84608 memory in use: 3233792 total memory: 4587520
z blockheap librb_fd_heap elements used: 568 elements free: 456 memory in use: 40896 total memory: 73728
z blockheap Total Allocated: 52943872 Total Used: 42500484
z Users 53042(1485176) Invites 4(48)
z User channels 226217(2714604) Aways 5895(209895)
z Attached confs 536(6432)
z Conflines 0(0)
z Classes 19(1064)
z Channels 25419(3991989)
z Bans 39259(1413324)
z Exceptions 4773(171828)
z Invex 14297(514692)
z Channel members 226217(2714604) invite 4(48)
z Whowas users 30000(840000)
z Whowas array 30000(6840000)
z Hash: client 131072(1572864) chan 65536(786432)
z linebuf 16(8448)
z scache 61(1779)
z hostname hash 131072(1572864)
z Total: whowas 7680000 channel 8119965 conf 0
z Local client Memory in use: 536(388064)
z Remote client Memory in use: 52564(14717920)
z TOTAL: 30908792 Available:  Current max RSS: 0
End of /STATS report
```

Figure 14-36. A small sample of available IRC server-side stats

Should you go this route, you'll have access to a wealth of information on community activity, as shown in Figure 14-36. You can also run clients or automated scripts to monitor traffic and alert you to content or suspicious behavior.

For most self-run real-time chat communities, however, IRC platforms are overkill. Several community platforms include chat capabilities, and you can get chat plug-ins for blogs. Third-party services like Firefly, shown in Figure 14-37, let you layer chat atop your website, but these haven't seen broad adoption.

Any real-time chat you run may become the target of drive-by spam and will require dedicated moderation to identify and block abusive users.

Running Social Networks

Given the lack of analytics within public social networks, you may want to run your own. You can use hosted services like Ning, Flux, KickApps, or Me.com to create private social networks for your community, but you're still dependent on these platforms to provide you with the means to implement your own page instrumentation or to supply you with reporting and analytics.

Figure 14-37. Firefly puts a chat page atop your website, where you can see what visitors are saying

Running Blogs

With thousands of new blogs appearing on the Internet each day, starting one is trivial. If you use a hosted service for blogging (such as Six Apart, WordPress, or Blogger), you won't have to worry about operating it. Because you can control the stylesheets and content of the blog, you'll also be able to embed many monitoring tools—JavaScript-based analytics and real user monitoring, voice of the customer surveying, and web interaction analytics tools—directly into the pages to capture a great deal of information on your blog's activity. Blogs also support plug-ins such as StatZen, shown in Figure 14-38, which will help you to identify popular content.

Running Wikis

You can run your own wiki using one of several platforms, including the MediaWiki software on which Wikipedia runs. You can populate your wiki with your own content. This is one way to build a community around a specific topic; if you can corral subject matter experts and get them to start populating a wiki, you may be able to convince your fans to continue the work.

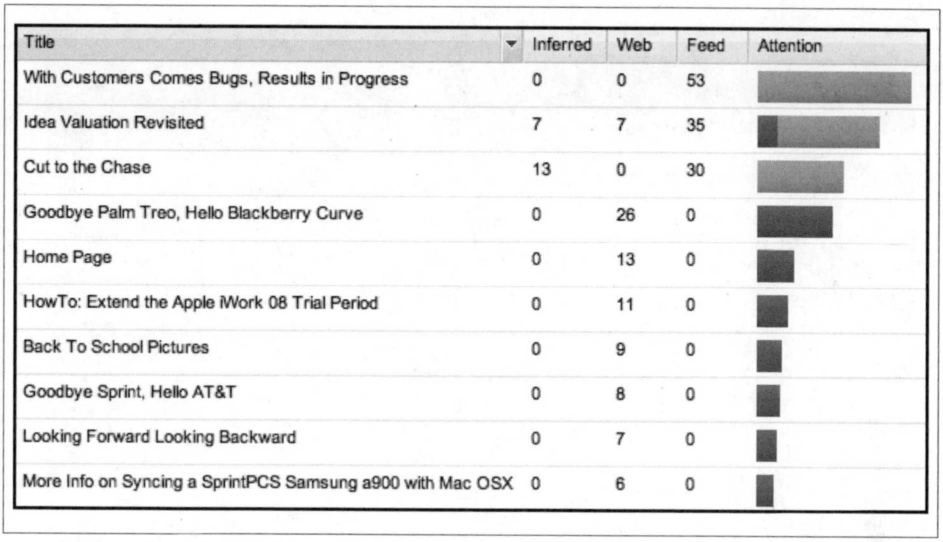

Title	Inferred	Web	Feed	Attention
With Customers Comes Bugs, Results in Progress	0	0	53	
Idea Valuation Revisited	7	7	35	
Cut to the Chase	13	0	30	
Goodbye Palm Treo, Hello Blackberry Curve	0	26	0	
Home Page	0	13	0	
HowTo: Extend the Apple iWork 08 Trial Period	0	11	0	
Back To School Pictures	0	9	0	
Goodbye Sprint, Hello AT&T	0	8	0	
Looking Forward Looking Backward	0	7	0	
More Info on Syncing a SprintPCS Samsung a900 with Mac OSX	0	6	0	

Figure 14-38. The StatZen reporting tool gives feed and post information

Incipient links

One of the most important metrics on a wiki is the *incipient link*. When members create wiki entries, they provide links to pages they anticipate writing in the future. These incipient links identify places where additional content is needed. Incipient link analysis provides several key KPIs for a wiki:

- A page with no links leading away from it, incipient or otherwise, is an orphan that's not connecting to the rest of the wiki.
- A page with many incipient links needs love and attention because it's incomplete.
- The average time that an incipient link exists before its destination page is complete is a measure of turnover and the performance of a content generation team.
- The most frequently clicked-on incipient links are those that should probably be completed first, since they show topics of interest that your wiki hasn't yet completed to its audience's satisfaction.

What operating a wiki can show you

When you run a wiki, you'll have access to much more information on which topics are being edited and where people are coming from. You will also be able to understand wiki-specific factors such as:

- *Top contributors*: you'll quickly see which members are generating the most content, as well as whose content is most disputed.

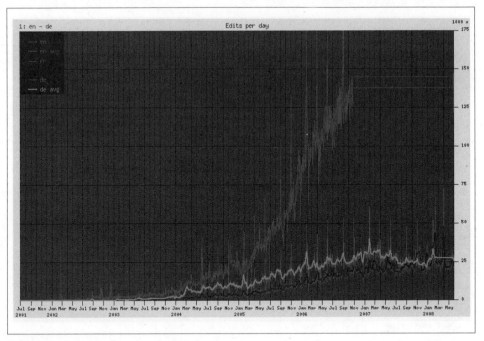

Figure 14-39. Wikipedia: new edits per day

- *Interconnectedness*: a wiki works best when content links to other content. Otherwise, it's simply a repository of articles. Analyzing how interconnected the wiki is overall is a measure of health and utility.

- *Edit patterns*: seeing repeated changes (such as a high number of edits by the same person in a small time period) can be an indicator of abuse or vandalism.

 For a good analysis of metrics and analytics surrounding wikis, see *http://www.socialtext.net/open/index.cgi?wiki_analytics* and *http://www.socialtext.net/open/index.cgi?wiki_metrics*.

Many of the projects and tools built for Wikipedia are available as open source code that you can repurpose for your own wiki. Figure 14-39 shows a Wikipedia tool that calculates the number of new edits per day on different language versions of the site.

If you don't want to run your own wiki, consider a hosted service you can embed in a website such as Wetpaint or Wikispaces. One advantage of running a wiki on your own site is that it shows search engines that you're a good source of authoritative information, because others will link to wiki content you create. If you create that content on Wikipedia, it will get a broader audience, but your domain won't get credit for it from search engines.

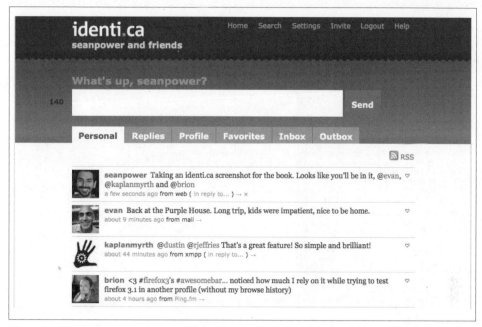

Figure 14-40. Identi.ca is an instance of the Laconi.ca open source microblogging platform

Running Micromessaging Tools

If you want to operate your own version of Twitter, you can use Laconi.ca, which is an open source alternative that resembles Twitter. The Laconi.ca team runs an instance of Laconi.ca, called Identi.ca, shown in Figure 14-40.

One of the advantages of Laconi.ca is its federated model, which allows you to join together several microblog instances so they can share messages between them. The micromessaging community is still defining what federation will do to the syntax of these platforms (for example, mentioning @seanpower might be limited to a local user with that name, but @@acroll might reference a user with that name on any federated node).

Running Social News Aggregators

If you want to run your own news aggregator, and need more visibility than what you'll get from a subreddit, you can run an open source service like Pligg (shown in Figure 14-41) and encourage your community to submit and vote on content.

Pligg has a variety of commercial plug-ins and templates that you can use to customize how it looks.

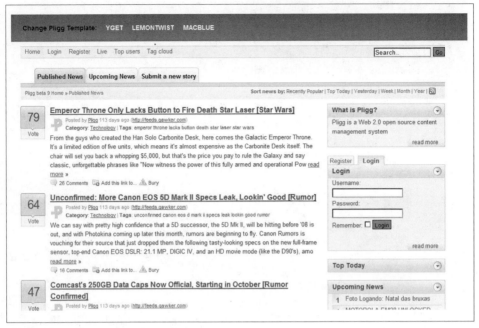

Figure 14-41. Pligg is a social news aggregator you can run on your own

Putting It All Together

We've seen that conversations are made up of people (the LTOF, the fanatics, and the occasional contributors), the topics they're discussing, and the eight types of community platform.

Your job is to search, join, moderate, or run each of these communities to get the visibility into your online presence that your organization needs.

- *To follow the people who drive your community*, use search tools and interact with community members on social networks. Subscribe via RSS to blogs they write and encourage them to join mailing lists and forums you moderate. Consider interviewing them for blog entries or showcasing things they're doing.

- *To track topics that matter to you*, rely on prospective and retrospective search. Set up a dashboard for topics that matter, featuring sites and search results for those topics. Aggregate data from micromessaging services. Create and use hashtags for the topics you care about, and reference them in your communications and on your site. If mailing lists don't exist, create and promote them.

- *To be sure you're looking at the right sites*, keep track of inbound traffic and the sites that come up in alerts, and aggregate their content using more targeted search and tools (such as Yahoo! Pipes) that can consolidate multiple feeds. Find the sites

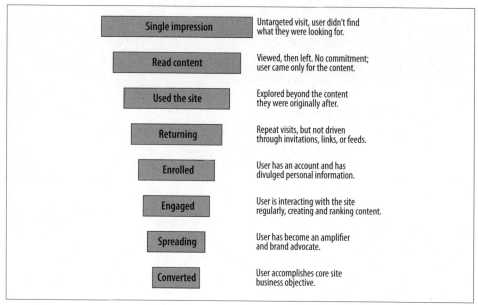

Single impression	Untargeted visit, user didn't find what they were looking for.
Read content	Viewed, then left. No commitment; user came only for the content.
Used the site	Explored beyond the content they were originally after.
Returning	Repeat visits, but not driven through invitations, links, or feeds.
Enrolled	User has an account and has divulged personal information.
Engaged	User is interacting with the site regularly, creating and ranking content.
Spreading	User has become an amplifier and brand advocate.
Converted	User accomplishes core site business objective.

Figure 14-42. The community funnel

where you're the subject of conversation, and if those sites don't exist, launch them yourself.

Measuring Communities and Outcomes

Your ability to track community engagement depends to a large degree on the nature of the community platform. As we've seen, when you run the platform you have far more insight into how your community engages with you. Because you control site design and the insertion of JavaScript, you can use many of the tools we've seen in earlier chapters, particularly web analytics, to understand the degree of engagement you're seeing within a community.

On the other hand, when you're merely a moderator or participant, measuring community engagement can be difficult.

As Figure 14-42 shows, you can think of a community's members as moving through eight levels of engagement, from their first impressions of the community all the way to the moment when they become active participants.

Single Impression

Any visitor who encounters a community gets an impression. If she doesn't see something of use, she will leave. Such visits typically last less than five seconds, and show up as bounces on your website (Figure 14-43). Of course, if you're not running the

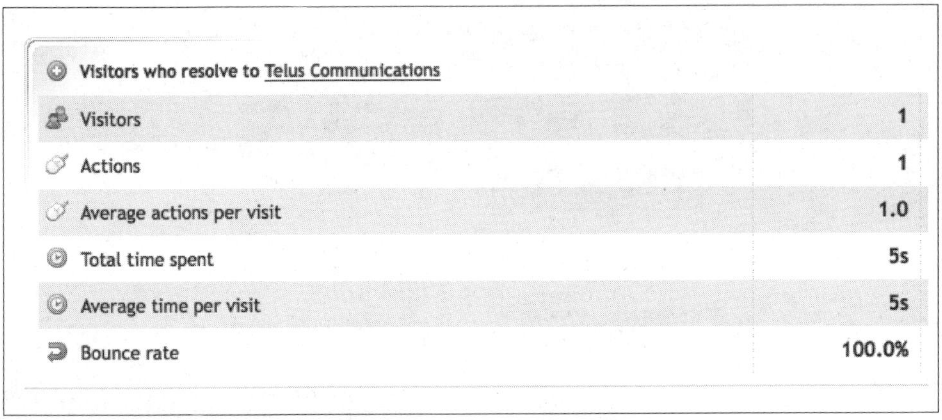

Figure 14-43. Comparing a visitor who failed to be impressed (and had only a single impression lasting five seconds) with the rest of your visitors

platform, you won't have any idea what the bounce rate of your messages is—you simply won't see activity, such as upvoting or comments, around the content.

Read Content

You've reached a visitor when he discovers your community and reads or interacts with content he sees there. Having found the content he wanted, there's little guarantee he will return. His visit includes only one page, but the average time on the page is greater than that of a bounced visit.

Used the Site

Visitors who navigated within the community website found a reason to stick around. They either had to navigate for their content or were intrigued by what they saw and wanted to see more. But they weren't interested enough to flag the site for future use (i.e., by bookmarking it) or to opt in for notifications about future content. These show up in traditional analytics tools as first-time visitors who saw more than one page (Figure 14-44).

Returning

Users who come back to the site over time are more committed than those who use it once. Browser cookies can measure this aspect of the community's visitors; if your blog uses a cross-blog commenting system, you may be able to spot returning visitors, too. Reports on new versus returning visitors show how likely community members are to return.

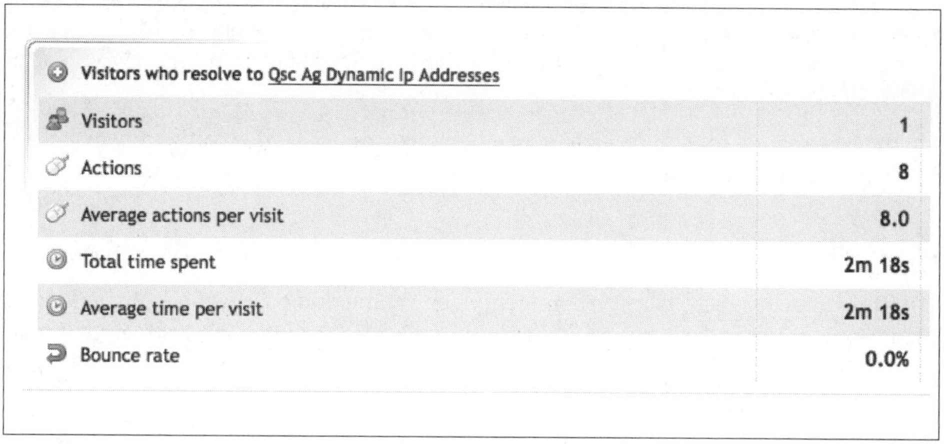

Visitors who resolve to Qsc Ag Dynamic Ip Addresses	
Visitors	1
Actions	8
Average actions per visit	8.0
Total time spent	2m 18s
Average time per visit	2m 18s
Bounce rate	0.0%

Figure 14-44. Visitors who performed multiple actions and used the site

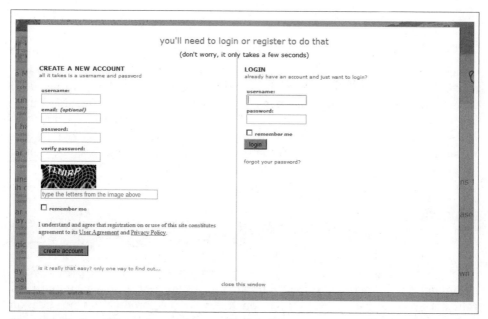

Figure 14-45. On reddit, users must first enroll before they can vote or contribute content

Enrolled

Enrollment happens when a visitor provides credentials or tries to establish a relationship with the community. This may mean that she creates an account (as shown in Figure 14-45), subscribes to an RSS feed, or joins a mailing list.

For social sites, enrollment occurs when a member connects with you. This may involve someone following you on Twitter or joining your group on Facebook. In these cases, you'll see who these users are, and they'll be added to your tally of friends or your social graph.

On the other hand, for sites such as social news aggregators, you won't know users are enrolled until they contribute or comment on content within a subreddit.

Engaged

Enrollment means establishing a connection; engagement means using it. A friend who sends you a message, a reader who comments on a blog, or a mailing list recipient who posts to the mailing list, is engaged. This may also be the act of upvoting or rating a submission (though this is harder to measure).

Users don't always have to enroll to engage you, however. They may interact with you in other ways—letting you know about inappropriate content, responding to a question via email, or responding to a survey, for example.

Spreading

Community members who help spread your message are further down the funnel. This may consist of discussing your content on other platforms, submitting it to a social news aggregator, reTweeting it (as shown in Figure 14-46), or mentioning it in a message or blog post. If your platform offers tools to help members invite others, members may also be willing to tell their social network about you directly by sending invites. This is a higher level of engagement that indicates brand trust, and is a sure sign of fledgling brand advisors.

Converted

As with all web analytics, community members are most engaged when they reach a business outcome. This is the return on community—it's where you get a financial benefit from your investment. It may be a customer getting help from a fellow community member rather than calling your support department. It may be a journalist contacting you because she heard about you. It may even be a product recommendation or a purchase. Whatever your conversion goals, this is the end of the funnel.

Reporting the Data

There's no standard way to track community interactions, and there's a wide variety of data to collect across the eight community types we've seen, so you're going to have to stitch together, clean up, and present community metrics yourself. There are paid

Figure 14-46. mollybenton reTweets an Akoha blog post

tools that can help you collect and analyze community activity, which we'll look at later in this chapter. First, however, let's consider what you need to report.

What's in a Community Report?

The first rule for reporting is to *remember why you have a community initiative in the first place*. If the goal is to increase brand awareness, you should be reporting on visibility and mentions. On the other hand, if it's to reduce support call volumes, you need to track inbound calls and use of the online support forum.

Remember all the promises you made to your executive sponsor when you launched the community initiative? Tracking your progress toward the goals you agreed upon back then is critical. This is why you needed to know what the purpose of the community was: it's the only way to know what to report. It's easy to get bogged down in statistics and metrics about communities, but you need to distill all of that data down to the few nuggets of insight that help you inch toward your goal. Your executive sponsor won't care how many friends you have—she'll care how many of them told one another to buy something.

Table 14-1 lists some of the data you may want to include in a community report for your organization.

Table 14-1. Suggested list of metrics that you should include in a regularly published community report

Topic	Detail	Statistics
Business outcomes	Conversions	Percent of visitors
		Dollar amount
	Resolutions	Number of queries resolved online
		User satisfaction with results
	Contributions	Numbers of posts, posters, and new first-time contributors
General information	Notes and excerpts	Specific comments worthy of attention
Internal community	Mailing list	New enrollments, bounce rate, reach
	Support forum	Top 10 topics
		Top 10 most frequently asked questions
		Answers per question
		Votes per question
		Question ranking feedback
		Tag cloud of key topics
		Analytics reports
	Chatroom	Volume of chat activity (messages, users)
		Analytics reports
	Facebook group	Number of members, number of messages, segmentation if needed
External community	Micromessaging	Topic activity, hashtag activity, tag cloud of inbound messages, number of followers
		Identification of key reTweets or first mentions
	Blogosphere	Number of post mentions, number of comment mentions, top 10 posts mentioning you ranked by Technorati ranking
	Social news aggregators	reddit, Digg, StumbleUpon, and Slashdot scores for any posting

You should always show reports against a historical graph (data is far more meaningful when it's delivered as a comparison) and you should annotate these reports with events such as mailouts, posts, outages, software releases, or marketing campaigns that might have triggered changes in community behavior.

The Mechanics of Tracking the Long Funnel

Tying business outcomes to the long funnel of a community is challenging. You can still understand the overall impact of communities on conversion, even if you're just overlaying the number of times your product is mentioned alongside website conversions at that time. This won't show you individual conversion rates, but it will show you what impact various conversations had on your business.

But what if you want to segment and analyze traffic more closely?

Custom URLs

An old marketing technique that works well is to create unique links for each social network, which will allow you to track their spread as you distribute them throughout the world.

If someone submits a URL from your website on reddit or broadcasts it through Twitter, that person will grab whatever string is in the address bar. Your site can dynamically modify this (for example, with a timestamp) and use this information to track when the link was first copied. This kind of integration isn't for the faint of heart, since it requires a coordinated effort from engineering and marketing, but if you look carefully at the URLs of some of today's biggest sites, you'll see it happening.

The three-step personal invite

Most of the analytics tools available for microblogging sites don't deal in business outcomes. They're more about popularity and frequency of communications, and less about which people or groups are driving customer behaviors you want. One reason for this is the inherent brevity of the messaging model. As an email marketer, you can send long custom URLs to each recipient and then track sessions by those URLs to assess whether a campaign was successful. However, it's hard to do that in 140 characters.

Traditional direct marketing relied on giving each prospect a unique identifier, allowing businesses to track prospects individually. This only works in a one-to-one model, however; on Twitter, everyone can read your status page and see the same URL.

One way around the constraints of one-to-many messaging is a three-step campaign model outlined in Figure 14-47:

1. Send a promotion to your audience, asking them to reply with a certain string if they're interested.
2. When users reply with that string, reply to them with a direct message that contains a unique URL, preferably one that embeds campaign and follower data.
3. Track referrals from that campaign and the rest of the conversion funnel using traditional analytics.

Today, this kind of tracking is relatively unsophisticated and isn't well integrated into analytics or other tools, but as micromessaging models become more commonplace we expect to see more marketing of this sort. This is a likely source of longer-term revenue for Twitter, too, since the company can dynamically rewrite URLs in transit.

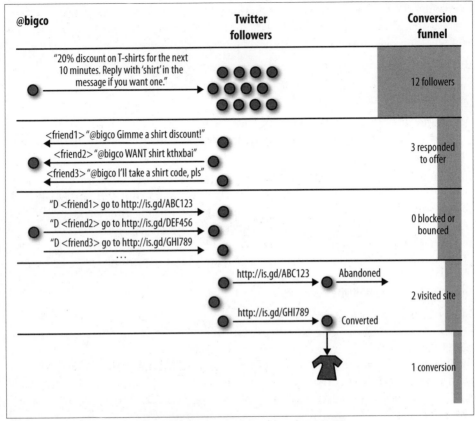

Figure 14-47. A three-step model for giving each follower a tracking URL in a promotion

Using campaign URLs and URL shortening

You can try out several messages to a community to see which works best by using URL parameters for marketing campaigns. This process is illustrated in Figure 14-48.

1. Create the marketing messages you want to test out.

2. For each message, create a custom URL using a campaign builder (such as Google's URL builder).

3. Using a URL shortening service, generate a shortened URL. This will contain not only the URL of your content, but also the campaign parameters encoded in the URL.

4. Publish the message to the community, along with its associated shortened URL.

5. When members of the community see a message that appeals to them, they'll click the shortened URL.

6. The URL service will redirect members to the full campaign URL.

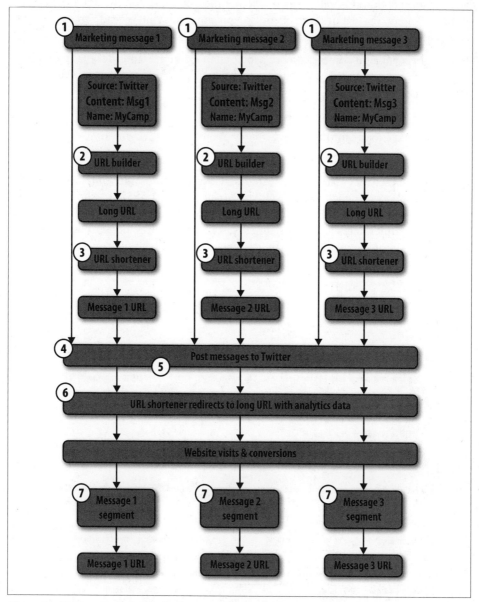

Figure 14-48. Test marketing multiple messages in microblogging with a URL shortener

7. When members are redirected to your content, the analytics tool will extract the campaign parameters and include them in your analytics reports as segments.

The best part is that this will happen whether members click on the link on the Web, within an email, or on a desktop client, so you can track viral spread across networks and tools.

This might seem a bit complex, so let's look at an example. Consider that we want to promote an upcoming event to our Twitter network. We first choose three message approaches:

- "Don't miss out on this exciting event"
- "We need experts for the event"
- "Please come to the event; I miss you!"

For each of these, we create a URL using the Google URL builder (*http://www.google .com/support/googleanalytics/bin/answer.py?hl=en&answer=55578*), shown in Figure 14-49. Note that we flag each of these messages with a unique "content" field, but we could use other fields for other segments, such as the social network we're posting the message on or the medium (email, text message, blog comment) we're using.

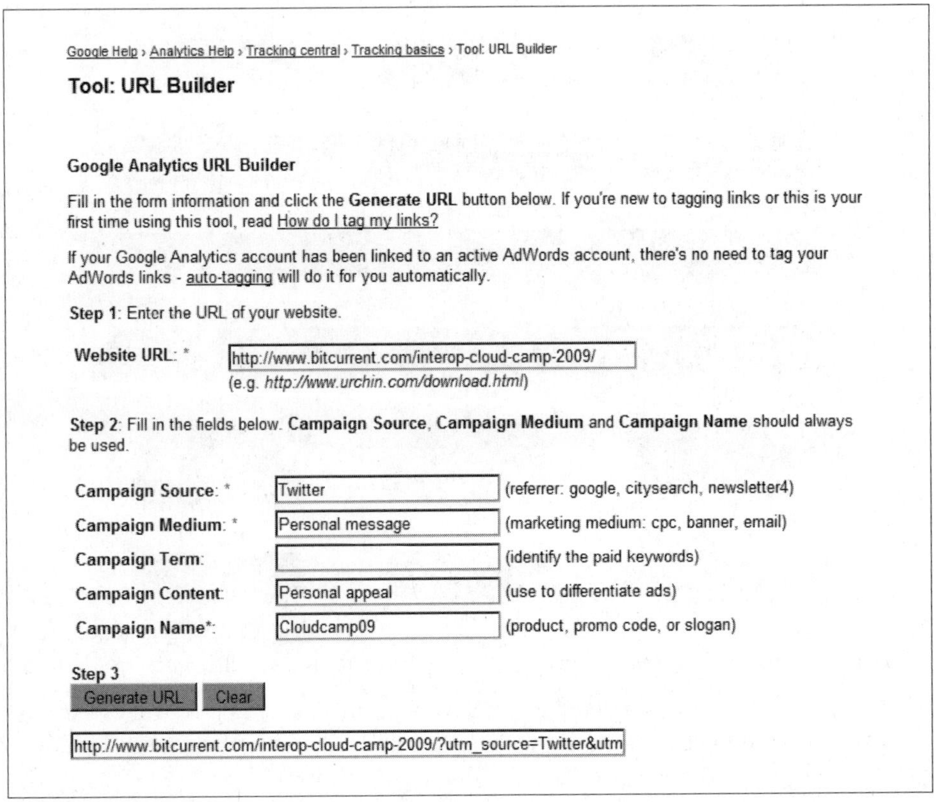

Google Help › Analytics Help › Tracking central › Tracking basics › Tool: URL Builder

Tool: URL Builder

Google Analytics URL Builder

Fill in the form information and click the **Generate URL** button below. If you're new to tagging links or this is your first time using this tool, read How do I tag my links?

If your Google Analytics account has been linked to an active AdWords account, there's no need to tag your AdWords links - auto-tagging will do it for you automatically.

Step 1: Enter the URL of your website.

Website URL: *   `http://www.bitcurrent.com/interop-cloud-camp-2009/`
                  (e.g. *http://www.urchin.com/download.html*)

Step 2: Fill in the fields below. **Campaign Source**, **Campaign Medium** and **Campaign Name** should always be used.

Campaign Source: *	Twitter	(referrer: google, citysearch, newsletter4)
Campaign Medium: *	Personal message	(marketing medium: cpc, banner, email)
Campaign Term:		(identify the paid keywords)
Campaign Content:	Personal appeal	(use to differentiate ads)
Campaign Name:	Cloudcamp09	(product, promo code, or slogan)

Step 3

[Generate URL] [Clear]

`http://www.bitcurrent.com/interop-cloud-camp-2009/?utm_source=Twitter&utm`

Figure 14-49. Using Google Analytics' URL Builder tool to generate a unique URL for the message

We now have a unique URL, which contains a web page and four parameters:

- *http://www.bitcurrent.com/interop-cloud-camp-2009/* is the destination page to which the link will lead.
- `?utm_source=Twitter` is the source of the traffic (where we intend to post the message).
- `&utm_medium=Personal%2Bmessage` is the format. If we'd used a banner or an email message, we'd mention it here.
- `&utm_content=Personal%2Bappeal` is the tone of the message (in this case, "personal message," but we will also generate URLs for "Don't miss out" and "Need experts").
- `&utm_campaign=Cloudcamp09` is the name of the campaign, which we use to distinguish it from other campaigns.

At 154 characters, that URL's a bit unwieldy—far too long to fit in a Twitter message, plus our community might notice that we've embedded a lot of information in there to track its members. We can easily shorten it into just a few characters, however, as shown in Figure 14-50.

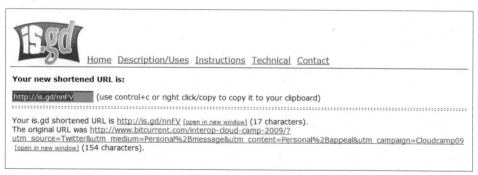

Figure 14-50. Creating the short URL for the campaign

Now we have a nice, short URL that we can embed in our message to Twitter (Figure 14-51). By combining the messages we've crafted with the shortened URLs that identify them, we can track which messages resonate best with our audience.

Figure 14-51. Posting the message in Twitter

After we run the campaign for a few days, we can see the results in Google Analytics, as shown in Figure 14-52.

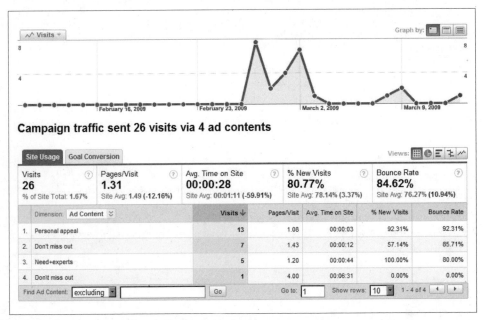

Figure 14-52. Examining the results in Google Analytics

Bear in mind that this technique is far from scientific: different messages go out at different times of day, when different numbers of visitors are online. More importantly, you'll get more attention for the first of the three messages you send, and a drop-off for subsequent messages, so try to pick a few tones of voice to stick with across several posts and transmit a different tone each time to remove this bias and learn what works best. Startup awe.sm and others are offering more automated, better integrated versions of this approach, too.

This is a good example of instrumenting the long funnel, tying together community analysis (discussions of a particular message) with analytical outcomes.

Facebook and other multisite trackers

In November 2007, Facebook announced a system called Beacon that would track users across partners' websites. The marketing advantages were obvious: with better visibility across the Web, marketers could understand how and why visitors did what they did. Facebook members were less than enthusiastic, and the company quickly relented.

Figure 14-53. Facebook Connect integrated with TechCrunch; the system "follows" a user across the blogs on which they comment on an opt-in basis, then posts their comments to their Facebook lifestreams for their social graphs to read

Since then, Facebook has introduced a more moderate system, called Facebook Connect, that lets Facebook users flag content elsewhere online and tell their social graphs about it. As Figures 14-53 and 14-54 show, a Facebook Connect account spans several websites. When users log into blogs with their Facebook accounts, comments they leave are associated with their Facebook personas. That means comments are shared with their social graphs as part of their lifestreams.

Cross-community sites like Facebook Connect are likely to become more popular, since they provide an easy way for community members to track their contributions and share with their friends. Sites that participate in Facebook Connect get increased visibility through the sharing of those comments, and Facebook gets analytical data it can monetize.

Responding to the Community

Every time you see a conversation online that's relevant to your organization, you need to consider following up. You might join the conversation, amplify what's being said, or directly engage with those who are talking. Your response may be as simple as visiting

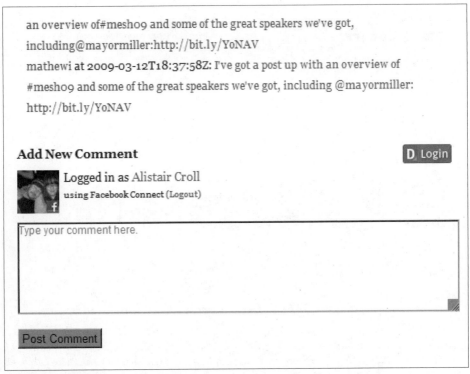

an overview of#mesh09 and some of the great speakers we've got,
including@mayormiller:http://bit.ly/YoNAV

mathewi at 2009-03-12T18:37:58Z: I've got a post up with an overview of
#mesh09 and some of the great speakers we've got, including @mayormiller:
http://bit.ly/YoNAV

Add New Comment   D Login

Logged in as Alistair Croll
using Facebook Connect (Logout)

Type your comment here.

Post Comment

Figure 14-54. Facebook Connect integrated with www.mathewingram.com/blog

a site and posting a comment or as complicated as mounting a dedicated marketing
campaign to address a problem.

The U.S. Air Force developed an assessment tool, pictured in Figure 14-55, that helps
it decide whether it should respond to comments and blogs. It's an excellent example
of turning community policies into clear, easy-to-follow directives that community
managers can follow.

Join the Conversation

The most obvious step to take in a community is to join the conversation. You can do
this to steer opinions, to meet others so that you can engage them personally, or simply
to improve your reputation within the community.

If you're adding information and opinions, make sure they're independently verifiable
and avoid spamming. Make sure you look at previous remarks to be sure you aren't
simply restating what someone has already said. Most importantly, start by listening:
it's a good idea to lurk for a while before jumping in.

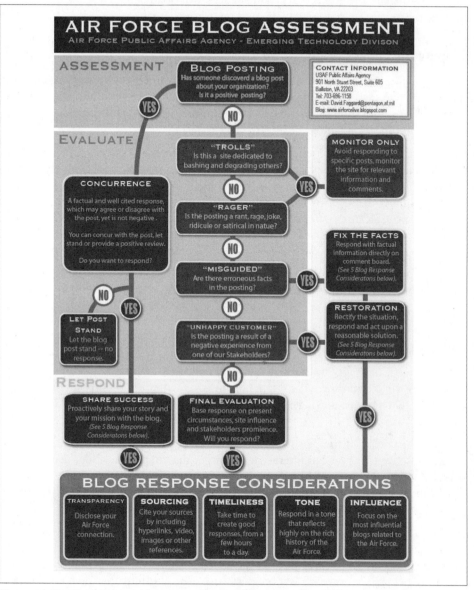

Figure 14-55. The Air Force blog assessment flowchart created by Capt. David Faggard, Chief of Emerging Technology at the Air Force Public Affairs Agency in the Pentagon

Remember that if you join the conversation, you may be unable to leave. Companies like Comcast and JetBlue, both of which have started to use micromessaging for support, have found that their communities expect them to always be around. These firms have had to share the community management workload across many employees to provide the always-on presence their customers now demand.

Amplify the Conversation

A second way to follow up on a mention is to try to encourage the spread of a message. If someone else says something about you, it will be far more genuine than if you say it yourself. But that shouldn't stop you from reinforcing it and pointing others to it, or from mentioning it on a blog or website.

Remember that there's a fine line between telling your friends about something and flat-out spamming. Many social news aggregators have sophisticated algorithms to try to weed out automated upvoting, but there are companies that specialize in pay-for-voting, resulting in an "arms race" between communities and gray-market, paid popularity. Jason Calcanis's offer to pay the top 50 users on any social networking site $1,000 a month further stirred up this controversy (*http://calacanis.com/2006/07/18/ everyones-gotta-eat-or-1-000-a-month-for-doing-what-youre/*).

The backlash you'll face for these kinds of practices often outweigh the value of the one-time burst of visits that occurs when your site makes it to the front page of reddit or Digg.

Make the Conversation Personal

You may want to follow up directly with someone who mentions you online. Many social sites will let you see what else a person commented on or submitted, and you can follow that person's messages on a variety of social sites, from FriendFeed to Twitter to Plaxo.

Avoid being creepy. If the person with whom you're interested in connecting has an online presence that you can interact with, target this first, rather than using a more personal medium like email. For example, if a user has a blog, start by commenting on something she has written and build the relationship gradually. Gentle cultivation of community members is an important skill.

Remember, too, that just because everyone has a voice on the Web doesn't mean everyone has a veto. These early days of community management have many marketers walking on eggshells, terrified of even the smallest sound from a plaintive customer. Communities will soon reach a more sustainable equilibrium that distinguishes between legitimate customer grievances and the whining of partisan entitlement.

Community Listening Platforms

Clearly, there's a tremendous amount of work required to find, engage, and report on communities. If you just consider the eight communities we've seen, and the four ways

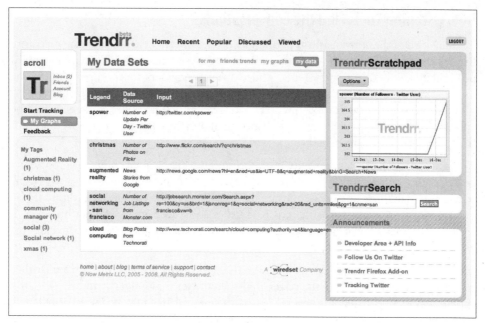

Figure 14-56. Trendrr is a general-purpose tool for tracking multiple sources of online data in a single place

you can get involved (search, enrollment, moderation, and running them), and the options to report, join, amplify, or contact members within a community, you have a daunting number of possibilities. Multiply this by the thousands of messages you might receive, and there's no way to keep up manually.

There are several tools and techniques for tracking your community presence. First of all, you can use consolidated reporting sites like Trendrr, shown in Figure 14-56, to collect and report on a variety of metrics.

Many commercial platforms, such as Radian6, Techrigy, ScoutLabs, BuzzMetrics, Sysomos, and BuzzLogic, provide reports and automate many of the search techniques we've outlined here (we're convinced that the term "buzz" has been forever claimed by this segment of the monitoring industry). Still other services and tools focus on bulk messaging and automated responses to handle a growing number of direct interactions between a company and its communities.

Regardless of the vendor, community monitoring tools do three basic things:

- *Find conversations* across many community platforms and websites
- *Aggregate the content* into digestible chunks and visualizations, sometimes detecting sentiment, demographics, and attitudes
- *Manage your responses* to what's happening in the community

How Listening Tools Find the Conversations

To collect all of the conversations happening on the Internet, social listening platforms use a variety of techniques. While they might rely on search engines like Google, many prefer their own collection systems. They run web crawlers that index websites and look for keywords. They subscribe to RSS feeds. And they harvest data from the APIs of micromessaging services like Twitter. Ultimately, they'll collect conversations and content from the many platforms we've looked at, but will usually stop short of joining those platforms on your behalf.

The use of crawlers for community monitoring is a controversial subject. To provide a real-time understanding of what communities are up to, these crawlers index sites more frequently than those of large search engines like Google, which can put additional load on servers. What really earns them the wrath of some website operators, however, is the fact that they crawl aggressively: they want to find any mention of a conversation, anywhere on the Web, even when told not to.

Most websites tell crawlers which pages on the site can be indexed through a file called *robots.txt*. This file specifies the pages that the crawler may traverse and those that it should leave alone. There are several reasons a site operator may not want crawlers to index the site. The site operator may want to force visitors to come to his sites to find content rather than finding it through an external search engine; he may not want the additional load; or the pages may be dynamic and subject to frequent changes.

Some social listening platforms' crawlers have been accused of ignoring *robots.txt*, or of using user agents that hide their true identities (*http://web-robot-abuse.blogspot.com/ 2008/05/wwwradian6comcrawler.html*). This is a violation of the site's terms of use, and some site operators go to lengths to block such crawlers through other means.

How Tools Aggregate the Content

Once they've collected the thousands of conversations that might matter to you, community monitoring tools try to make sense of them so that the important interactions rise to the top. This takes several forms, including rivers of news, keyword summaries, influencer scores, threading and drill-downs to individuals, and time graphs.

River of news to tap into the feed

A river of news is a lifestream for the topics you've flagged as important, as shown in Figure 14-57. You can segment the river by platform, most popular topics, or most influential participants. This serves as a good real-time visualization of how people are discussing your brand or topics online, as well as where those conversations are happening. You can often drill down to individuals participating in the conversations.

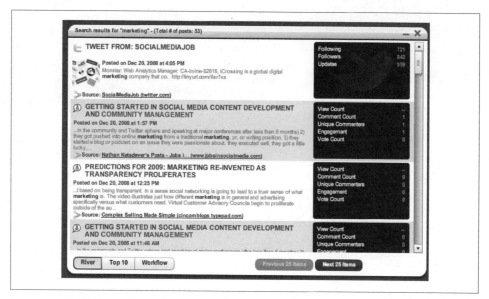

Figure 14-57. The river of news report in Radian6 shows recent activity on a topic

Keyword summary to see what's being discussed

Rather than viewing community activity as a feed of news, you may want to aggregate conversations by keywords over time. You can compare specific terms within a certain topic for popularity, as shown in Figure 14-58.

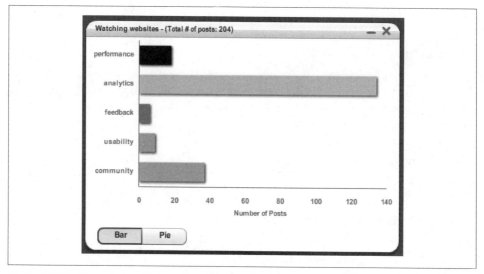

Figure 14-58. An analysis of keywords across blog postings within Radian6

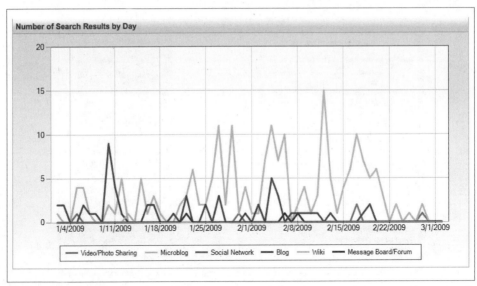

Figure 14-59. The Share of Voices report in Techrigy's SM2 product shows the number of times people have searched for a particular subject

Choosing the right words within a segment, such as product names, helps you estimate the relative popularity of particular topics or assess overall mood across communities. Be careful not to read too much into such results: without context, the results may be misleading. It's much better to visit the places you're being discussed and get a sense of what's happening, rather than relying on how often a particular word is mentioned.

You can also view graphs that show you the distribution of keywords across social media channels (Figure 14-59).

Influencer scores to see who's got clout

Every mention on a community has someone behind it, and some community members are more influential. This is the "who's talking" part of things. Perhaps they have more followers, a better Technorati ranking, more page views according to Compete.com, more influence on Klout.net, or a bigger social graph in OpenSocial. Whatever the case, you need to identify key influencers who are discussing you.

Figure 14-60 shows an influencer report that lets you compare the people contributing to topics you care about.

Threading and drill-down to look at the people

Once you've identified an important influencer, regardless of his social network or the platform he's using, you'll want to learn more about him. Many of the community

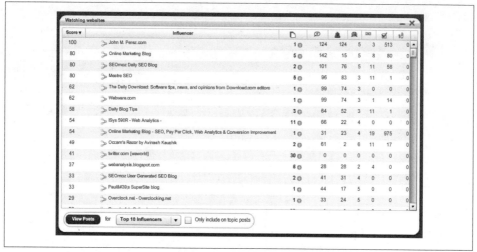

Figure 14-60. A Radian6 report on the influencer scores of various members of a conversation

management tools available today will mine multiple social networks, as well as public social graphs like OpenSocial, to build a profile of your influencers (Figure 14-61).

Figure 14-61. An individual's profile in Techrigy's SM2 platform

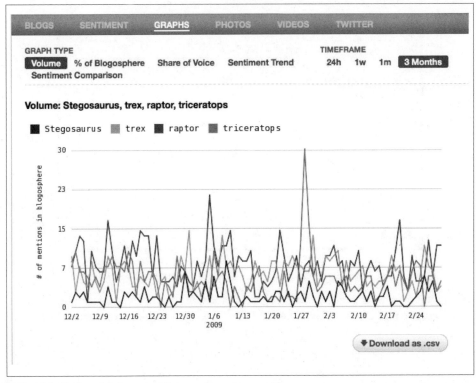

Figure 14-62. Scoutlabs' report of keyword mentions on blogs

Graphs over time to understand how much

It's not enough to know which topics are important and who's talking about them. You will also want to understand the trends of particular topics within a community or the blogosphere at large, as shown in Figure 14-62.

Historical trend analysis of this kind can be overlaid with analytical data to see whether a spike in community traffic corresponds to a rise in site traffic or goal conversions. You can also use this kind of analysis to compare competitors and see who's pulling ahead.

Sentiment analysis to understand tone and mood

While it's fine to see mentions of a topic, it's just as important to understand the context in which those mentions occurred: was the sentiment positive or negative?

Evaluating sentiment requires that a machine parse online mentions and assign it a score—positive, negative, or neutral. If a blog post reads, "*Complete Web Monitoring is awesome!*" it would be scored as a positive response. On the other hand, "I thought *Complete Web Monitoring (http://oreilly.com/catalog/9780596155131/)* was rubbish!" would clearly be a negative response (and, of course, patently false).

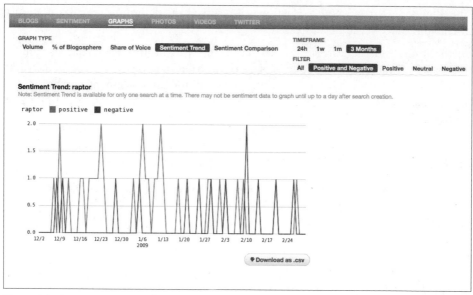

Figure 14-63. A sentiment graph in Scout Labs, based on a semantic analysis of language

But sentiment analysis isn't easy. Consider, for example, posts that contain irony, sarcasm, or colloquialisms. The sentence "Great, I'm so happy I just got Rickrolled again" isn't positive, but a computer would score it as such because of the tone of the keywords it contains.

Margaret Francis, VP of products at Scout Labs, a community management tool that includes sentiment analysis features says, "If [a sentiment analysis] algorithm ever gets good enough to [natively] classify that as negative, I will start stockpiling the explosives needed to take down Skynet"—in other words, such a tool would be frighteningly close to true artificial intelligence and might pass a Turing Test.

Understanding whether people like, dislike, or disregard certain features in your product helps you set product road map priority and anticipate support issues. Figure 14-63 shows an example of a sentiment analysis report.

How Tools Manage the Response

The third function of community management tools is handling responses. We've looked at how to do this on a personal level, and we believe that individual interactions from your market deserve individual responses—to do otherwise is disingenuous at best and disrespectful at most.

Some community management platforms help ensure you can follow up with community interactions by treating mentions as "tickets" that can be assigned to someone within your organization. CRM (customer relationship management) vendors, in particular, are integrating community interactions with customer support applications.

This allows you to track how a situation was resolved and determine which community interactions have yet to be addressed.

On the other hand, there are some tools that let you craft an automated response similar to a form letter and send it to followers based on what they say. Some companies may choose to adopt this functionality to reduce community management overhead or to deliver mass-marketing messages. You do so at your own peril, and you'll likely be called out for doing it.

Community Monitoring Maturity Model

Here's how organizations develop their community monitoring as they become more comfortable with and sophisticated about communities. Note that this applies to external communities only—we'll look at a maturity model for internal communities in Chapter 15.

	Level 1	Level 2	Level 3	Level 4	Level 5
Focus	Technical details	Minding your own house	Engaging the Internet	Building relationships	Web business strategy
Emphasis	Flags, alerts, and basic awareness	Brand and message update	Community mindset, viral spread, and rankings	Nurturing communities, measuring ROI	Interactive marketing, multivariate community testing, moderation
Questions	Am I being discussed?	Do I have a presence? Are people engaging with me on my own site?	Are they listening & amplifying me? What's sentiment like? Which messages work best with which communities?	How does the "long funnel" affect my revenues? What's the lifetime engagement of a visitor across all social networks? Who are my most vocal/supportive community members?	What will the community buy? How do I automatically match the right messages to each audience? How does my community help itself? Is virality a part of business planning?

CHAPTER 15
Internally Focused Communities

The communities and tools we've looked at so far face the public. Communities are flourishing within enterprises too, driven by a need to communicate freely and transparently with coworkers. Internal communities are a recent innovation in the field of knowledge management (KM). Ideally, KM lets organizations maximize their information resources and optimize collaboration.

That's easier said than done. It's hard to get employees to share what they know, since much of their knowledge is tacit, informal, and hard to extract. A significant amount of a company's information assets is buried in email messages and attachments as unstructured data. Employees spend their time digging through inboxes and duplicating efforts; wisdom walks out the door every night.

Internal communities provide one way to change this. "KM 1.0" consisted of static intranets that were little more than a place to store information. Recently, however, organizations have realized that for KM to work well, it has to be integrated with the way people work, rather than a separate system that expects workers to change. By giving employees tools like chat, instant messaging, wikis, and forums, a company can make more of its inherent knowledge accessible in ways that employees adopt quickly, and that don't require them to change how they behave.

The methods we've looked at for monitoring online communities also apply well to internal communities. There are some important differences, however: employees, unlike anonymous web users, are required to use KM tools as part of their jobs, and the metrics that we use to define success of internal communities are different from those for outside ones.

Internal communities have different goals than their externally facing counterparts. They need to:

- *Capture knowledge* from many different internal sources, many of which are integrated into existing chat, email, and messaging platforms.
- *Improve knowledge retrieval* by users so that employees can find the right information faster.

- *Communicate what's important* so coworkers can see which topics and issues are salient to a particular project or timeline and also to see what other employees are working on.

- *Provide feedback to managers* so they can identify top contributors and use employees' productivity for performance reviews.

Knowledge Management Strategies

Not all companies organize their information in the same way. Consider two leading analyst firms, Accenture and McKinsey.

Accenture focuses on the storage of information—employees are rewarded for documenting and storing knowledge in a way that makes it easy for others to retrieve, using consistent formats and metadata. Those employees are less likely to be subject matter experts. By contrast, McKinsey is all about expertise: the company emphasizes making it easy to find the person who is the authority on a particular topic.

This difference has far-reaching consequences, as documented in "Analysis of Interrelations between Business Models and Knowledge Management Strategies in Consulting Firms" (Sven Grolik, available at *http://is2.lse.ac.uk/asp/aspecis/20030056.pdf*). The study found that Accenture has younger employees and a higher rate of employee turnover, but because its KM emphasizes the codification of knowledge, this is acceptable. By contrast, employees at McKinsey, who are rewarded for service inquiries, tend to be older, and work there longer. Accenture has a standardized, centralized approach, while McKinsey allows less standard forms of communication and more decentralization. There's no right or wrong strategy here, but you want the KM you employ should align with the structure of the organization in which it will be used.

When crafting a strategy for monitoring internal communities, consider not only the goals of the community—knowledge capture, findability, communication of what's important, and management feedback—but also the organization's KM strategy. This will dictate what you measure, as well as which tools or platforms fit the way your employees work.

There are some other big differences between internal and external communities. First, employees may be expected to use the internal community as part of their jobs, so you may have a higher percentage of contributors than you would have on a public site. Second, you can contact your internal community directly—you have the email addresses of all the members, so you can announce changes and encourage interactions.

Just because employees may be obligated to use an internal community doesn't mean they'll want to. If you're in charge of an internal community, your focus will be on quality of contributions and making the application something your users *want* to use, instead of just something they have to use.

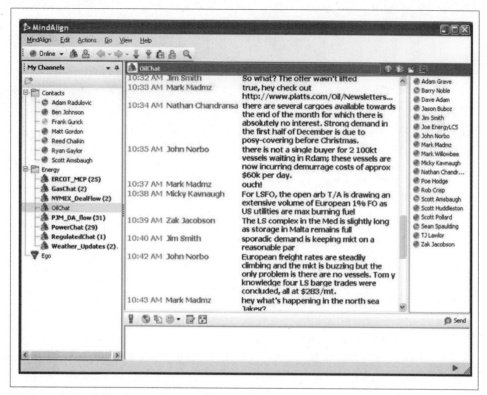

Figure 15-1. MindAlign by Parlano (now part of Microsoft's Office Communication Server)

Internal Community Platform Examples

Most companies already have the building blocks of a community. With email systems, organizational charts, and company directories, they have social graphs. Community platforms extend these systems rather than being standalone platforms. They incorporate several community models under a single umbrella.

Chat

With a known social graph, chat systems can tie into existing messaging servers. For example, Parlano's MindAlign (now part of the Microsoft Live Communication Server) works with an employee's existing contacts, as shown in Figure 15-1.

Hosted solutions, such as Campfire (shown in Figure 15-2) are geared toward smaller businesses that are looking for ways to reach out to one another in real time. These types of solutions are much easier to install and manage than enterprise platforms.

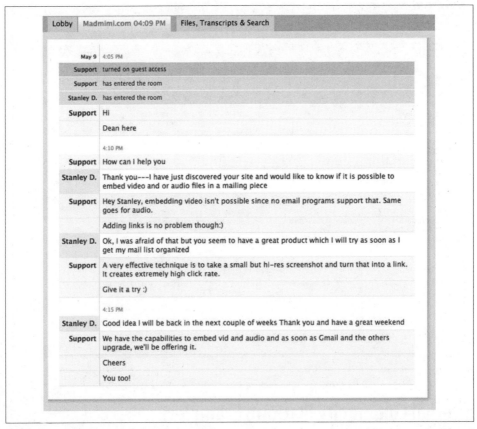

Figure 15-2. Hosted internal chat platforms like Campfire are aimed at smaller, more distributed organizations

Large enterprises will run chat internally in part because of reporting and compliance legislation. Some industries require that all interactions be archived and available for discovery in the event of legal disputes. Also, by using an internal chat system, a company can block outside chat and filter content that reaches outside organizations to reduce malware breaches and slow down corporate espionage.

Social Networks

Enterprise social networks are the rebirth of corporate intranets. They're often tied to social graphs and the company directory, and they provide analytics on contributions. There are some open source alternatives to commercial in-house social networks. Elgg.org, shown in Figure 15-3, is one such example.

Other firms, such as Socialtext (Figure 15-4), tie together profile data, status updates, and content repositories across an organization.

Figure 15-3. Elgg.org is an open source social network platform

Figure 15-4. Socialtext is a commercial social network that includes wikis, micromessaging, and collaboration

Wikis

Some organizations use wikis for technical information, particularly when it's tied to engineering tools such as bug tracking and documentation. There are a variety of commercial and open source packages available, including MediaWiki, Twiki, Confluence, and extensions to trouble ticketing systems such as Jira. Unfortunately, because they're not tied back to social graphs and employee databases, it can be hard to monitor employee productivity with such systems in ways that a company may require.

Many small organizations are turning to Google Sites or other SaaS-based document management tools that can be used as a sort of wiki. Unfortunately, purely hosted models like these lack many of the detailed analytics you get from running your own site, and won't let you embed analytics into them.

Micromessaging Tools

Twitter's rapid growth has given birth to many competitors, several of which have focused on the enterprise as a way to make money. In particular, Present.ly and Yammer (Figure 15-5) offer solutions for internally focused micromessaging.

Figure 15-5. A Capgemini employee in Yammer

These tools create ambient awareness about your coworkers and what they're up to, and perhaps this is better than the traditional water cooler model. But while the basic functionality of micromessaging is simple, the real value of these solutions is in the way they collect and interpret what's being said so that it becomes a resource for the rest of the company.

When considering micromessaging tools, it's more important to look at how they aggregate and visualize content than at how they transmit messages. For example, social collaboration tools such as Brainpark analyze what's being said, then try to offer relevant searches, documents, or coworkers, based on the task at hand.

Social News Aggregators

There's a close parallel in internal communities to reddit, Digg, and Slashdot: *prediction markets*. Prediction markets aim to harness the wisdom of crowds by letting people try to guess an outcome. Call them "Suggestion Box 2.0."

There are dozens of companies working in field of prediction markets, which has been around for over a decade. These companies include Consensus Point, NewsFutures, Xpree, Nosco (shown in Figure 15-6), QMarkets, Exago, Prokons, Spigit, and Inkling. With mainstream acceptance of wisdom-of-the-crowd models and broader adoption of enterprise communities, prediction markets are becoming an integral part of the way that some companies make decisions.

Figure 15-6. A predictive market discussion on Nosco

You'll need to factor predictive market monitoring and reporting into the rest of your internal community monitoring strategy as the technology becomes more commonplace.

The Internal Community Monitoring Maturity Model

As we've done with other kinds of monitoring, let's look at how internal community monitoring becomes more mature over time. Note that because your audience is in-house, you'll be more concerned with productivity and contribution by the organization and by its ability to create useful content that becomes an asset.

	Level 1	Level 2	Level 3	Level 4	Level 5
Focus	Technical details	Minding your own house	Engaging the organization	Building relationships	Web business strategy
Emphasis	Volume of content	Usefulness of content	Contribution patterns	Collaboration patterns, ROI from predictions, productivity gains	Community data mining, employee performance reviews based on contribution
Questions	How much information have we generated?	How good is the information we're generating? What's most and least used? How easily can employees find it?	How integrated is KM with corporate culture? What are people sharing most and least?	What's the payoff? How does KM improve per-employee contribution or reduce per-employee cost? How are people connecting with one another? Are our collective guesses good?	What does our organization know? What is our organizational knowledge worth? How can we use better KM posture to improve competitive position? Are we hiring and firing based on employee knowledge contribution?

CHAPTER 16

What Are They Plotting?: Watching Your Competitors

This book is about getting a complete picture of every website that affects your business. You will know the most about websites and communities you run, and you can learn a significant amount by moderating and joining communities. You can also gain important insights into your competitors through many of the techniques we've seen so far.

- You can use *community search* approaches to eavesdrop on your competitors and see what people think of them.
- You can use *synthetic testing* to compare your competitors' performance and uptime to your own.
- You can use *search and ranking tools* to understand how well your competitors are viewed by the rest of the Web.
- Some *usability tools* may even let you analyze how test users interact with competitors' websites, helping you to improve your own.

Before you embark on any competitive analysis, you need to consider whether you care if your competitors know what you're up to. If you perform competitive research from your own domain, their analytics will show your visits. What's more, they may have specific terms of service on their websites that prohibit things like synthetic testing or crawling by others.

Let's look at how you can keep an eye on your competitors.

If you haven't already done so, review Chapter 14. All of the approaches we describe there apply equally well to competitors.

How Much Should You Worry About Competitors?

Guy Kawasaki is a vocal proponent of smart, information-driven startups. In his most recent book, *Reality Check* (Portfolio Hardcover), he talks about how companies can build and market great products. We asked him for his thoughts about getting too caught up in competitive analysis, particularly for startups.

Complete Web Monitoring: How has the availability of free online competitive analysis changed the product planning process?

Guy Kawasaki: My guess is that the product planning process is not too affected by the wealth of competitive information that's available. The basics are obvious: you need to know what your competition's product does, how much it costs, when it will be available, and what people are saying about it.

But the truth is that there is so much to do that's internal and, ironically, out of real control, that I would bet that a product manager would like to know what his engineers are doing more than what the competition is.

CWM: How much time and energy should a startup devote to this process versus other things, and when should they do so?

Guy: It's hard to quantify, but my response is "not much." A startup should just create a great product that it loves and let the chips fall where they may. It should not assume a world of perfect information where the whole marketplace understands exactly what you or the competition is doing.

The greatest competition startups face is ignorance—that is, people don't know why they should use a product at all as opposed to why they should use products that do the same thing from company A or company B.

CWM: How does a company know when it's too obsessed with competitive monitoring and not focused enough on making a product or service that rocks?

Guy: Generally, a company should be obsessed with creating a great product or service. It should never be obsessed with the competition. If a young company has anyone with a title like "competitive analyst," then it's gone too far. Max, competitive analysis should take up about 5% of a company's mindshare.

Watching Competitors' Sites

The wonderful thing about the Internet is that everyone's online. Your competitors have gone to great lengths to tell you all about themselves, putting product specifications, lists of senior employees, and in some cases, pricing and technical specifications within your reach.

Web crawlers index these sites. Search engines analyze inbound and outbound links to them. They're ranked for popularity and monitored by testing services. Thousands of people search for them, using all kinds of terms. All of this information is collected and stored, and much of it is free for the taking. For more detailed analysis, there are paid services that rank popularity and estimate traffic levels.

Before you start gathering information on your competitors, however, ask yourself why you care. Only then can you decide what to collect and how to report it. Some possible questions you want to answer include:

- Do I have competitors I don't know about?
- Are others getting more traffic than me?
- Do competitors have a better reputation than I do?
- Is a competitor's site healthier than mine?
- Are others' marketing and brand awareness efforts more successful than mine?
- Is a competitor's site easier to use or better designed?
- Have competitors made changes that I can use to my advantage, such as changes in products, funding, or staffing?

All of these questions, and more, can be answered easily. To report competitive data on others, you'll need to have comparable data on yourself. For example, if you want to know whether competitors' sites load faster, you need historical measurements of your own website's performance.

Do I Have Competitors I Don't Know About?

The Web is full of directories and groups that try to make sense of all the sites you can visit. Some are manually created directories, such as those from Yahoo! and Google (Figure 16-1). One of the easiest ways to ensure you're aware of your competitors is to find yourself in a directory and see who else is listed there.

But directories are built by fallible humans, and sometimes your competitors won't be included. You should search for specific terms and see what sites come up, and use Google Sets (*http://labs.google.com/sets*) to enter your known competitors and see if there are others you don't yet know of.

Once you've got a list of competitors, it's time to determine who's a threat.

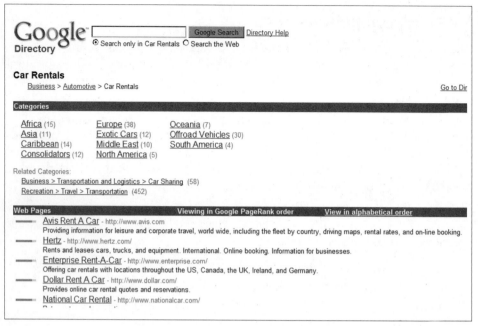

Figure 16-1. Car rental companies in Google Directory

Are They Getting More Traffic?

You can't know exactly how much traffic a competitor has on its site, but some services such as Compete.com (Figure 16-2) provide rough estimates.

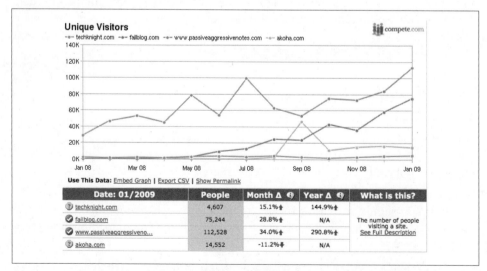

Figure 16-2. Comparison of estimated traffic on Compete.com

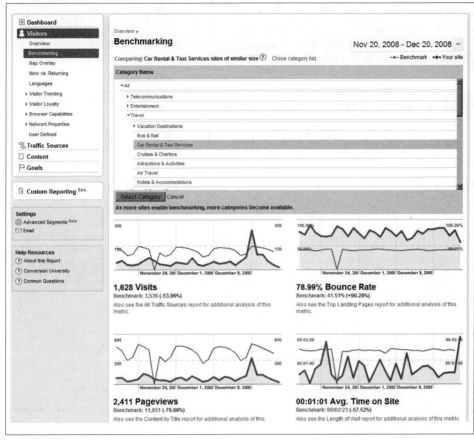

Figure 16-3. The Benchmarking function of Google Analytics compares key traffic metrics to similarly sized sites in your industry

As we've seen, traffic to a site is no guarantee of outcomes: you may have a competitor with far less traffic than yourself—but with a much higher conversion rate—who's making far more money online. Without knowing what others' conversion rates are like, there's no way to know exactly how well their businesses are running. Traffic volumes are merely a starting point.

There is, however, a way to determine how well you're doing against your competitors in general: benchmarking in Google Analytics (Figure 16-3).

To use Google Benchmarking, you first specify the business category you're in. Google Analytics will then show you how you compare to others in that category in terms of page views, retention, and so on.

You can multiply estimates of competitors' traffic (from Compete.com and others) by the Google Analytics benchmarks for a category as shown above, and get a "best guess" of what your competitors are experiencing. Be aware, however, that this is only a guess:

bounce rate, conversion, time on site, and so on will vary widely across websites. This kind of analysis also won't bring you closer to conversion or goal attainment, since Google Analytics' Benchmarking doesn't share this data, and conversion goals vary widely from site to site depending on how analytics are configured.

Do They Have a Better Reputation?

At this point, you have a list of competitors that you care about and an idea of how you're doing against them in terms of traffic and some other basic analytical metrics. But what do Internet users think of them?

You can apply community listening tools to your competitors just as easily as you can apply them to your own online presence, and most of the tools we've seen in earlier chapters can compare your products or services to those of competitors. You can also measure what the Web's authorities, such as Google and Technorati, think of you.

One of the basic elements of online reputation is the referring link. It's the basis for Google's PageRank algorithm. You can find out how many sites link to a particular competitor with the link *prefix*. To use this, simply search for `link:<weburl.com>` in Google (Figure 16-4). You'll get a list of all sites that link to the URL you enter.

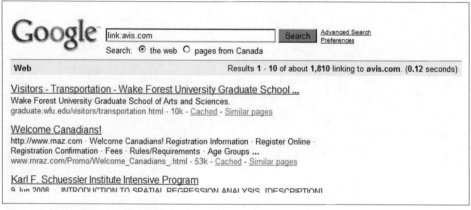

Figure 16-4. Analysis of sites that link into Avis.com

Comparing the number of inbound links in this way can show you a measure of a site's popularity. Inbound links affect everything from Google PageRank to Technorati scores, and a sudden change in the number of inbound links shows that a site is getting more attention and growing the early stages of its long funnel.

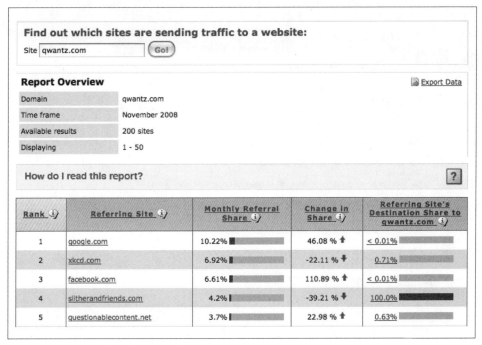

Figure 16-5. Referral breakdown in Compete.com

Compete.com also offers referral analysis tools (Figure 16-5), but free accounts are limited to only a few referring pages, so if you want a more complete link analysis (and other features) you'll have to pay for it.

PageRank

To understand what Google thinks of a site, you can look at its PageRank scores within the Google Toolbar when you visit the site. If you don't have the Google Toolbar installed, you can use several websites that will report PageRank for you (Figure 16-6). This isn't a book on search engine optimization, but if you're tracking your competitors, knowing how they compare with you over time should be part of a competitive comparison.

SEO Ranking

Google's search algorithm isn't the only one on the Internet. Your site and your competitors' sites vary with respect to how well search engines can index them. This is a combination of metadata, title structure, and so on. Several services can compare web pages and tell you how well optimized they are (Figure 16-7), and this can be a good baseline for your web design and SEO teams. You can track your own site and those of key competitors over several months to understand where you stand.

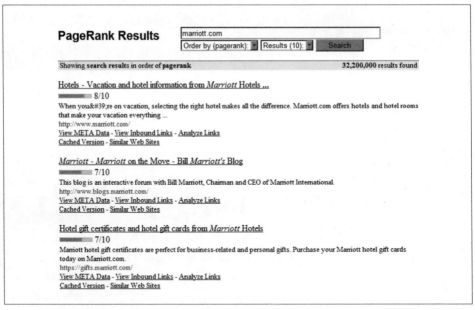

Figure 16-6. Search results including PageRank and links on SEOchat.com

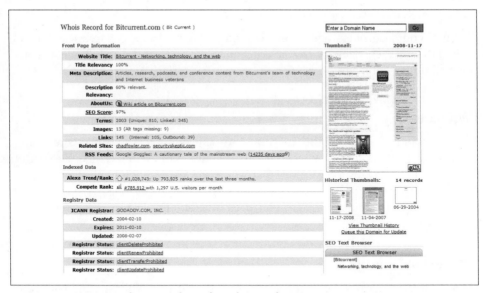

Figure 16-7. DomainTools.com analysis of a website with SEO score

Figure 16-8. Technorati blog ranking with reactions from other sites

Technorati

If your site is a blog, Technorati ranking is an alternate tool for measuring popularity and reputation. Technorati looks at inbound blog links in recent months as a way of determining which blogs are most respected—what Technorati calls a blog's "authority" (Figure 16-8).

In addition to authority, Technorati will show reactions about a site from elsewhere on the Internet.

Are Their Sites Healthier Than Mine?

You should visit your competitors' sites regularly using browser add-ins like Firebug or hosted test tools like WebPageTest.org, as described in the sections on performance management and monitoring. Each time you notice a change to their sites, measure

Airlines Flight Search Benchmark

November 01, 2008 - November 30, 2008/0:00 - 24:00 EST

Response Time Rating			Availability Rating			Consistency Rating		
Rank	Site	Response (sec)	Rank	Site	Availability (%)	Rank	Site	Consistency (sec)
1	Delta	5.48	1	Alaska Airlines	99.75	1	Frontier	3.86
2	Southwest	6.83	2	USAirways	99.68	2	Delta	4.46
3	AirTran	7.78	3	Continental	99.64	3	USAirways	4.99
4	Alaska Airlines	8.20	4	Travelocity Air	99.61	4	CheapTickets Air	5.17
5	JetBlue	10.04	5	Delta	99.60	5	AirTran	5.64
6	United	12.21	6	Southwest	99.58	6	United	5.79
7	Spirit Airlines	12.72	7	NWA	99.55	7	Continental	6.08
8	Travelocity Air	13.19	8	Orbitz Air	99.46	8	Alaska Airlines	6.26
9	American Airlines	13.40	9	CheapTickets Air	99.23	9	JetBlue	6.55
10	Continental	13.97	10	JetBlue	99.19	10	Midwest Airlines	6.59
	Average	**14.15**	11	Frontier	99.13	11	Priceline Air	6.76
11	USAirways	15.30	12	United	98.89		**Average**	**6.83**
12	Midwest Airlines	15.69	13	Priceline Air	98.83		Orbitz Air	6.83
13	Priceline Air	16.39		**Average**	**98.57**	13	Travelocity Air	7.24
14	Frontier	16.63	14	American Airlines	98.27	14	Southwest	7.88
15	NWA	20.87	15	Midwest Airlines	98.24	15	Spirit Airlines	8.21
16	CheapTickets Air	21.57	16	Expedia Air	98.08	16	American Airlines	8.81
17	Expedia Air	22.11	17	AirTran	94.34	17	NWA	9.22
18	Orbitz Air	22.31	18	Spirit Airlines	93.23	18	Expedia Air	12.64
o	ATA	--	o	ATA	--	o	ATA	--

Figure 16-9. Gomez performance benchmark for airline websites

how long their pages take to load, as well as how many third-party components they're using.

If you want to keep a closer watch on competitors, use synthetic testing tools to run regular tests, providing you're not violating their terms of service.

Several of the commercial synthetic testing services provide comparative measurements of leading sites in vertical industries (Figure 16-9). Even if you don't subscribe to a service's detailed research, you can use the data as a guideline for acceptable responsiveness and uptime.

Is Their Marketing and Branding Working Better?

If your marketing team is doing its job, it will get more coverage. Web crawlers are the perfect tools for comparing the effectiveness of marketing. If you're spending any money in AdWords, you should understand which ads are being taken, how much they're worth, and how you compare in terms of ad-spend against your competitors. Some of this information is available through competitive analysis sites like Spyfu (Figure 16-10).

Google Trends is another good resource for comparing competing products or services, and it will show you which news stories occurred in concert with changes in a particular trend line (Figure 16-11).

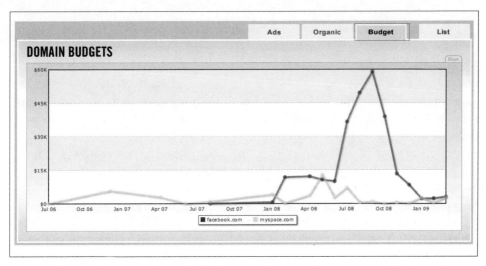

Figure 16-10. Spyfu.com provides a detailed analysis of ad-spend by company or by keywords

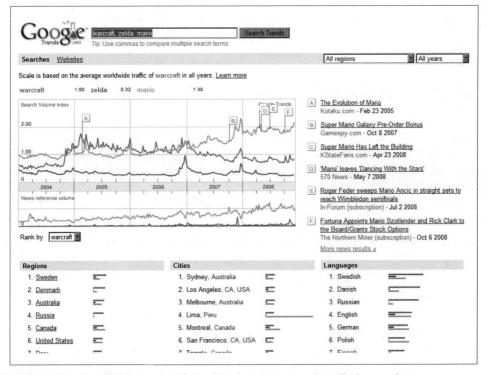

Figure 16-11. Google Trends, part of Google Labs, compares web traffic for specific terms

Google Trends is available at *www.google.com/trends/*. See *http://adlab .microsoft.com/alltools.aspx* for similar tools from Microsoft.

Don't overlook competitive questions when designing your VOC surveys. Ask respondents which other products they considered or ask them to describe how they feel about you and your competitors.

Finally, here's a dirty trick that exploits an information leak in CSS to let you test whether your visitors have also visited competitors. First, include links to your competitors' sites on one of your web pages in an iframe the visitors can't see, but that their browser's JavaScript can parse. When visitors arrive, these links will be marked as visited or unvisited in their browsers (but they won't see this). Then use JavaScript— which can see the links—to check the state of those links and determine whether they've been visited. (See *http://www.azarask.in/blog/post/socialhistoryjs/* for more details on this approach.)

By sending this data back to your analytics platform, you can track how many of your visitors also frequented your competitors' sites. You need to decide whether this information is worth the potential backlash you'll face if you're caught collecting data from visitors without their consent. It may be simpler—and more ethical—just to ask them.

Are Their Sites Easier to Use or Better Designed?

There aren't any tools that will automatically show you whether a competitors' site is better than your own. You can, however, use paid panels or usability testing to watch a recruited test group complete tasks on your site and a competitor's site to see where they get stuck. This kind of subjective feedback can be invaluable for design groups, particularly when you choose a task—such as reserving a room or adding a second passenger—that your analytics tools tell you is a common point of abandonment during a conversion process.

First, identify the places within your website where abandonment is common. Then identify similar processes within your competitors' sites. Write task descriptions for your test subjects, asking them to accomplish the same tasks on both sites, and observe their reactions.

Have They Made Changes I Can Use?

Several of the change monitoring services we've seen, such as WatchThatPage.com, will tell you whether a competitor's site has changed. You should monitor pages that give you insight into the health of your competitors' business. These typically include:

- *Staffing* pages, listing key employees and advisors
- Pages that mention *investors* or financial backing
- *Product* listings
- *Press releases* (you should also be tracking press content through RSS feeds if it's available)

Preparing a Competitive Report

If you're in charge of competitive research, you'll want to publish a competitive brief at regular intervals to your product and sales teams, as well as members of your executive management team. You'll also want to send real-time updates when important news breaks.

What's in a Weekly Competitive Report

The purpose of the weekly report is to give your internal audience an at-a-glance understanding of key performance indicators you're tracking. Information should include:

- A list of competitors you're tracking, with basic business information, such as stock price and number of employees.
- Any newcomers to the list, as well as any who have been removed.
- Charts for each metric you're comparing, such as traffic volumes, Google PageRank, Technorati ranking, uptime, performance, and number of mentions. Include one series line for each competitor, and clearly show your own organization for comparison.
- A table showing the score for each competitor, as well as the change from the previous reporting period.
- A summary of activity for each competitor, including any changes detected in each one's online content as well as links to any important news picked up through alerts or feeds.
- Community reports for any online profiles your competitors operate, such as the volume of Twitter messages, contributions to groups or forums, and so on.

Once you start to publish regular competitive reports, you'll train the company in what to expect.

Communicating Competitive Information

Your weekly updates can be sent via an internal mailing list. You may also want to publish competitor-specific information to an internal wiki so that it can be accessed by others doing research. Depending on how important competitive intelligence is to your company, you may even want to capture and store competitive data within a community platform automatically.

Breaking news, such as changes to a website, product or marketing updates, or other alerts that can have a material impact on your company's business, should be communicated through a real-time model, such as a micromessage feed or a mailing list.

Competitive Monitoring Maturity Model

Here's how organizations mature their competitive monitoring strategy:

	Level 1	Level 2	Level 3	Level 4	Level 5
Focus	Technical details	Minding your own house	Engaging the Internet	Building relationships	Web business strategy
Emphasis	Basic awareness	Site-to-site comparison	Web-wide comparison, identification of new competitors	Turning fans into information sources, testing competitive strategies	Data mining informs long-term strategic planning
Questions	What are my competitors doing?	How does my site compare to competitors'? Is it as fast? As easy to use?	Am I more popular than competitors? Better ranked? More adopted by certain segments?	How loyal are my customers? What can they tell me about my positioning versus others?	Is competitive analysis pulling data from VOC, community, and EUEM? How should competitors' strategies affect my own?

PART V

Putting It All Together

We've looked at a tremendous number of data sources you can track in order to understand how your audience interacts with you online. Putting all of this information into sensible, understandable formats that can be shared with others is challenging and time-consuming. In Part V, we look at how you can integrate all of this data for your various audiences. Part V contains the following chapters:

- Chapter 17, *Putting It All Together*
- Chapter 18, *What's Next?: The Future of Web Monitoring*

Putting It All Together

So far, we've looked at the many kinds of data you need to monitor, both on your site and elsewhere on the Internet. It's a deluge of data, and you can spend your entire day analyzing it, leaving you little time for anything else.

The final step in complete web monitoring is putting it all together. This is hard work: while many organizations are now recognizing the need for complete web monitoring, few understand how to assemble it into a single, coherent whole.

But assemble it you must. Silos of web visibility lead to the wrong conclusions. Imagine, for example, a website redesign. If conversions go down following the design, you might assume that the design was a failure. If your service provider had a problem that made pages slow at the same time, however, you can't tell whether the redesign was at fault or was, in fact, better.

With many moving parts behind even a simple website, having one part affect others is commonplace: conversions are down because people can't understand a form field; a page isn't usable because third-party content isn't loading; a synthetic test is pointed at part of the site visitors never frequent; a survey gets no responses because the SSL certificate is broken.

In the end, you need to tie together performance, usability, web analytics, and customer opinion to get a complete picture. Here are some of the reasons for doing so.

Simplify, Simplify, Simplify

With so much data at your disposal, your first instinct may be to collect and aggregate as much as you can in hopes that some sort of pattern will emerge, or that someday, it will all make sense. Thankfully, the places & tasks model should protect you from wasting your time chasing useless numbers. By using the questions that you've written down as a metrics blueprint, you'll go on missions to find relevant numbers within any arbitrary tool that might provide it. This differs greatly from using a tool's number to try and figure out what's important to you. Places and tasks helps you avoid information overload by seeking out only the metrics that matter (and avoiding those that distract).

Drill Down and Drill Up

Many of the tools we've seen start with an aggregate view of what's happening on a site. This makes sense, because with so much happening on a website, you should begin with a view of traffic as a whole.

When viewing aggregate data, however, you want to drill down quickly to individual visits. If you're seeing a pattern of clicks on a page, you may want to replay mouse movements for that page to better understand what's happening. If you're seeing errors, you will want to replay visits to see what visitors did just before arriving at that page. When you notice a performance problem, you will want to analyze individual visits that experienced the problem to see what caused it.

You will also want to drill up. Given an individual visit, you'd like to know what was happening to the rest of your visitors. When someone calls the helpdesk with a web problem, the first question should be, "Are others affected?"

Drilling up is part of the diagnostic process. If you have a visitor who's experiencing slow page loads, you want to analyze all other visitors who experienced similar conditions to see what they have in common. For example, if 10 percent of all users are having very slow page load times, analyzing all of their visits together can show what they have in common. You might notice that they're all in the same city, or all have the same browser type, or all bought more than $500 in products.

Drilling up means creating a custom segment based on attributes of an individual visit (for example, "All visits from Boston on Sprint using Firefox"). It may also mean moving across several data sources in the process.

Visualization

Even if you're not trying to drill down or up, you may want to see everything on a single screen. This is easier to do than actually tying individual visits to aggregates, because you can just pull aggregate information from many sources and display it on a single chart with the same time range. Typically, graphs and data will be grouped by time, site, network, or region, since these attributes are universal across almost all kinds of monitoring.

For example, you could display performance, community mentions, page views, and conversion rates over time. You could also show them by region, network, or web property for a given time period (e.g., the last week). You'd quickly see correlations and be able to compare traffic sources.

Segmentation

As we've discussed, web monitoring isn't just about collecting visit data, it's about segmenting those visits in useful ways to see what's working better or worse. Many of the dimensions along which you want to segment are contained in a single tool—you can segment goals by referring sites within an analytics tool directly. But what if you want to segment data in one tool by a dimension that's only available in another?

Let's say you'd like to know whether people who submit feedback on your site through a VOC survey are less likely to buy something. You can approach this in several ways. You can store both VOC and analytics data in a central data warehouse, then analyze it later. Or you can modify your VOC tool to send a message to the analytics service that marks the visit as belonging to someone who has been surveyed, then segment conversion rates by this marking.

Measuring the impact of page load times on conversions is another obvious example that requires cooperation between several systems. You may want to answer questions that require even subtler segmentation—for example, do people who scroll all the way to the bottom of the page have a faster page load time?

To do this after the fact, you need to compute the new segments from individual visits and pages that your systems monitored, which means storing all visit data somewhere.

Efficient Alerting

Another reason for consolidating your monitoring data is alerting. You can't watch everything all the time. You'd like to feed monitoring data into a system that can baseline what "normal" is and let you know when something varies.

Sometimes the first clue that there's a problem is a drop in sales—indeed, anecdotal evidence suggests that sales per minute is one of the key metrics huge online retailers like Amazon look at to determine whether something's wrong (though none of them would confirm this for us).

Complete web monitoring implies complete web alerting, bringing all of the problems detected by all of your monitoring tools into a single, central place where you can analyze and escalate them appropriately.

Getting It All in the Same Place

The amount of consolidation you're going to get depends on how much work you're willing to do and how much you want to invest. You can buy a solution from a single provider that collects many kinds of data, or you can build your own data warehouse and generate your own reports. Even if you don't want to invest a lot of time and effort, you can still herd all the web monitoring data you're collecting into a few sensible views.

Unified Provider

The simplest way to get a consolidated view is to use a vendor that provides it. Since much of the data we've seen is collected in the same way—through JavaScript—it makes sense that vendors are broadening their offerings to include a more comprehensive perspective.

Here are some examples:

- Omniture's Genesis offering is a central tool for aggregating data, and many third-party products can push their information into Genesis, where you can analyze it. You can also use Omniture's segments to query Tealeaf sessions.
- Gomez and Keynote offer both synthetic testing and JavaScript-based RUM in a single service.
- Google Analytics includes a site overlay view that shows WIA click analysis.
- AT Internet combines many WIA and analytics functions in a single offering.
- Tealeaf captures WIA data for analysis and replay, and also captures page performance information (RUM) and allows you to specify goals (analytics).
- ClickTale records page timings of the visitors whose interactions are recorded, combining RUM with WIA.
- Analytics provider WebTrends is partnered with community monitoring tool Radian6.
- Coradiant is primarily focused on RUM and user performance, but can extract custom fields (such as checkout price) from pages for segmentation later. It can also export page records to analytics packages as a form of inline collection.
- Clicky includes FeedBurner and Twitter follower information within its display.

We expect to see considerable consolidation and collaboration among vendors in this manner as a complete web monitoring model is more broadly adopted. While unified provider tools often cost more money, they may have a smaller impact on page loading because they require only a single monitoring JavaScript on the client.

Data Warehouse

The other way to get a complete picture is to paint your own. Web analytics is really a subset of a much broader category known as business intelligence (BI), which is dominated by firms like IBM (Cognos), MicroStrategy, and Oracle (Hyperion). Companies use BI to crunch numbers, from employee performance to which products are selling best in which stores.

You can think of BI as a gigantic version of the pivot tables you see in spreadsheets, where any piece of information can be segmented by any other. BI tools query data from a data warehouse for analysis. The data warehouse is essentially a very large database that's specially optimized so the BI tool can segment and query it along many dimensions.

Every time you make a purchase with a grocery store loyalty card, book a hotel with a frequent flier card, or use a bookstore discount card, data about your transaction winds up in a data warehouse for analysis by a BI tool.

If you put web monitoring data into a data warehouse, you can use a BI tool to generate custom reports that span multiple data sources. If you load aggregate data (visits per hour, comments per hour, performance for an hour) into the BI tool, you can get aggregate reports. But the real value of a BI tool comes from loading individual visit data into the tool, which allows you to segment it in new and creative ways.

Merging visitor data: The mega-record

Before you can load anything into a data warehouse, you have to understand the structure of the information you're inserting.

Imagine that you have a visit to your site, and you're running five kinds of visitor tracking. After that visit, you have four different vantage points from which to consider the visit:

- The web analytics system knows each page that was visited, as well as all the analytical metadata about pages and the visitor (such as browser type, ad campaign, and whether the visitor abandoned a transaction). It also knows which online community referred the visit, if any.
- The WIA system knows about pages it watched, such as a home page's clickmap and a checkout form's field completion.
- The RUM appliance knows the performance of each page the visitor saw, as well as any errors that occurred. We have excluded synthetic data from this list because it doesn't offer per-visit information; it's used for aggregate views instead.
- The VOC tool keeps a copy of the survey the visitor completed, along with any per-page feedback rankings.

Some of this data, such as browser type, applies to the visit as a whole. Other data, such as RUM performance or VOC ranking, applies to individual pages. There might even be information on individual objects (such as an embedded Flash element) and actions taken on those objects (such as clicking play or pause). You can even think in terms of the visitor, because this visit may be one of many that the visitor has made to your site.

In the end, the data you can collect is hierarchical. Different metrics from different perspectives apply to different parts of this hierarchy, as shown in Table 17-1.

Hierarchy	Analytics data	WIA data	RUM data	VOC data
User	Username, number of visits	Where visitors most often click on arrival	Percentage of visits that have errors	Number of surveys completed
Visit	Number of pages seen, whether a goal was achieved, entry page, browser	Events that happened during the visit	Total network time on session, total bytes downloaded, number of errors encountered	Whether the visitor provided feedback this visit
Page	Time on page, previous page visited	Mouse movements on page	Total page load time	VOC page vote
Object	Whether component rendered in browser	Whether component was clicked	Object size	-
Action	In-page events or labels	On-page movements and clicks	Refresh latency	-

This user-visit-page hierarchy is just one of many that are useful when defining how to structure data on web activity. We've found it works well for problem diagnostics, helpdesk, conversion optimization, and replay. There are many other hierarchies, however. For example, a SaaS company that has to report SLA performance might have a hierarchy that organizes information first by each subscribing customer, then by section within that site, then by page. This would make it easier to provide SLA reports.

Choosing the right hierarchy for a data warehouse is one of the reasons BI is such a complex topic: how you organize data affects how efficiently you can analyze it afterward.

Once you have specified a hierarchy, you need to determine the unique elements of visit records that you'll rely on to assemble the many data sources. For example, all four monitoring tools can record three basic facts: the time of the visit, the visitor's IP address, and the URL of the page the visitor was on. You can use this information to merge the data from each system into one mega-record. Then you can pick a metric at any level of the hierarchy (i.e., the VOC page-level information on whether a visitor provided feedback on a particular page) and tie it to another metric (i.e., analytics page-level information on whether that visit yielded a conversion).

Unfortunately, relying on only time and IP address is seldom enough. You need to consider many other factors, such as unique cookies or data within the DOM that all tools can collect.

Taken together, the hierarchy you choose and the rules you use for creating a gigantic mega-record of visits make up the schema for your data warehouse. Figure 17-1 shows one way this can happen:

1. A visitor surfs the website, visiting four pages. Each page has a session cookie that identifies it as part of the same visit, as well as a referring URL and timestamp from

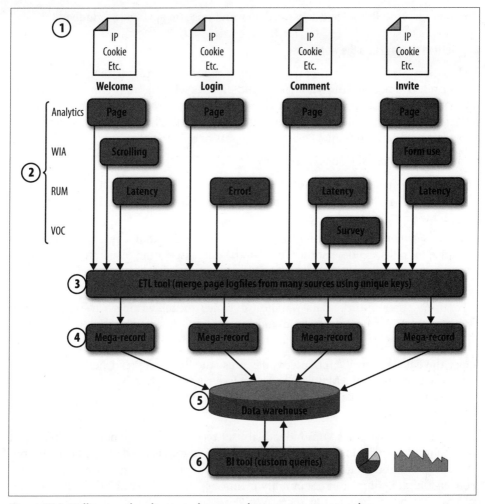

Figure 17-1. Pulling together disparate data according to common unique keys

the previous page, from which we can understand the sequence of the page requests.

2. The various monitoring tools collect their information. While analytics and RUM collect data on every page, some other monitoring tools may only collect information on certain pages—a survey on the comment page, for example, or form analytics on an invitation page.

3. At regular intervals, log data from all of the tools is merged according to the unique session cookies. This is a process known as *Extract, Transform, and Load* (ETL).

4. The result is a mega-record of each page.

5. The records are stored in a data warehouse, where they're indexed in useful ways (by time, geography, page URL, or visitor name, for example).

6. You can use a BI tool to query the data warehouse for aggregate or individual data.

ETL: Getting the data into the warehouse

ETL is the most complex part of this operation, since it requires merging many sources of nonstandard data for further analysis and dictates which kinds of analysis you'll be able to do. The ETL task is performed by a software package that gathers, modifies, and inserts data into a data warehouse according to the hierarchy and schema you provide, using certain unique keys to merge sets of data together.

As its name implies, ETL has three steps:

1. *Extracting* data from various sources (analytics, RUM, WIA, VOC, and communities) through RSS feeds, export tools, or APIs. Your ability to do this will depend on the product you choose.

2. *Transforming* data by normalizing it (for example, one tool may measure performance in milliseconds, another in seconds) and associating disparate data sources (for example, determining which VOC feedback goes with which analytics data by looking at unique cookies the two data sources have in common).

3. *Loading* data into the data warehouse, which means writing it to some kind of storage in a structured fashion and indexing it according to the dimensions by which you want to group and analyze.

Once the data is in the warehouse, you can use the BI tool, spreadsheets, or even advanced analysis tools like Maltego (*www.paterva.com/maltego/screenshots/*) to generate reports. Merging records into a mega-record like this is the only way to perform drill-down and drill-up across multiple data sources.

The example above deals with individual user records. If you're not concerned with drilling into individual records or segmenting based on visitor data, you can still put aggregate data into your data warehouse. For example, you might store site latency, conversion rate, number of community mentions, and number of VOC surveys completed every hour, then overlay them. While useful, this won't let you ask many questions of your data.

Be warned: BI is a complex field, and your organization probably has other priorities for its very expensive BI systems. Even if you get access to the BI tool and a data warehouse in which to store every visit, you're effectively building a new analytics tool of your own.

The Single Pane of Glass: Mashups

You can get a long way by simply displaying many information sources on the same screen. Where the BI approach first merged data, then reported it, this approach first generates a report within each monitoring tool, then merges those reports into a single view.

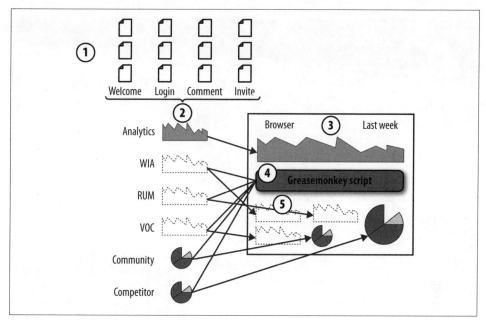

Figure 17-2. Using Greasemonkey and a browser mashup to display data from many sources at once

You won't be able to create your own segments easily—rather, you'll have to settle for the segmentation your tools offer—but you will be able to see the big picture and drill into individual tools. This is a good start, and often enough to satisfy executives who want regular updates. There are several ways to accomplish this.

Browser mashup

You can bring together several data sources within a browser using client-side scripting such as Greasemonkey (Figure 17-2). This is how Kampyle, which collects VOC data on whether visitors like a page, interacts with Google Analytics.

1. Visitors visit your site.
2. The various monitoring tools collect relevant information from their activities, both on your site and off. They may also collect data on competitors. Anything that can be correlated by time is a candidate.
3. You visit the analytics portal (such as Google Analytics) and select a particular time range or segment.
4. The Greasemonkey add-in runs a script that's triggered by your original request to Google Analytics. This script reads the time range from the Google Analytics page (e.g., "last week"), then rewrites the contents of the page that has just loaded so that it contains elements from other sites.
5. The data from each service—charts and graphs all aligned for the same time range—display in a single view.

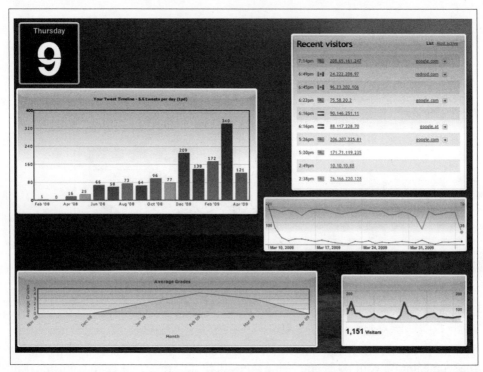

Figure 17-3. A desktop mashup of TweetStats, GetClicky, FeedBurner, Kampyle, and Google Analytics data displayed through the Mac console. Note the lack of consistent time ranges across components

You can create your own Greasemonkey scripts to consolidate data like this. One of the advantages of this approach is that it runs on your browser, where you probably have already saved logins to the various tools that are required. A third-party website can't sign in on your behalf and retrieve all of the data from all of the tools, but in this case it's your browser, not the third-party site, that's retrieving it.

Desktop mashup

Some desktops allow you to embed page snippets from several sources (Figure 17-3). On the Mac, for example, the Console view can display data from many sources by querying several websites and displaying specific DIV tags.

On Windows, you can use Active Desktop or widgets to embed data like this. Desktop mashups work well for a wall display or other visualization that won't change much.

Site mashup

A number of websites let you pull together snippets of other sites, RSS feeds, and social network traffic. Social network consolidators like FriendFeed and Plaxo (Figure 17-4) subscribe to several data sources and build a timeline of activity across all of them.

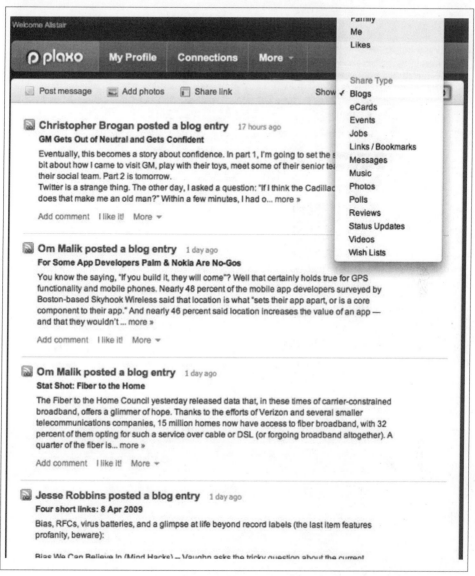

Figure 17-4. A Plaxo feed filtered for blogs; status update aggregators can combine notifications across many different message sources into a single river of news

Another class of consolidated view is the personal dashboard provided by sites like Pageflakes or Google's iGoogle home page (Figure 17-5).

Many of the web mashup platforms that are available for free on the Internet limit what you can embed, so for more control, you may want to build a consolidated view of

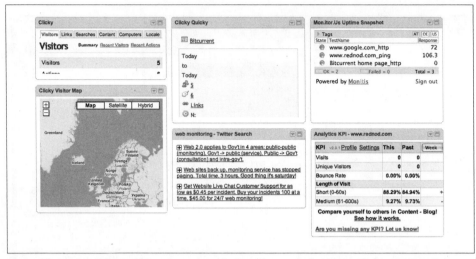

Figure 17-5. A combination of widgets and RSS feeds from searches assembled in iGoogle helps put many views in one place

what's going on either in your own web page or on a shared intranet server such as SharePoint.

Some vendors' dashboards also behave like personal dashboards, letting you add third-party dashlets from other sources to their displays (Figure 17-6).

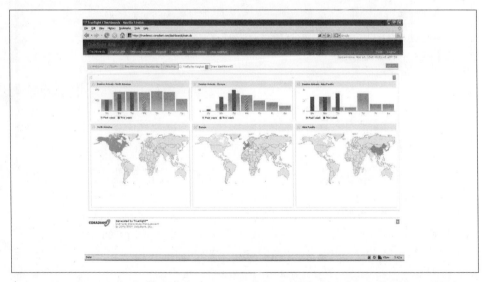

Figure 17-6. A TrueSight dashboard showing RUM data above dashlets from a third-party source (in this case, maps from Google Charts)

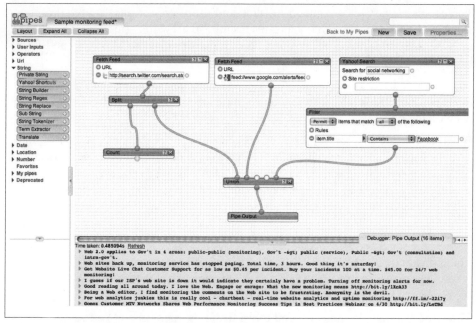

Figure 17-7. A Yahoo! Pipes example blending the RSS feed of a Twitter search, a Google Alert feed, and filtered Yahoo! Search results into a new feed

One of the advantages of a site-based single pane of glass is that it's accessible to anyone and available from anywhere. You're not tied to a particular desktop or browser installation. On the other hand, some monitoring services may not let other sites pull in data on your behalf, and each component will stand alone, without being able to take time ranges or segmentation cues from other components.

Roll your own mashup

An increasing number of sites and tools are integrating APIs into their products. Google Analytics is opening up its data to third-party tools, and social networks like Twitter have wide-open APIs that have created the wide variety of Twitter analysis sites we saw in Chapter 14.

Building a feed with Yahoo! Pipes

Yahoo! Pipes and Microsoft Popfly let you link feeds of data, text processing, filters, and searches together to consolidate information. For example, you can pull several RSS feeds into one master feed, then filter that feed for certain keywords or authors (Figure 17-7).

These tools are easy to configure using drag-and-drop connections between processing steps. When you consider that you can subscribe to a Twitter search term, a competitor's website, and a blog, all through RSS, the power of tools like Pipes and Popfly becomes quickly apparent.

As these APIs become more common, it'll be easier than ever to pull in raw data from many sources and render it in your own application, should you have the time and resources to do so.

Whether you use a client, browser, or site mashup, or decide to roll your own, make sure the data you display is clickable—either to drill down into more detail or to link back to the service from which the data came. That way, when you see something that's broken, you can quickly link to the individual tool that generated that part of the consolidated display and investigate further.

Comparing apples to apples

The individual tools from which you're collecting data generate reports in their own ways. Your goal is to line up multiple data sources to display them sensibly so you can see the big picture. You can do this based on data they all share.

Time, for example, is something all tools' data has in common. You can get graphs from several monitoring tools for a particular day. There are other dimensions you may be able to segment by: page name or URL, visitor name, campaign name, region, carrier, and so on.

If you're creating a browser-side mashup, your "anchor" data source will likely be web analytics. With the Greasemonkey approach outlined above, the analytics application dictates the segmentation—a particular time, a certain URL, or a single city, for example. The JavaScript you execute with Greasemonkey finds this segment by parsing the page, then it queries all the other tools' reports for the same segment. If you're viewing a week-long graph of conversion rates, the client-side JavaScript could send a request to a synthetic monitoring tool and retrieve a graph of performance that week, then insert it as a component within the analytics page container.

Desktop mashups can't do this. Each component of the desktop mashup stands alone because the various components can't "take their cues" from data on the parent page. Every chart and graph is unrelated to the others on the screen.

If you're using a website mashup such as Pageflakes or iGoogle, you probably won't have this much programmatic control. The best you can do is to define tabs that make sense and then collect related reports from each tool. Here are some examples to get you started:

- Your "regions" tab might show the site's performance by region alongside the analytics tool's view of traffic by region.

- Your "performance" tab might blend DNS uptime from a synthetic testing service with checkout page host latency from a RUM product and show conversion rates to help tie it back to business outcomes.

- Your "awareness" tab might include a chart of Twitter followers, new visits, Google Insights search trends, FeedBurner subscriptions, and Technorati ranking.

- Your "URLs" tab could show the top URLs in terms of poor performance, arrival and departure, bounce rate, and time on page.

- Your "virality" tab could show the top search results for certain terms, the number of people who have completed an invitation goal, bounce rate data from an email system, and which social networks and messages are being forwarded most.

Alerting Systems

You don't just want to consolidate for visualization. If you want to be sure your PDA isn't constantly vibrating with alerts, you need a way to bring together alerts and metrics from many sources, detect problems, and let you know about the important ones.

IT operators deal with hundreds of servers and network devices, and over the years vendors have built sophisticated algorithms for baselining and detecting problems. Their jobs revolve around a queue of incidents flowing in, each of which needs to be triaged and repaired. To handle the deluge, many monitoring tools include algorithms that dynamically baseline measurements, then detect problems when those baselines are exceeded.

The good news is that many of these systems don't care what kind of data you feed into them. Companies like Netuitive, ProactiveNet (now BMC), and Quantiva (now NetScout) can take in lots of different data sources and form an opinion about what's normal.

You can pump in conversion rates, the number of feedback forms visitors have submitted, page satisfaction rankings, and network latency. Monitoring tools will chew on all this information and let you know when anomalies occur. Some RUM vendors, such as Coradiant, build this technology into their products.

Consolidating information for alerting involves four steps:

1. *Normalize the data.* These systems work best when they receive numeric information at regular intervals. If you're using RUM data, for example, you don't just send web requests to the analysis tool. You need to feed a metric such as "95th percentile host latency for the last minute," which can take some computation.

2. *Collect the data.* These days, simple protocols like RSS feeds make it relatively easy for one system to pull data from another. You can even use web services like Yahoo! Pipes to pull many sources together, or you can build a collection service using a cloud platform like Google's App Engine.

3. *Do the math*. Even if you don't have access to sophisticated algorithms, you can still derive relatively simple ones. We know of one web operator whose main metric is the average rate of change in host latency—a gradual slowdown or speedup isn't a problem, but a sudden spike is easy to identify. You could do the same thing with checkout prices on the site.

4. *Send alerts*. This may involve sending an email or SMS (Short Message Service) message, turning a light red, or creating a trouble ticket. Much of this already exists in an organization's Network Operations Center (NOC), but the things they're monitoring generally don't include business data like conversions or form abandonments.

The point here is that you may need to pull in data from various monitoring services at regular intervals through whatever APIs or RSS feeds those services provide, then decide how to treat that data as a set of signals about the overall health of the website.

Tying Together Offsite and Onsite Data

One of the most difficult types of data integration involves linking what happens elsewhere in the world to your web analytics data. Whether it's a poster on a bus that contains your URL, a mention on a social network, or a call to your sales department that began on your website, it's hard to monitor the long funnel.

There are two main ways of tying offsite activity back to your website. The first is to gather information from social network sites where visitors opt in and share their identities. The second is to mark your offsite messages with custom URIs that you can track with your analytics tool.

Visitor Self-Identification

The first approach, which is the basis for systems like Open Social, Beacon, OpenID, myBlogLog, and Disqus, requires that you enroll yourself in the social network somehow. Then visitors who want to share their identity with you can do so.

This is the simplest way to track users. You may even be able to analyze inbound links from certain social networks to see who's sending traffic. For example, referrals from *www.twitter.com/acroll* indicate that the visit came from the profile page of Twitter user "acroll." In practice, traffic of this kind is only a fraction of Twitter traffic, as most people view links in their own Twitter feeds and with third-party desktop clients like Tweetdeck and Twhirl.

Many social networks want to monetize the analysis of their traffic; indeed, "selling analytics" is the new "ad-driven" business model. It's therefore unlikely that you'll get visitor analytics for free at the individual visitor level unless you ask visitors to log in. But don't write this possibility off: in many situations, your market will gladly tell you who it is. Musicians, for example, have legions of fans who crave the attention and

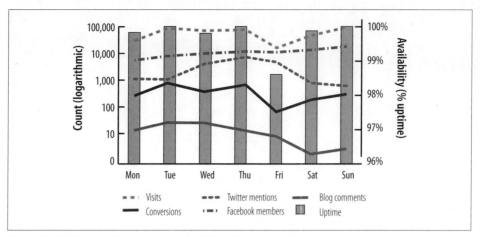

Figure 17-8. The most basic way to consolidate data manually is to display multiple data sources in a spreadsheet using time as a common key

acclaim of the artists they support, and may be willing to divulge their identities in return for access to a celebrity or early notification about tickets or appearances. There are often ways you can encourage visitors to self-identify if you're creative.

Using Shared Keys

Any data analysis requires a key of some kind. The more specific that key, the more closely you'll be able to tie together the data. All interactions have some basic attributes: where they happened, when they happened, any keywords that were mentioned, and in some cases, the person who said it.

The easiest way to compare two sets of data, then, is by time. Lining up graphs of mentions on Twitter, Facebook friends, web traffic, and so on, is an easy way to see which factors are related (Figure 17-8).

If keys like date or geography aren't enough to consolidate your information—particularly if you want to link individual actions to individual outcomes, rather than just comparing aggregate data over time—you'll need to make your own keys. These may be a unique URL on printed materials, a URI parameter hidden in a shortened URL, a unique phone number for each visitor to call, or some other way of linking together the many offsite interactions your visitors have with you.

In the end, you'll have to take whatever data you can get, assemble it according to common keys like time or geography, and track it over time to determine how your many sources of web monitoring information are related.

CHAPTER 18

What's Next?: The Future of Web Monitoring

> *"Tis some visitor entreating entrance at my chamber door ... I betook myself to linking fancy unto fancy, thinking what this ominous bird of yore—what this grim, ungainly, ghastly, gaunt, and ominous bird of yore meant in croaking 'Nevermore.'"*
>
> —Edgar Allan Poe, "The Raven," 1845

In the first chapter of this book, we said that smart companies make mistakes faster. It's perhaps more accurate to say that smart companies *adapt* faster, learning from their mistakes and forever tightening the feedback loop between what they attempt and what they achieve.

The monitoring tools and measurement techniques we've covered in this book are your eyes and ears, giving you an ever-improving view of your entire online presence. We've tried to lay out some fundamental principles within these pages: determine your goals, set up complete monitoring, baseline internally and externally, write down what you think will happen, make some changes, experiment, rinse, and repeat.

If writing this book has taught us anything, however, it's that the technology of web monitoring is changing more quickly than we can document it. New approaches to improving visibility surface overnight, and things that were nascent yesterday are mainstream today.

In the course of researching and writing the text, we've talked to over a hundred companies building monitoring technologies—some still stealthy—and this has afforded a tantalizing glimpse of what's coming.

Accounting and Optimization

All of the monitoring we've seen in this book serves one of two clear purposes: accounting or optimization.

In many ways, analytics is simply how we account for the online world. After all, we're collecting metrics, such as daily sales, average shopping cart size, and advertising revenue. Accounting-centric analytics is likely to become the domain of the finance department as that department pushes for more visibility into the online business.

Today, analytics is an afterthought in most application development efforts, thrown in as JavaScript tags just before a site launches. This is unlikely to continue. Expect accountants and controllers to define standards for data collection from web systems, and expect investors to demand more real-time data on the progress of the online business. It won't be long before Generally Accepted Accounting Principles (GAAP) are part of the requirements documents that web analysts need to incorporate into their analytics strategies.

The other half of analytics focuses on continuous improvement. This is the domain of product management, interface design, and copywriting. The goal is to get more visitors to do more things you want for more money, more often. Every tweak to your site is an attempt to edge ever so slightly toward that goal, and every visit a hint at whether the attempt was successful.

While much of this optimization can be done manually, at least at first, organizations with sites of any substantial size will soon find this an overwhelming task. They'll turn to multivariate testing, as well as the automation of usability analysis and visitor feedback, allowing web platforms to adjust their content and layout dynamically to maximize conversions based on what works best for a particular segment.

From Visits to Visitors

It's relatively easy to consolidate monitoring data by time and to segment it by shared metrics. For example, conversions by country can be compared to historical performance within that country, and periods of poor performance can be aligned with bounce rates to see if there's a correlation.

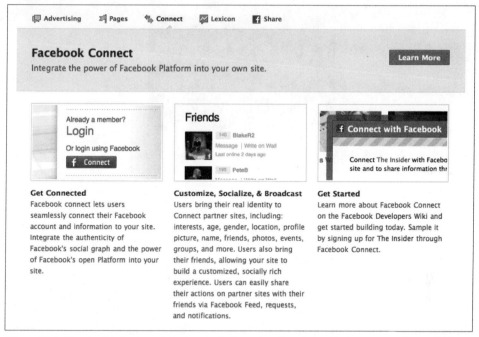

Figure 18-1. Facebook's business solutions include onsite advertising, the Lexicon analytics platform, paid pages for companies, and tools such as sharing and Connect to encourage users to share their actions

A far bigger challenge, however, is linking individual visits to individual outcomes. In essence, the web monitoring industry is moving from a focus on visits to a focus on visitors. The individual visitor—either named or anonymous—must be tracked across social networks and the conversion process in order to understand the return on any marketing investment. It's a return to the days of database marketing.

Identifying an individual visitor is critical. This happens in several ways:

Visitor self-identification

Visitors sign up for a federated social model, such as Facebook Connect, Microsoft Passport, or OpenSocial, which shares their activities with others (Figure 18-1). The information may be anonymized by the owner of the social platform, but already, tools like Radian6 rely on OpenSocial to identify key influencers in a social network.

Grammatical links

Text from comments and blog posts are analyzed using language-parsing algorithms that try to identify a single person's online identity from the content she creates. This works for prolific web users for whom a considerable amount of content exists.

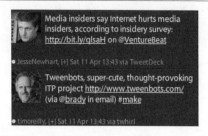

Figure 18-2. Social messages, such as these ones on Twitter, contain clues about the spread of social conversations, in some cases even documenting when a social message changes platforms

Metadata and fuzzy association

Many social networks include basic information about users, such as their first names or the cities in which they live. By combining this information, analytics systems can make educated guesses about who's who and try to join multiple online profiles into one metaprofile.

Understanding the flow of messages and referrals

Most social conversations carry threads within them that can be tracked. On Twitter, words like "via" and special characters such as "RT" (reTweet) signify information that's being passed down the long funnel. Tracking tools can follow these threads to examine the social spread of a message (Figure 18-2).

Embedding tracers in online activity

For years, analytics companies have relied on cookies to persist information on browsers so they can track new versus returning users. As we've seen, with the emergence of short message interactions like Twitter messages and Facebook status updates, there's nowhere to easily hide tracking information other than within the URL.

Asking visitors to log in

Sometimes the simplest solutions are the best. Asking visitors to sign in so that they can be uniquely identified and tracked is the surest approach to tracking. It's also less risky from a legal and ethical standpoint, since site operators have the visitor's permission to know who they are.

Web analytics companies that focus on tracking visits are quickly retooling to track individual visitors, either anonymously or as known individuals, linking user logins to Twitter accounts and blogs. Of course, harvesting and mining every individual's interaction across the Internet has serious privacy concerns, but it also steers web analytics toward the realm of CRM and a one-to-one relationship with a company's customers.

Personal Identity Is Credibility

A corollary to the move from visits to visitors is that personal identity becomes currency. Companies and individuals will start to manage—and defend—their online reputations as personal currency that gets them jobs and gets their voices heard. Online reputation is the twenty-first century's reference letter.

Expect to see certified profiles and recommendations provided by the community. Already, highly ranked contributors on SAP's community site use their community-granted awards on their Linkedin profiles, leveraging them as proof of their knowledge and ability to work with others when seeking new employment.

These reputations will have a strong influence over which links communities follow, and personal credibility scores will factor into community analytics. Metrics such as friend count, number of followers, and other metrics of popularity are already giving way to more sophisticated measurements on sites like Twitalyzer, such as the reach of a reTweet or the number of community members with whom a person converses regularly.

From Pages to Places and Tasks

We've looked at the concept of places and tasks several times in this book: as a way to think about your website, as a means of defining analytics metrics, and as a tool for understanding the end user experience you need to measure.

The shift from document-centric monitoring to a places-and-tasks model is inevitable. We simply don't use the Web as a set of pages any more. Startups like Kissmetrics are poised to take advantage of this shift in the fundamental structure of web monitoring.

This will change what we monitor. For places, we'll care about efficiency and productivity; for tasks, we'll focus on effectiveness and abandonment. We'll also use new metrics, such as the number of times a user undoes a particular action, to measure efficiency and effectiveness.

Instrumenting places and tasks requires the collection of many new metrics, often from sources that can't simply be instrumented with JavaScript: email campaigns, URL shorteners, RSS feeds, site crawlers, dynamic ad content, and inbound calls. The collection infrastructure needed to launch a complete web monitoring service will become far more significant and will require the integration of many different disciplines.

Mobility

The majority of devices that can access the Internet today are mobile. That means they're being used in a wide range of locations in conditions that are strikingly different from the desktops most developers assume. iPhone surfers pinch and tap their way through websites, posing unique usability challenges. Wireless devices suffer from sluggish performance, spotty coverage, and high levels of packet loss.

It's not just the surfing conditions that change when the Web goes mobile. Now, location is an important factor. Knowing where a user is physically located will determine which content they see and which ads they're served. Advertisers will want to know which shopping mall a visitor is in so they can tie online traffic to real-world outcomes. Some stores are even launching free Wi-Fi tied to Facebook Connect—a pact in which the visitor reveals his identity in return for Internet access.

Blurring Offline and Online Analytics

Communities will form around locations, much as they coalesce around hashtags today. Transient groups will appear and dissolve as web visitors move from physical location to physical location, and this data will be useful not only for segmentation, but also for mapping social graphs based on physical proximity.

Real-world activities are increasingly tracked online, through devices like the Fitbit or games like Akoha, and they're changing the way we think about analytics. Networks of sensors will soon collect and share our lives with one-time, tacit approval, leading to a new understanding of audiences—and new privacy concerns.

This blurring of online and offline worlds is aided by real-world tracking tools such as RFID (radio frequency identification) tags, barcodes that can be scanned by PDAs, VOIP services that link calls to analytical data, caller ID systems, and so on.

Standardization

One of the reasons it's hard to get a complete picture of your web presence is the lack of standards. With the exception of rudimentary logging (like CLF), some standard terms, and some industry initiatives for scoring (such as Apdex), there's no consistent way to collect, aggregate, or report web monitoring data.

This is particularly true for performance monitoring. It wouldn't take much to improve the state of the art with a few simple additions to the technologies we use today. If, for example, the world's browser makers decided to cooperate on a single DOM object that marked the final click on the preceding page, it would dramatically improve the usefulness of JavaScript-based RUM. Unfortunately, such standards are woefully absent.

Synthetic testing vendors differentiate their offerings through their instrumentation (browser puppetry versus scripting), the number of test points they have, the detail they offer, and the reports they generate. There's no reason why we can't have a standard language for defining scripts, or one for annotating page load times—this wouldn't undermine any vendor's uniqueness, but it would be a boon for web operators who need to manage and interpret test data.

Similarly, a common format for associating analytics, WIA, RUM, and VOC data across service producers would make it far easier for web analysts to do their jobs; yet data is seldom available for export. We need a common format for visitor records that can work across products.

The same is true of community monitoring. There's no standard measurement of ranking across social network analysis tools, and it's up to customers to demand openness and interoperability.

Increased concerns over data privacy and online activity may spur such standardization. It's unlikely, however, that web monitoring companies will voluntarily standardize the log formats and metrics that lock in customers and make migrating from provider to provider more difficult.

Agencies Versus Individuals

Public relations (PR) has traditionally been an outbound business, controlling the flow and the message. Today, the new PR is community interaction. Lobbying and advocacy are so passé; we'd rather they change sentiment and tweak the zeitgeist.

While many companies have started community management as an in-house effort—often part of user groups, support communities, or blogging—PR agencies are taking notice. Seeing their roles as keepers of the corporate message at risk, agencies want to own the interaction.

This heralds an arms race between agencies and the organizations they serve. On one side of this pitched battle will be agencies, which have the ear of the marketing department and can spread worries of liability and spin control. They'll employ mass-messaging tools—many of which we've seen but have yet to be launched publicly—to send personalized mass messages to followers. On the other side will be the righteously indignant individuals claiming the moral high ground of the genuine and devoting themselves to nurturing real relationships with their markets and followers.

How this will play out is uncertain. Several agencies we spoke with are building their own proprietary tools to mine communities on behalf of their clients; meanwhile, conversation listening platforms are turning support teams into skunk works PR agencies.

The outcome of this battle is in the hands of the communities themselves, who will either embrace the new communications channels even if they're disingenuous, or will reject them, exposing agencies as an insulating layer that impedes interaction to an organization and its audience.

Monetizing Analytics

Silicon Valley is littered with business plans based on ad revenue. In the era of free software, companies expected to make their money from advertisements. Yet pay-per-click advertising margins are notoriously thin, and often insufficient to support a business model.

Knowing which segments are most likely to do what you want—and how to reach them—is critical for any online business. Much of this information flows across Internet connections, social networks, search engines, and websites, and the keepers of these sites see analytics as the new AdWords: in the world of free, analytics is the cash register, and many companies seek to monetize their services by selling insights into online visitor behavior.

Let's look at some of the industries hoping to cash in on analytics and web intelligence.

Carriers

Mobile phone carriers had a great business. They provided services such as voicemail, text messaging, videoconferencing, newsfeeds, and more, for a fee. With the advent of the consumer Web on smartphones, however, they've lost the ability to unbundle individual services and charge for each of them. Apple finally capitulated and allowed Skype on the iPhone, which will have a material impact on long-distance revenues. Carriers have lost control of their original business models.

Instead of just becoming "dumb wires" for data applications, one way carriers can reclaim their revenues is to watch what mobile users do. Already, some firms (like Neuralitic) make analytical tools that capture mobile user traffic, aggregate it based on the applications and services those users access, and help the carriers analyze their businesses. The data could easily be sold to third-party clients who want to better understand mobile usage patterns.

Search Engines

Search engines are the most obvious candidates for analytics revenues, since paid search is their primary source of revenue, and giving customers tools to understand the effectiveness of their campaigns is a part of the sales cycle. Search engines can do more than just analyze keywords, however: they have data on trends, geographic spread, and more. We've covered many of the free tools that search engines—particularly Google—make available, and most of them are subsidized by advertising revenues. Search engines do have premium offerings for some of their tools, however.

URL Shorteners

The many URL shortening sites—dominated by bit.ly and Tinyurl—are well poised to make money from analytics. They sit between a web user's intentions (clicking the link) and the outcome (delivering that user to a site) and can build a great deal of information on the spread of social messages, as well as which content works best for which demographics, social networks, times of day, or geographies.

These URL shorteners also threaten social news aggregators. Sites like Digg, reddit, and Slashdot rely on surfers to submit stories they find interesting, and despite the efforts of these companies to make the submission process easy—through toolbars, embedded buttons, and so on—it still requires effort to submit a story.

bit.ly, awe.sm, and Tinyurl require no effort beyond the shortened URL, however, to learn what people find interesting. They can see which content is spreading most quickly, and segment it by user demographics, simply because they're acting as middlemen between the links and the content.

While today many of these sites offer analytics for free, bit.ly is already forming business relationships with marketers who want to track the spread of their messages. As organizations seek to do multivariate testing of community messages, they'll turn to these URL shortening companies for metrics and reporting data.

Social Networks

There's a war brewing between open, opt-in tracking through sharing and participation across many community sites, and the all-inclusive walled gardens of social network giants.

When you receive a message from a friend on Facebook, you're notified by email. To respond to that message, however, you need to click on the link provided in the message and respond using Facebook's tools. The same is true, to a large degree, for other walled garden social networks. LinkedIn is somewhat more open—email messages can be taken out of the platform and into traditional email—but social networks in general have been reluctant to integrate with systems outside their walls. It's only in the face of competitive pressure from sites like Twitter that social networks have been willing to put even carefully guarded openings in their garden walls.

Walled garden social networks want to be a one-stop social platform whose users don't need to venture elsewhere. Even "open" features like Facebook Connect are aimed at pulling content from elsewhere on the Web inside those walls, where members can further discuss them.

It's far easier to track—and monetize—community activity when that activity all happens in one place. A picture, leading to a Wall post, leading to a message, resulting in an invite, can all be tracked back to a user and segmented by demographics. As soon as the user's online experience transcends a single site, tracking him becomes far more difficult.

The tension between walled gardens and open communities isn't just technical—it's cultural. Early adopters and the web-savvy favor an archipelago of online communities loosely connected through APIs and hyperlinks. More mainstream consumers, however, want a unified set of tools built around their social graphs, a group hug with the friends they know. The social makeup of communities makes this distinction clear: *a real-world meeting of Twitter friends feels like speed dating; a Facebook meet-up is more like a class reunion.*

Whether a site is a lone island in an archipelago of social sites or a one-stop continent, it's still got to make money. To do so, the site's operators will deploy monitoring tools that help advertisers understand the relationship between social network activity and business outcomes. This will happen either by keeping them in the walled garden—the Facebook model—or by finding ways to connect community members' many online interactions.

This will in turn cause (justified) concern over the tracking of personal data online. Our network of families, friends, and acquaintances will become increasingly valuable to marketers who seek to optimize online activity, but we'll be increasingly concerned about sharing that network with others.

This picture of social networks might seem bleak: rather than them becoming a tool for collective human consciousness, they'll slide inexorably toward their role as yet another marketing tool. But this has happened throughout much of the Web already, from websites to blogs to email, without entirely tragic consequences.

Online communities will continue to thrive and connect, but when organizations want to monitor their roles within their communities and tie this information back to business outcomes, they'll start tracking them.

SaaS Providers

One area in which analytics has been glaringly absent is within hosted applications. SaaS offers an unprecedented opportunity for employers to measure the productivity of their employees, yet many SaaS offerings lack even rudimentary visibility into who is using the application and how they're doing so.

Sales is a numbers game, and sales force automation products, such as Salesforce.com, already give managers the ability to see which employees are meeting their goals. Call center tracking applications have similar analytics. With more and more enterprise applications available in SaaS formats, it's likely that we'll see "productivity analytics" become a more commonplace feature of SaaS sites.

Today, however, it's hard to implement this type of tracking yourself. Most modern SaaS platforms won't let customers embed tracking tools within their pages, so you're at the mercy of the SaaS platform to find out how much your employees are using the service, where they're getting stuck, or where they're pressing Delete the most.

As businesses start to expect the transparency and accountability of analytics, we expect them to demand this kind of visibility from all of their software tools.

A Holistic View

What's most obvious from today's monitoring industry is that siloed perspectives will not last. Vendors are rushing to broaden their feature offerings, either by building all the components themselves, acquiring other companies, or using creative mashup approaches to integrate their reports into entrenched solutions, such as Google Analytics. As Figure 18-3 shows, the distinctions of analytics, WIA, EUEM, and VOC are fuzzy at best, with many firms offering strong products in several segments.

Analytics companies are positioning themselves as the anchors of all this analysis. They are the cornerstones of monitoring strategies because they are the tool that sees the payoff. Whether it's an improvement in performance, positive visitor sentiment, a better page design, or a groundswell of community support, everything that happens online has an outcome that ties back to the organization's goals, and the analytics tool is where those outcomes are defined.

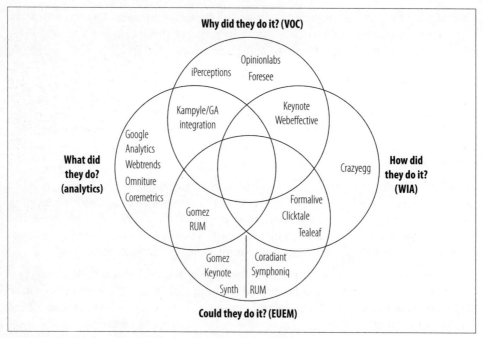

Figure 18-3. Examples of blended product functionality as the monitoring industry consolidates and vendors merge

Web analytics tools, however, are really a subset of the broader category of BI. For larger enterprises, it's likely that a central BI repository will include not only the online outcomes tracked by analytics platforms, but also call center traffic and retail transactions. In other words, the analytics giants of today may be consumed by even bigger companies that are the central repositories of a firm's interactions with all of its customers.

The Move to a Long Funnel

Increasingly, web users learn about destinations from their peers and their friends. Referrals have been the basis of much of the Internet economy, driving online advertising and paying for many of the websites we take for granted. This, in turn, has fueled the analytics industry. A referral from a website, however, isn't as genuine as one from a trusted friend or an admired celebrity. Already, Twitter mega-users rival Digg and reddit in their ability to send thousands of visitors to a site in an instant, bringing it to its knees.

Community monitoring tools will have no choice but to partner, as marketing executives demand more details about how their money is being spent. The community monitoring industry today is where the analytics industry was only a few short years ago (Figure 18-4).

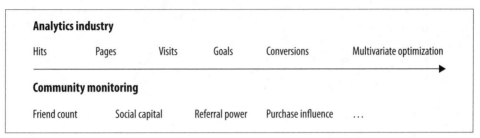

Figure 18-4. The community monitoring industry lags behind web analytics by a few years, and will undergo many of the same changes

Then, analytics focused on hits and pages; today, it's about conversions, segmentation, lifetime visitor value, viral coefficient, and optimization. Similarly, today's community monitoring tools focus on popularity and sentiment, but are quickly moving toward an analysis of users' long-term impact on the business.

The shift from web referrals to social referrals has far-reaching consequences for the analytics industry. While analytics continues to be important for onsite activity and optimization, such as reducing abandonment and deciding which content works best, it's becoming less useful for offsite analysis. Expect to see a flurry of partnerships between established analytics vendors, as well as changes to the offerings of leading advertising network providers who try to insert themselves into the social graph.

A Complete Maturity Model

As we've looked at various forms of monitoring, we've considered a maturity model that begins with basic technical visibility and moves toward a more integrated, comprehensive perspective of your entire online presence. Table 18-1 shows the questions an organization is trying to answer at each level of maturity.

Table 18-1. *The questions an organization answers at each level of web monitoring maturity across various forms of monitoring*

	Level 1	Level 2	Level 3	Level 4	Level 5
	Technical details	Minding your own house	Engaging the Internet	Building relationships	Web business strategy
Analytics	How many hits did I get?	How many conversions have I had?	What segments or sites are sending me the most, best traffic?	How loyal are my users, and what's their lifetime value? What content works best for them?	What KPIs should I use to manage my organization? How are online and offline channels related?
WIA	What did visitors see and click?	Where in a form or a page do users get stuck or abandon? How can I improve my design?	How do different visitor segments interact with my site?	How long does it take for people to learn the application? Where do they get stuck? Can I resolve their problems by having their sessions in my CRM?	Can I automatically serve optimal content and navigation for different segments and levels of visitor experience? How efficient/productive are users?
VOC	What do emails and "contact us" forms say about visitor satisfaction?	When I intercept visitors, what do they tell me about my site, their perception of me, and their loyalty to my products or services?	When I engage my audience elsewhere on the Internet, what do they say? What's their sentiment? How does word of mouth spread?	How does visitor feedback correlate with conversions? What do visitors tell me about my competitors?	How do I tie visitor sentiment to employee performance, software release success, and product road maps?
Synthetic monitoring	Is my site up? How fast is it?	Are transactions working? When things are slow, which tier of the infrastructure is guilty?	How are the communities and third parties on which I rely working? How's my CDN? Which parts of the Internet are slowest? Did I make my overall SLA?	How does my performance and availability affect my business? What are my aggregate scorecards?	What's the best balance of performance, capacity cost, and revenue? What's the business plan for uptime? Does IT get bonuses based on delivery? How cost-effective is our elastic capacity?

	Level 1	Level 2	Level 3	Level 4	Level 5
RUM	What are the traffic flows? When a problem happens, what does the trace show me?	How fast are pages loading for actual users? Which pages, servers, and objects are slowest? How healthy are transactions for actual users?	Which segments of visitors are most correlated with performance? Who are my worst-served customers? For whom were SLAs violated?	How did the latest release affect user experience? What's a given user's lifetime SLA?	How do online incidents tie back to CRM and support response? Have I automated the refund and dispute resolution process with RUM data?
External community	Am I being discussed?	Do I have a presence? Are people engaging with me on my own site?	Are they listening to & amplifying me? What's my sentiment like? Which messages work best with which communities?	How does the "long funnel" affect my revenues? What's the lifetime engagement of a visitor across all social networks? Who are my most vocal/supportive community members?	What will the community buy? How do I automatically match the right messages to each audience? How does my community help itself? Is virality a part of business planning?
Internal community	How much information have we generated?	How good is the information we're generating? What's most and least used? How easily can employees find it?	How integrated is KM with corporate culture? What are people sharing most and least?	What's the payoff? How does KM improve per-employee contribution or reduce per-employee cost? How are people connecting with one another? Are our collective guesses good?	What does our organization know? What is our organizational knowledge worth? How can we use better KM posture to improve competitive position? Are we hiring and firing based on employee knowledge contribution?
Competitors	What are my competitors doing?	How does my site compare to competitors'? Is it as fast? As easy?	Am I more popular than competitors? Better ranked? More adopted by certain segments?	How loyal are my customers? What can they tell me about my positioning versus others?	Is competitive analysis pulling data from VOC, community, and EUEM? How should competitors' strategies affect my own?

Ultimately, your complete web monitoring strategy needs to span both your site and the other sites and communities that affect your business. It needs to track visitors through their offsite activity, arrival, usage, and long-term engagement. It must monitor how they help your business, and how they refer others to you.

A Complete Perspective

Figure 18-5 is a quick reference for many of the things you need to consider when formulating your web monitoring strategy.

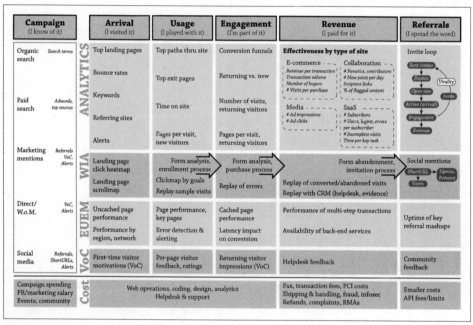

Figure 18-5. What a complete web monitoring strategy looks like

Complete monitoring includes analytics, WIA, EUEM, and VOC. It also includes looking at the many stages of visitor engagement, from initial arrival and use of the application through revenue growth and referral management. It encompasses effectiveness and efficiency measurements, as well as the instrumentation of virality and social spread. Finally, monitoring must be considered in the context of the many costs that are involved.

The Unfinished Ending

We're hoping this text provides a good background on web monitoring and that it has armed you with enough detail to dramatically improve the visibility you have into every website that affects your business. But it doesn't end here; by the time you read this book, the information is out of date.

We hope you'll join us at *www.watchingwebsites.com* for updates and perspectives on this dynamic topic. Together, we can continue the conversation.

KPIs for the Four Types of Site

At the start of this book, we looked at the many metrics you can track in order to understand your online presence. Those metrics you track and how you weigh them is your *monitoring mix*. Recall that there are four site archetypes: media, transaction, collaboration, and SaaS. The one that you're running dictates which metrics and KPIs matter most to you. Table A-1 shows how important each element of the monitoring mix is to each type of site.

Table A-1. The monitoring mixes for the four major site types[a]

	Media	Transaction	Collaboration	SaaS
How well did visitors benefit my business?				
Conversion and abandonment	1	4	1	3
Click-outs	4	1	3	1
Offline activity	1	3	1	3
Subscriptions	4	1	2	1
Billing and account use	1	1	1	4
Where is my traffic coming from?				
Referring URLs	3	4	3	1
Inbound links from social tools	4	3	3	1
Visitor motivation	2	4	3	1
What's working best (and worst)?				
Site effectiveness	2	4	2	1
Ad and campaign effectiveness	1	4	1	1
Findability and search effectiveness	4	3	3	1

What's working best (and worst)?				
Trouble-ticketing and escalation	1	1	2	4
Content popularity	4	2	4	1
Usability	2	4	4	3
User productivity	1	1	3	4
Community rankings and rewards	3	2	4	1
How good is my relationship with my users?				
Loyalty	4	3	3	1
Enrollment	4	2	2	1
Reach	3	4	4	1
How healthy is my infrastructure?				
Availability and performance	2	3	3	4
SLA compliance	1	1	1	4
Content delivery	4	1	1	1
Capacity and flash traffic	4	3	2	1
Impact of performance on outcomes	2	4	1	1
Traffic spikes from marketing efforts	3	4	1	1
Seasonal usage patterns	4	4	2	1
How am I doing against the competition?				
Site popularity and ranking	4	2	3	2
How people are finding my competiitors	3	4	2	1
Relative site performance	4	3	2	4
Competitor activity	1	3	1	3
Where are my risks?				
Trolling and spamming	3	3	4	1
Copyright and legal liability	1	1	3	1

Where are my risks?				
Fraud, privacy, and account sharing	1	3	1	4
What are people saying about me?				
Site reputation	4	1	4	1
Trends	4	3	1	1
Social network activity	4	1	3	1
How is my site and content being used elsewhere?				
API access and usage	4	3	4	2
Mashups, stolen content, and illegal syndication	4	2	3	1
Integration with legacy systems	1	3	1	4

[a] Key:
  1 Not Important
  2 Somewhat Important
  3 Very Important
  4 Primary Metric

Tailoring the Monitoring Mix to Media Sites

Media organizations care about the volume, loyalty, and interests of their visiting population, as well as the rates of click-through for advertising and their ability to cost-effectively handle traffic during peak load. They also watch the rest of the Internet for plagiarism and content theft, as well as incoming links from news aggregators that can foreshadow traffic spikes. Finally, they need to monitor comment threads for abusive behavior.

How Much Did Visitors Benefit My Business?

Primary metrics: click-outs; subscriptions

A visitor helps a media site by clicking on advertising, which generates revenue. If you have a premium subscription model, this is a second source of income, although you should treat the part of your site that converts users from "free" to "subscription" models as a transactional site.

This is especially important when comparing your own web analytics with those of advertisers or sponsors who owe you money.

Where Is My Traffic Coming from?

Primary metric: inbound links from social sites and search engines

Secondary metrics: referring URLs; visitor motivation

Since the site's job is to deliver content to others, it's important to reconcile where someone came from, her demographics, and where she went. This allows you to identify segments of visitors who are more likely to click on ads, so you can focus on attracting those segments.

On media sites, traffic will spike when news breaks or when content becomes popular. When this happens, it's important to identify the referring organization or social network that started the traffic and to encourage additional upvoting by the members of that network to make the most of your short-lived visibility. You also need to look at what was on visitors' minds that caused them to visit, which you can collect through VOC surveys.

Since many visitors to media sites arrive in search of some content—a recent TV clip, an interview with a celebrity—the search systems by which they find that content are important to track.

What's Working Best (and Worst)?

Primary metrics: findability and search effectiveness; content popularity

Secondary metrics: site effectiveness; community rankings and rewards

Media sites are all about content. To maximize visits, it's important to show visitors content that will grab their interest or is related to what they initially came for, so that they stay on the site longer and see more ads. To lower your bounce rate, it's essential to track which content is most popular and display it prominently on landing pages.

A lesser but still important concern is community ranking. In many cases, you will allow your visitors to rate and rank content. This is vital not only for identifying popular media and encouraging people to enroll in order to vote, but also for harnessing the wisdom of the crowds in flagging inappropriate material.

How Good Is My Relationship with My Users?

Primary metrics: loyalty; enrollment

Secondary metric: reach

The Web is increasingly dependent on permission-based marketing. Occasional visitors benefit you when they become loyal, returning subscribers or enroll via a mailing list or RSS feed, as this makes the site more attractive to advertisers. The number of people who have given you permission to communicate with them is a critical metric for media sites.

However, you can't just measure how many people have enrolled in your mailing list or subscribed to your RSS feed. You need to track your reach—the number of subscribers who act on what you send them by clicking on a link or returning to your site

How Healthy Is My Infrastructure?

Primary metrics: content delivery; capacity and flash traffic; seasonal usage patterns

Secondary metrics: availability and performance; impact of performance on outcomes; traffic spikes from marketing efforts

While all web operators need to know their sites are working properly, media site operators care about specific aspects of availability and performance.

Your site will often be a reference that's cited by others. If your site's content is updated often, you need to be sure caching is properly configured and data isn't stale, so that news gets out to returning visitors. You also need to keep archives available if you continue to be an authoritative source for other sites that link to your content.

Your site needs to load quickly, particularly for new visitors who aren't sure you have what they want. Returning users might tolerate occasional latency, particularly if they know they're going to get what they're looking for, but a first-time visitor won't. This is where internal service level targets and comparative synthetic test results are useful.

Performance and availability aren't just about delivering your pages to users. They're also about making sure ads and rich media reach visitors. Many ad networks do this for you by embedding one-pixel images before and after ads to confirm that they were properly displayed. For rich media such as video or interactive advertising, you may need to track other metrics, such as rollover and playback progress.

How Am I Doing Against the Competition?

Primary metrics: site popularity and ranking; relative site performance

Secondary metrics: how people are finding competitors; visitor motivations; brand recognition in surveys

Web users can get their news from many places. Knowing how you're doing against other sites that cover your news or offer similar media is an essential business metric, but one that only shows you how you're doing—not the cause of differences in popularity between your site and your competitors'.

This difference might be due to site performance—faster sites can lead to more active visitors. It may be because of your relative ranking in search engines. Or it may be that your competitors' brand awareness and engagement are leading potential visitors to them instead of you (or vice versa).

You also need to look at how people are finding your competitors. Certain keywords may be useful for you to start bidding on as well, in order to take a share of their market. You can also gauge the effectiveness of your brand.

Where Are My Risks?

Secondary metric: trolling and spamming

Media sites publish content they create. When reusing content, however, they need to be sure copyrights and terms of use are respected. Since most of the content is their own, copyright is less of an issue than it is for collaborative websites.

But the modern media site is a dialogue between visitors and content creators. These sites give visitors places to comment and respond. As a result, many popular media sites are plagued by comment spammers and trolls looking for a fight. If you're running such a website, you need to track problem users and quickly remove offensive content (or let the community do it for you) even as you strive to make it easy for your visitors to join the conversation.

What Are People Saying About Me?

Primary metrics: site reputation; trends

Media sites get visitors based on their reputations, which includes how well search engines trust their content and how well-known their brands are. If your media site is particularly topical, knowing which trends and topics are on the rise is also important if you want to stay relevant to your audience and adjust your coverage accordingly.

How Are My Site and Content Being Used Elsewhere?

Primary metrics: API access and usage; mashups, stolen content, and illegal syndication; traffic volume in bytes sent; top URLs in logfiles

If you're seeing huge amounts of outbound traffic but relatively few site visits, it's a clear sign that someone is embedding your media into a web page without your approval. This is a common form of illegal content syndication; it's bad for your media site because you don't get to show (and get paid for) the accompanying advertising.

If you're running a media site, you need to watch traffic levels to see who's putting your content elsewhere on the Net. When content becomes popular on a social network site, it's common for spammers to submit a second copy of the story, which links through to their sites first so that they can benefit from some of the traffic—a practice known as *linkjacking* (see Figure A-1).

Figure A-1. Linkjacking on reddit

Sometimes this second submission can even outstrip the original one. If you watch social news aggregators when your content first becomes popular, you can often detect this and report it to the site's operators—though the best defense is to make it easy for people to upvote your authentic copy of the content through better site design so your original story rises to the top.

You should also search the Web for strings of text within your popular articles to see if others are reprinting them as their own.

Tailoring the Monitoring Mix to Transactional Sites

Transactional sites make their money from outcomes. These might be subscriptions, purchases, or enrollments—whatever the case, the company's revenues are directly driven by the percentage of visitors who complete a particular process.

How Much Did Visitors Benefit My Business?

Primary metric: conversion and abandonment

Secondary metric: offline activity

If you're running a transactional site, you need to constantly compare different site designs, offers, and pricing strategies to find the ones that have the highest conversion rates. Page views mean nothing unless you can turn visitors into buyers.

If the end of your transaction happens offline—talking to a sales rep, getting contact information from a site, or starting an online chat—treat this as your goal and try to tie the online portion of the transaction to its offline outcome. At a minimum, provide a unique phone number on the website and track call volumes to that number.

Where Is My Traffic Coming from, and Why?

Primary metrics: referring URLs; visitor motivation; traffic volume by segment

Secondary metric: inbound links from social tools

The second big question for transactional sites is how many visitors come from where. You get and pay for your business by understanding which search terms, campaigns, social networks, and referring sites have the highest conversion rates. You should also understand *why* visitors come to your site so you can be sure you're satisfying their needs and putting appropriate offers in front of them. VOC surveys can reveal visitor motivations.

When looking at advertising, distinguish qualified from unqualified visitors. If you sell cars, young children who visit the site aren't likely to be buyers and will reduce the ROI of your ad campaigns. Consequently, you need to eliminate unqualified leads from your conversion analysis and adjust your advertising spend to ensure that such leads don't poison otherwise healthy conversion numbers.

What's Working Best (and Worst)?

Primary metrics: site effectiveness; ad and campaign effectiveness; usability

Secondary metrics: findability and search effectiveness; content popularity; community ranking and rewards

When you're monitoring conversions and segmenting traffic, you need to focus on site effectiveness. Transactional sites thrive by upselling visitors. As more and more users buy through onsite search tools, it's also important to monitor the effectiveness of searches and determine which search terms lead to purchases most often. The key to maximizing site effectiveness is constant experimentation in the form of A/B testing and multivariate analysis.

Since peer recommendations ("Other people who bought this also bought…") are one of the strongest influences on purchase upselling, it's also critical to monitor the user community for comments, ratings, and other feedback, and to borrow a page from collaborative sites by encouraging visitors to comment and recommend products.

Finally, you need to look at usability: examining how users interact with the site, particularly at places where abandonment occurs, is essential if you're going to improve usability. Do users scroll down? Does your offer appear above the fold? Do buyers click on the button or the text? Does the page take a long time to load? All of these factors can impact how effectively you maximize each visit.

How Good Is My Relationship with My Users?

Primary metric: reach

Secondary metrics: enrollment; loyalty

Transactional site operators care about their ability to send messages to their users and have them act on those messages. Email campaigns, RSS feeds of promotions, and similar forms of enrollment are all useful, but they only count when someone clicks on the link.

You also want to know the lifetime value of a customer. You may find that a particular segment of the market purchases considerably more over months or years, and it's wise to cater to that segment in your marketing, positioning, and offers. As retailers move away from broadcast and toward community marketing, enrollment and loyalty may overtake reach as the key measurement of relationship strength.

How Healthy Is My Infrastructure?

Primary metrics: impact of performance on outcomes; traffic spikes from marketing efforts; seasonal usage patterns; consistency of performance and availability

Secondary metrics: availability and performance; capacity and flash traffic

Transactional site operators care about performance primarily for the way in which it affects conversion. If performance degrades, conversion rates will fall. The site may be slow because of sudden spikes in traffic, peak seasonal periods (such as holiday shopping), content changes, or modifications to code and infrastructure.

Of particular interest is the impact of marketing campaigns on performance. Your marketing efforts must be tightly tied to capacity planning and performance monitoring to ensure that a successful marketing campaign doesn't backfire and break your infrastructure.

How Am I Doing Against the Competition?

Primary metric: how people are finding competitors

Secondary metrics: site popularity and ranking; relative site performance; competitor activity

If you're running a transactional site, you care how people are finding your competitors, and how they can find you instead. This is often a battle of keywords and search terms, and in terms of organic search it is also a matter of how relevant Google and others think you are.

You also care whether you're fast enough. It's not necessary to be as fast as possible, but you should compare your performance to relevant industry indexes to ensure you're not falling behind what users consider acceptable. On the Web, your competition may not be who you think it is. In addition to other sites that offer the same products and services, you're competing against the expectations set by other websites your target market frequents. If those sites constantly improve and innovate while you don't change, your audience will eventually grow disenchanted, even if those sites don't compete with you in the traditional sense.

Where Are My Risks?

Secondary metrics: trolling and spamming; fraud, privacy, and account sharing

For transactional sites, most risks come from password and credit card leaks, which are a matter for your security team. But if you're letting visitors rate and rank products, you need to be on the lookout for abusive behavior. For example, in an effort to improve the quality of ratings, Apple's App Store chose to limit reviews to only those visitors who had purchased an application (*www.alleyinsider.com/2008/9/apple-flexes-even-more-muscle-at-the-iphone-app-store-no-reviews-till-you-pay-up-aapl-*).

You may also care about account sharing—if multiple users, all with unique tastes, share one account, your suggestions and recommendations will be less accurate and will undermine upselling attempts.

What Are People Saying About Me?

Secondary metric: site reputation

Your site's reputation figures in both word of mouth and search engine ratings. But as a transactional site operator, you care mostly about which online conversations are leading to conversions, rather than reputation for its own sake.

How Are My Site and Content Being Used Elsewhere?

Secondary metrics: API access and usage; mashups, stolen content, and illegal syndication; integration with legacy systems

If you're running a transactional site, you care less about your content being used elsewhere, particularly if it helps spread the word about your products and your brand. Travel site Kayak.com, for example, compares flight prices across many airline portals, but it makes its money through affiliate fees from the airlines from which the visitors ultimately buy tickets.

On the other hand, if somebody is scraping pricing data from your site for price comparisons, it can undercut your margins and lead to price wars, so you need to identify hostile crawlers that harvest content from your site. You can then set up a *robots.txt* file to block well-behaved crawlers from those parts of the site, then identify those that ignore it and block them by user agent, source IP address, or a CAPTCHA test.

If your transactional site is a large-scale marketplace, you may have an ecosystem of buyers and sellers who've built tools around your application. There are hundreds of tools for eBay sellers, for example. You need to monitor these interactions so you don't alienate power users, but also so they don't violate terms of service.

If you're selling online, you may also have backend connections to payment systems (such as PayPal) that need to be monitored as part of the overall site health.

Tailoring the Monitoring Mix to Collaborative Sites

If you're running a collaboration site, you want to be sure users engage with your application; create, edit, and rank content; and spread the word. You also want to mitigate bad content and stop users from disengaging.

You're in a unique position: compared with a transactional or media site operator, you have much less control over your destiny. You're dependent on your visitors and your community to generate content and build buzz. You also need to walk a fine line between rewarding a few extremely active participants and making sure that content is open and democratic.

How Much Did Visitors Benefit My Business?

Secondary metrics: click-outs; subscriptions

While the advertising side of your collaborative site is run like a media business, from a collaboration standpoint you care about users that are creating content, whether through uploads, writing, ranking, or editing. You also care whether this content is valuable—are others reading it?

A second factor is how much users are engaging with the site. Do they track comments on items they've created? Are they building social networks within the site and rating one another?

On many collaborative sites, a small population of users will generate the majority of content. This can actually be a liability for site operators: a big attraction for collaborative sites is that they harness the long tail of public opinion and provide more targeted content than the mainstream media. Sometimes, the focus on the long tail has casualties. On September 24, 2008, Digg announced that as part of its new financing it would be banning its biggest users, saying it could "not have the same 1 percent of users generating 32 percent of visits" to the company's site (*http://socializingdigg.word press.com/2008/09/24/diggs-new-biz-model-ban-top-users-and-hit-300m/*).

Also, because much of the growth of collaborative sites comes from invites, you should treat invitations as a form of conversion. Facebook, for example, lets users share their content with friends who don't have Facebook accounts. Those invited friends can see the shared content immediately, but must sign up when they try to browse elsewhere on the site.

Where Is My Traffic Coming from?

Secondary metrics: referring URLs; inbound links from social tools; visitor motivation

As the operator of a collaboration site, you care less about where visitors are coming from than the operator of a transactional site might. But knowing about the social groups and referring sites helps you to tailor content to their interests. Similarly, visitor surveys can reveal why people are coming to the site and what other collaborative sites they frequent.

What's Working Best (and Worst)?

Primary metrics: content popularity; usability; community ranking and rewards

Secondary metrics: site effectiveness; findability and search effectiveness; trouble ticketing and escalation; user productivity

This is the most important set of metrics for a collaborative site. With so much riding on your visitors, your site has to be usable and it must be easy for visitors to find and rate popular content. You need to reward active contributors and make them feel a part of the community, showcasing their work. Also, find out what causes visitors to invite their friends, and make it easy for them to do so and for their friends to get immediate gratification from the invitation.

Your site won't succeed if there are a large number of complaints and problems that make it hard for users to create. You also want the collaboration site to become a reference for users. After all, you'll eventually make your money by turning the content your users provide into media to which you can attach advertising, so the site must be searchable and properly indexed.

If you're using a wiki model, you need to track incipient links—essentially, links to pages that haven't yet been created. If a page has too few incipient links on it, it's an orphan. If it has too many, related material hasn't yet been created. You should identify incipient links that are frequently clicked by visitors and flag them so that their destination pages are the next to be created.

How Good Is My Relationship with My Users?

Primary metric: reach

Secondary metrics: loyalty; enrollment

You need contributors to keep coming back. Informing them that others are interacting with their content—essentially giving your contributors their own analytics—is one way to accomplish this, as are updates and friend feeds. For all of these to work, you need permission to contact your visitors via email or RSS feeds, and you need them to follow the links you send them. As a result, you need to track reach, loyalty, and enrollment and encourage users to engage with the community to maximize collaboration.

How Healthy Is My Infrastructure?

Secondary metrics: availability and performance; capacity and flash traffic; seasonal usage patterns

Collaboration sites may experience sudden growth, particularly when viral marketing kicks in. Slideshow producer Animoto, for example, went from 25,000 to 250,000 users in three days.[*] While availability and performance should always be monitored, your primary concern is that they do not interfere with collaboration and that you can quickly detect growth in traffic so your systems engineers know to add capacity.

How Am I Doing Against the Competition?

Secondary metrics: site popularity and ranking; how people are finding my competitors; relative site performance

Competition isn't as important with collaboration sites as attention is. Because the long-term goal is to make money on media contributed by others, however, you do need to track your site ranking to be sure that the content your users are creating is relevant to advertisers and is gaining the attention of search engines. How are Internet

---

[*] *http://mashraqi.com/labels/animoto.html.* For a more detailed look at Animoto's use of elastic computing resources, see Werner Vogels' presentation on Amazon Web Services at *http://www.cca08.org/files/slides/w _vogel.pdf.*

users finding out about topics you cover? Can you better mark your pages so they get the attention of search engines and you rise above competitors in organic search?

Where Are My Risks?

Primary metric: trolling and spamming

Secondary metrics: copyright and legal liability

The biggest risks for a collaboration site are bad content and the addition of illegal, inappropriate, or copyrighted material to the site. Trolls will deter visitors from returning, and spammy content will reduce the value of the site in the eyes of both users and search engines.

You also need to watch how quickly content is rejected, which can be a sign of abusive behavior or an attempt by spammers to downvote other users' content in order to bring theirs to the forefront.

Depending on the type of collaboration site you're running, you may need to monitor for illegal uploads. If users post content that could expose your site to litigation, you must be able to demonstrate effective tools for flagging the content, investigating it, and removing it quickly. Such actions need to be backed by terms of use and community management policies.

What Are People Saying About Me?

Primary metric: site reputation

Secondary metric: social network activity

Your site's reputation in the eyes of both search engines and users is key. In the early stages of a collaborative application, you need to watch social networks to track buzz and manage complaints by addressing user concerns. Because your site is so dependent on the contributions of others, its ranking and the attention it receives from microblogs and news aggregators can make or break you.

If you're focused on a specific segment of the Internet, you need to be sure you're reaching that community directly. Imagine you have a website where people contribute plans for paper airplanes: are aeronautical engineers discussing you? How about paper companies? Science teachers? Where can you go to find them?

How Are My Site and Content Being Used Elsewhere?

Primary metric: API access and usage

Secondary metric: mashups, stolen content, and illegal syndication

Your content, and that which your users create, is valuable. If it winds up on others' sites without you being able to insert advertising, you'll never make money from the community you're nurturing. At the very least, content should be attributed to you so you'll rise in search engine rankings and gain visibility. So you need to watch APIs and automated retrieval of content, particularly the embedding of rich media for acceptable use.

Many multimedia collaboration sites embed their advertising directly into the media as preroll messages, interstitial advertising, or overlaid logos that link users back to the site itself. This is one of the main attractions of Flash- or Silverlight-based encoding of video and audio. If you're using this approach, you don't mind that others embed your content in their sites as long as you tie your rich media content back to your analytics systems so you can see when it's played elsewhere.

Tailoring the Monitoring Mix to SaaS Sites

SaaS companies want to replace desktop software. Their products must be at least as fast and as dependable, and ideally more convenient than the desktop alternatives they're replacing. They should also offer features, such as sharing and group scheduling, that aren't easily available on standalone desktop applications. So most of the metrics a SaaS provider cares about focus on performance, usability, and end user productivity.

How Much Did Visitors Benefit My Business?

Primary metric: billing and account use

Secondary metrics: conversion and abandonment; offline activity

SaaS companies make money when subscribers pay for access to the application. Monthly billing is based on the number of users and seats, and should be the basis for revenue reporting. Usually, this kind of accounting is done within the application itself and tied back to payment systems directly.

You should view the parts of your site that sign up new subscribers or convert free users to paid services as a transactional site using traditional conversion-and-abandonment monitoring.

The interaction your subscribers have with your helpdesk, however, needs to be tracked because it is a direct reflection of performance, availability, and usability issues with the SaaS site itself. Helpdesk calls are negative outcomes that need to be traced back to the pages, processes, or outages that caused them.

Where Is My Traffic Coming from?

No primary or secondary metrics

Traffic sources aren't particularly important to a SaaS provider, with one exception. You may care how many users are accessing the application from home (rather than from an office) or from a nonstandard device (like an iPhone instead of a PC). Tracking this kind of data can warn you about usage patterns you may need to address in future releases, such as stronger privacy for home PCs or a different resolution for mobile devices.

Of course, the part of a SaaS provider that tries to attract customers cares a lot about traffic sources, but it's a transactional site, not a SaaS site.

What's Working Best (and Worst)?

Primary metrics: trouble ticketing and escalation; user productivity

Secondary metric: usability

Everything on your SaaS application ties back to productivity. If users are more productive with the hosted application than they were with a desktop alternative, your business subscribers will be happy.

Tracking productivity metrics is key. Identify which actions are at the core of your application—filing an expense report, sending a client an estimate, looking up a contact, and so on—and track them ruthlessly.

Every time a user's session goes to a helpdesk, you should flag that session and be able to replay it. If the problem was a user error, you need to fix the page and make it more usable. If it was a technical error, the steps needed to reproduce it will have been captured and you can add them to your testing of future releases.

You also care whether users are embracing new features and enhancements. If they're only using a subset of the functionality you offer, you may find yourself competing against other services simply because your customers aren't aware of your entire product.

How Good Is My Relationship with My Users?

No primary or secondary metrics

Your subscribers pay to use the application. You aren't as concerned with engagement or reach as you would be with other types of site, though you do care about bounced mails if email notification is an important part of your application.

For most SaaS firms, customer relationships are a sales issue, but you need to arm the sales force with salient data about the customers' experiences with the site so they can sell and renew subscriptions.

How Healthy Is My Infrastructure?

Primary metrics: availability and performance; SLA compliance

Infrastructure health is especially important for SaaS companies. Not only do you need to be as fast as an in-house alternative, you may have to offer refunds to customers if you fail to meet SLA targets. If you're unable to deliver the application to users, you need to know why. Was it the network or the application? Is network delay due to large pages or poor networks?

SaaS companies are some of the strongest adopters of End User Experience Management (EUEM) technology because they need to end the finger-pointing that typically accompanies service outages. Knowing what's your fault and what's your subscribers' fault can mean the difference between cutting a belligerent customer a refund check and finally getting him to admit that he was wrong.

Because SaaS application use is often part of the workday, you also need to know daily usage patterns by time zone. As different parts of the world wake up and use your application, you'll see different spikes: logins in the morning, reporting in the afternoon, and a lull before night owls access the application from home. You need to ensure that you can handle this rolling traffic across the day and detect any changes that might signal a problem before it interferes with your SLAs.

How Am I Doing Against the Competition?

Primary metric: relative site performance

Secondary metrics: competitor activity; site popularity and ranking

While it's important to know how you're faring against the competition in any industry, you care a lot less about social networks and search engine rankings in your operation of the SaaS application itself. Search engine rankings are, of course, important to your acquisition of new customers, but that process is treated as a transactional site.

When it comes to the SaaS portal, you should look at industry indexes or tests of your competitors to ensure that your performance is on par with the industry. Are your competitors using content delivery networks to reach far corners of the Internet? Have they embraced on-demand infrastructure in an effort to scale? Are they paying a premium for high-end managed hosting or faster broadband connections?

Where Are My Risks?

Primary metric: fraud, privacy, and account sharing

When you're running an application, your main fraud concerns are about protecting your customers' data. You're responsible for private financial information that may be subject to specific legislation—HIPAA for healthcare, PCI for credit card transactions, OSHA for human resources, and so on. And this legislation may affect what you're allowed to collect and monitor, and who has access to that monitoring data.

Be particularly careful around account management and termination. Since your customer is the enterprise that subscribes to the SaaS application, but your end user is the employee, you may have to remove employees from the system when they leave the company. SaaS applications need much more powerful on-demand forensics to manage user accounts and assist with security investigations when employees defraud their employers.

One other area of concern for SaaS companies is account sharing. If you generate revenue per subscriber, you need to monitor account usage to detect when several people are sharing a single account.

What Are People Saying About Me?

No primary or secondary metrics

As a SaaS provider, you don't care much about what the Internet thinks of you. Your sales department may, but as the operator of the application itself, most of your attention is turned inward or toward your customers.

How Are My Site and Content Being Used Elsewhere?

Primary metric: integration with legacy systems

Secondary metric: API access and usage

The main concern for SaaS companies is how their hosted applications integrate with the enterprise's in-house data. You may have to monitor communications with your customers' CRM, HR information, or Enterprise Requirements Planning (ERP) systems. This means monitoring a variety of APIs, some of which will be based on older protocols, to ensure that the application works properly. For example, a login process on your hosted application may need to talk with a company's LDAP server or an OpenID service to verify users. Any slowdown in that connection will affect the login process on the site, so it needs to be part of your monitoring strategy.

Index

Numbers

80/20 rule, 433

A

A/B comparisons
  VOC, 204
A/B testing
  experimentation with, 141
abandonment
  about, 16
  and bounce rate, 107
  form analysis, 160
Accenture, 540
accounting
  future trends, 582
accounts
  sharing, 45
  tracking use, 21
add-ons
  browsers, 297
address translation, 285
administrators
  blogs, 464
  communities, 430
  forums, 456
  mailing lists, 455
  real-time chat rooms, 458
  social networks, 463
ads
  about, 95
  effectiveness, 25
  online, 72
  tracking clicks, 109
advocates

communities, 429–432
affiliate IDs, 97
affordance
  web site usability, 156
agent-based capture
  RUM, 373
agents
  server agents, 117
aggregate reports
  RUM, 380
aggregation
  content, 532
  synthetic testing, 346
alerts
  about, 565, 577
  Google Alerts, 42
  synthetic testing, 343
Altimeter Group's model of community
        participants, 432
analytics, 57
  (see also web analytics; WIA)
  monetizing, 588–591
  online versus offline, 586
Apdex (Application Performance Index)
  alternative to ITIL, 247
API calls
  tracking, 49
APIs
  access and usage, 49–51
  contributors of, 437
  performance, 51
  Twitter, 472
applications
  desktop, 305
  embedded in web pages, 86

We'd like to hear your suggestions for improving our indexes. Send email to *index@oreilly.com*.

on life with Twitter, 469
functional tests
    defined, 337
funnels
    communities, 405, 406, 515
    conversion optimization, 167
    forms as, 161
    goal funnels, 79, 84, 105
    social referrals, 592
    stage in web analytics, 90–100
    tracking business outcomes, 520
    typical e-commerce conversion funnel, 16

G

Genesis, 566
geographic distribution
    synthetic testing, 340
GET command, 263, 324
Get Satisfaction, 421
GetStats, 70
Gibson, J.J.
    on affordance, 156
Global Server Load Balancing (GSLB)
    EUEM, 275
Gmail, 497
goals
    segmentation, 160
    tracking visits and outcomes, 78, 105
    VOC studies, 202
    web analytics implementation, 137
    websites, 16, 112
Gomez, 566
Google, 72
Google Alerts, 42
Google Analytics, 134, 135, 551, 566
Google Apps, 87
Google Insights, 47, 442
Google PageRank, 47, 553
Google Trends, 556
graphs
    community conversations, 536
Greasemonkey, 297, 571, 576
group administrators
    social networks, 464
group creators
    social news aggregation, 475
group/hashtag creators
    micromessaging, 474
groups

joining, 497
    moderating, 502
    running, 507
    searching conversations, 482
GSLB (Global Server Load Balancing)
    EUEM, 275

H

hard errors
    defined, 253
hashtags, 470, 472, 474
hashtags.org, 505
HEAD command, 263
heatmaps, 158
helpdesks
    WIA, 169
histograms
    versus averages, 58
history
    online communities, 391–394
    web analytics, 69–89
hits
    defined, 73
    tagging content, 82
Homepage portal
    referring URLs, 23
host latency
    databases, 289
host time, 267
hosted tools
    versus in-house tools, 146
Hotmail, 411, 415
HTTP
    commands, 263
    counting requests, 73
    ELF, 69
    referring URLs, 22
    status codes, 264

I

ICMP ECHO
    testing network connectivity, 322
identifying segments, 130
iGoogle, 573
IM (instant messaging)
    history of, 397
    online communities, 399–403
images

messages, 471
    open API, 472
type-in traffic, 92

U

V

About the Authors

Alistair Croll is an analyst at research firm Bitcurrent, where he covers emerging web technologies, networking, and online applications, and a principal of startup accelerator Rednod, where he advises companies on product and marketing strategy. Prior to Bitcurrent, Alistair cofounded Coradiant, a leader in online user monitoring, as well as the research firm Networkshop. He has held product management positions with 3Com Corporation, Primary Access, and Eicon Technology.

Alistair contributes to industry events such as Interop (where he runs the Cloud Computing and SaaS tracks), Structure, Enterprise 2.0, Mesh, Velocity, eMetrics, and Web2Expo. He writes for a variety of blogs (including *www.bitcurrent.com*) and is the author of numerous articles on Internet performance, security, cloud computing, and web technologies. Alistair is also the coauthor of *Managing Bandwidth: Deploying QOS in Enterprise Applications* (Prentice-Hall).

Sean Power spends way too much time on the computer and needs to get out more. He has worked as a web systems administrator since the mid '90s, has worked with online communities for companies such as MTV Northern Europe, and helped users reduce the headaches of managing and monitoring web infrastructures through Coradiant, a web performance monitoring vendor. Prior to working at Coradiant, he was a technical reviewer for the Addison-Wesley book *Troubleshooting Linux Firewalls*.

Sean last worked as community gardener for Akoha, a company pioneering the industry of "social games," where he handled all things community and analytics. He is currently a cofounder of Watching Websites.

He completes his full plate by supporting the companion website to *Complete Web Monitoring*.

In his spare time, Sean makes sure that servers stay up and curses spammers in the EFnet IRC community. When he's not writing web optimization articles, he occasionally updates his personal music-related blog.

You can hear him ramble away on his Twitter account at *http://twitter.com/sean power*, or read about other stuff he's thinking of at the website he shares with Alistair, *http://www.watchingwebsites.com*.

Colophon

The animal on the cover of *Complete Web Monitoring* is a raven. The raven *Corvus corax* is a member of the family *Corvidae*, which includes crows, jays, and magpies. They are one of the most widespread, naturally occurring birds worldwide. While they can be found throughout most of the Northern Hemisphere in many types of habitats, they are permanent residents of Alaska, where they nest anywhere from the Seward Peninsula to the mountains of southeast Alaska. Ravens prefer open landscapes such as seacoasts, treeless tundra, open riverbanks, rocky cliffs, mountain forests, plains, deserts, and scrubby woodlands. There is no mistaking the raucous call of the raven; its deep, resonant caw is its trademark, yet the bird can produce an amazing assortment of sounds.

The raven is distinguished from other *Corvus* species by their massive size and is the largest all-black bird in the world. In Alaska, the raven is sometimes confused with a hawk or crow. The birds have have large, stout bills, thick necks, shaggy throat feathers called "hackles" that they use in social communication, and wedge-shaped tails, which are most visible when the birds are in flight.

Most ravens first breed at three or four years of age; once a raven finds a parter, it mates for life. Ravens begin displaying courtship behavior in mid-January, and by mid-March, adult pairs roost near their nesting locations. The female lays three to seven eggs and then incubates them; the male contributes to the birth of his young by feeding the mother-to-be while she nests. The chicks hatch after about three weeks and leave the nest about four weeks after hatching. Both parents feed their young by regurgitating food and water stored in their throat pouches. Ravens are omnivores, but most of their diet is meat, they are known to consume a wide variety of both plant and animal matter. They scavenge for carrion and garbage and also prey on rodents and on the eggs and nestlings of other birds.

Ravens are excellent fliers and often engage in aerial acrobatics as they soar to great heights. During the day, ravens form loose flocks, but by night, many of them will roost together. As many as 800 ravens have been seen in one roost. Unlike many birds, ravens do not undertake long migrations, but they do relocate locally for nesting each year.

The raven has played important roles in many cultures, mythologies, and writings. Ravens disobeyed Noah during the great flood by failing to return to the ark after being sent to search for land. In Norse mythology, the god Odin ordered two ravens named Thought and Memory to fly the world each day so they could inform him of what was happening. The spiritual importance of the raven to Alaska's Native people is still recognized today.

The cover image is from Cassell's *Natural History*. The cover font is Adobe ITC Garamond. The text font is Linotype Birka; the heading font is Adobe Myriad Condensed; and the code font is LucasFont's TheSansMonoCondensed.

Related Titles from O'Reilly

Web Authoring and Design

ActionScript 3.0 Cookbook

Ajax Hacks

Ambient Findability

Creating Web Sites: The Missing Manual

CSS Cookbook, *2nd Edition*

CSS Pocket Reference, *2nd Edition*

CSS: The Definitive Guide, *3rd Edition*

CSS: The Missing Manual

Dreamweaver 8: Design and Construction

Dreamweaver 8: The Missing Manual

Dynamic HTML: The Definitive Reference, *3rd Edition*

Essential ActionScript 3.0

Flex 8 Cookbook

Flash 8: Projects for Learning Animation and Interactivity

Flash 8: The Missing manual

Flash 9 Design: Motion Graphics for Animation & User Interfaces

Flash Hacks

Head First HTML with CSS & XHTML

Head Rush Ajax

Head First Web Design

High Performance Web Sites

HTML & XHTML: The Definitive Guide, *6th Edition*

HTML & XHTML Pocket Reference, *3rd Edition*

Information Architecture for the World Wide Web, *3rd Edition*

Information Dashboard Design

JavaScript: The Definitive Guide, *5th Edition*

JavaScript & DHTML Cookbook, *2nd Edition*

Learning ActionScript 3.0

Learning JavaScript

Learning Web Design, *3rd Edition*

PHP Hacks

Programming Collective Intelligence

Programming Flex 2

Web Design in a Nutshell, *3rd Edition*

Web Site Measurement Hacks

Our books are available at most retail and online bookstores.
To order direct: 1-800-998-9938 • *order@oreilly.com* • *www.oreilly.com*
Online editions of most O'Reilly titles are available by subscription at *safari.oreilly.com*